Nanotechnology for Dentistry Applications

IPEM–IOP Series in Physics and Engineering in Medicine and Biology

About the Series

The Series in Physics and Engineering in Medicine and Biology will allow IPEM to enhance its mission to 'advance physics and engineering applied to medicine and biology for the public good'.

Focusing on key areas including, but not limited to:

- clinical engineering
- diagnostic radiology
- informatics and computing
- magnetic resonance imaging
- nuclear medicine
- physiological measurement
- radiation protection
- radiotherapy
- rehabilitation engineering
- ultrasound and non-ionising radiation.

A number of IPEM–IOP titles are published as part of the EUTEMPE Network Series for Medical Physics Experts.

Nanotechnology for Dentistry Applications

Edited by
Ashutosh Kumar Shukla
Ewing Christian College, Prayagraj, India

IOP Publishing, Bristol, UK

ISBN 978-0-7503-3671-0 (ebook)
ISBN 978-0-7503-3669-7 (print)
ISBN 978-0-7503-3672-7 (myPrint)
ISBN 978-0-7503-3670-3 (mobi)

DOI 10.1088/978-0-7503-3671-0

Version: 20211201

IOP ebooks

British Library Cataloguing-in-Publication Data: A catalogue record for this book is available from the British Library.

Published by IOP Publishing, wholly owned by The Institute of Physics, London

IOP Publishing, Temple Circus, Temple Way, Bristol, BS1 6HG, UK

US Office: IOP Publishing, Inc., 190 North Independence Mall West, Suite 601, Philadelphia, PA 19106, USA

To my parents

Contents

Preface

This collection highlights the applications of Nanotechnology for dentistry applications. There are 13 chapters contributed by the experts from different laboratories/regions. Nanotechnology applications in different fields of wide interest including dental caries and erosion and dental implants have been covered in detail. Chapters covering the nanotechnology applications for dental therapeutics and periodontal disease management are undoubtedly meant to connect the reader to the promising trends and alternative therapies. In addition, individual chapters with focus on nanotechnology applications in specialties like orthodontics, endodontics and prosthodontics have been included that are of interest to novice researchers in addition to professionals. A chapter on neurotoxicity of different nanomaterials used in dentistry needs a special mention as this emphasizes the need to proceed cautiously while exploring the new dimensions of 'Nano'. Contributors have mostly adopted a bottom-up approach. I have tried my best to avoid intrusion in the plan of contributing experts and this is probably the reason that individual chapters have come up with a sense of completeness in them.

I sincerely thank Michael Slaughter, Senior Commissioning Editor—Bioengineering, Medical Physics, Biophysics, IOP Publishing for giving me an opportunity to present this volume. I wish to thank Phoebe Hooper, ebooks Editorial Assistant, IOP Publishing for extending all the support during the different publication stages of this project. I sincerely thank the expert contributors for contributing to this volume in a timely fashion and helping me to meet the schedule.

Ashutosh Kumar Shukla, Prayagraj, India
June 2021

Editor biography

Ashutosh Kumar Shukla

With two decades of physics teaching and research experience, publications in peer reviewed journals, review articles, books and many edited volumes prepared in collaboration with experts from different countries, I intend to continue in academics.

List of contributors

Ra'ed Al-Dboush
The University of Alberta, Edmonton, AB, Canada
Jordanian Royal Medical Service, Amman, Jordan

B Alekhya
Postgraduate Student, Department of Oral Pathology and Microbiology, GITAM Dental College and Hospital, Visakhapatnam, India

Nafiu Aminu
Department of Pharmaceutics and Pharmaceutical Microbiology, Faculty of Pharmaceutical Sciences, UsmanuDanfodiyo University, Sokoto, Nigeria

Meena Ajay Aras
Department of Prosthodontics, Goa Dental College and Hospital, Bambolim, Goa, India

Heitor Ceolin Araujo
São Paulo State University (Unesp), School of Dentistry, Araçatuba, Department of Preventive and Restorative Dentistry, Araçatuba/São Paulo, Brazil

Laís Salomão Arias
São Paulo State University (Unesp), School of Dentistry, Araçatuba, Department of Preventive and Restorative Dentistry, Araçatuba/São Paulo, Brazil

Papia Basuthakur
Department of Applied Biology, CSIR-Indian Institute of Chemical Technology (CSIR-IICT), Uppal Road, Tarnaka, Hyderabad, Telangana, India

Shatarupa Biswas
Department of Microbiology, School of Life Science and Biotechnology, Adamas University, Kolkata, India

Anupama Bonam
Graduate Student, GITAM Dental College and Hospital, Visakhapatnam, India

Wilmer Ramírez Carmona
São Paulo State University (Unesp), School of Dentistry, Araçatuba, Department of Preventive and Restorative Dentistry, Araçatuba/São Paulo, Brazil

Saptarshi Chatterjee
Department of Microbiology, School of Life Science and Biotechnology, Adamas University, Kolkata, India

Anne Caroline Morais Caldeirão
Graduate Program in Dentistry (GPD—Master's Degree), University of Western São Paulo (UNOESTE), Presidente Prudente/São Paulo, Brazil

Cora Abigail Coutinho
Prosthodontics, Dental Concepts Dental Clinic, Margao, Goa, India

Ivy Fernandes Coutinho
Department of Prosthodontics, Goa Dental College and Hospital, Bambolim, Goa, India

Alberto Carlos Botazzo Delbem
São Paulo State University (Unesp), School of Dentistry, Araçatuba, Department of Preventive and Restorative Dentistry, Araçatuba/São Paulo, Brazil

Dongmei Deng
Academic Center for Dentistry Amsterdam (ACTA), Department of Preventive Dentistry, Amsterdam, The Netherlands

Kathleen Manuela D'Souza
Department of Prosthodontics, Goa Dental College and Hospital, Bambolim, Goa, India

Tarek El-Bialy
The University of Alberta, Edmonton, AB, Canada

Rob Exterkate
Academic Center for Dentistry Amsterdam (ACTA), Department of Preventive Dentistry, 1081 LA, Amsterdam, The Netherlands

Beatriz Díaz Fabregat
São Paulo State University (Unesp), School of Dentistry, Araçatuba, Department of Preventive and Restorative Dentistry, Araçatuba/São Paulo, Brazil

Amir I O Ibrahim
Faculty of Dentistry, Department of Restorative Dentistry, Nile University, Khartoum, Sudan

Arundeep Kaur
Department of Periodontics, Maulana Azad Institute of Dental Sciences, B.S. ZafarMarg, New Delhi, India

Monalisa Mishra
Neural Developmental Biology Lab, Department of Life Science, NIT Rourkela, Rourkela, Odisha, India

Douglas Roberto Monteiro
São Paulo State University (Unesp), School of Dentistry, Araçatuba, Department of Preventive and Restorative Dentistry, Araçatuba/São Paulo, Brazil
Graduate Program in Dentistry (GPD—Master's Degree), University of Western São Paulo (UNOESTE), Presidente Prudente/São Paulo, Brazil

Desigar Moodley
Smile Center, Wolfurt, Austria

Neeharika Mortha
GITAM Dental College and Hospital, Oral Pathology and Microbiology, Visakhapatnam, India

Eman Osman
Chemical Metrology Division, National Institute of Standards, Egypt

Chitta Ranjan Patra
Department of Applied Biology, CSIR-Indian Institute of Chemical Technology (CSIR-IICT), Uppal Road, Tarnaka, Hyderabad, Telangana, India

Juliano Pelim Pessan
São Paulo State University (Unesp), School of Dentistry, Araçatuba, Department of Preventive and Restorative Dentistry, Araçatuba/São Paulo, Brazil

Anju Rana
Senior Resident, Maulana Azad Institute of Dental Sciences, B.S. Zafar Marg, New Delhi, India

Banashree Roy
Department of Microbiology, School of Life Science and Biotechnology, Adamas University, Kolkata, India

Caio Sampaio
São Paulo State University (Unesp), School of Dentistry, Araçatuba, Department of Preventive and Restorative Dentistry, Araçatuba/São Paulo, Brazil

Divya Uppala
Department of Oral Pathology and Microbiology, GITAM Dental College and Hospital, Visakhapatnam, India

Muhammad Sohail Zafar
Department of Restorative Dentistry, College of Dentistry, Taibah University, Al-Madina Al-Munawara, Saudi Arabia

Igor Renan Zen
São Paulo State University (Unesp), School of Dentistry, Araçatuba, Department of Preventive and Restorative Dentistry, Araçatuba/São Paulo, Brazil

Chapter 1

Use of phosphate-based nanoparticles to enhance the effects of fluoride against dental caries and erosion

Juliano Pelim Pessan, Caio Sampaio, Igor Zen, Dongmei Deng, Rob Exterkate, Alberto Carlos Botazzo Delbem and Douglas Roberto Monteiro

Fluoride therapy is considered one of the main strategies for dental caries control and is responsible for the substantial decrease in the prevalence of this disease worldwide. Nonetheless, despite the large body of evidence on the effects of fluoride on caries dynamics, concerns on its potential side-effects, the limited action against dental erosion, and the polarization of dental caries, have stimulated the study of alternative strategies to enhance the effects of conventional fluoridated products on dental caries and erosion. In this sense, the association of fluoride and polyphosphate salts has shown to be effective in enhancing enamel remineralization and reducing enamel demineralization and erosive wear, both in professionally-applied and self-application products; such effects have been recently reported to be enhanced by the use of nanoparticles. In addition to polyphosphates, nanoparticles of hydroxyapatite, silver, calcium fluoride, calcium sodium phosphosilicate, and bioactive glass, have demonstrated promising effects on variables related to dental caries and dental erosion. This chapter will present the main data on the association of fluoride and different nano-sized compounds, from studies with different levels of evidence, assessing the effects on enamel de- and re-mineralization, erosive tooth wear, and on dental biofilm. Insights on the mechanisms of action and future steps for research will also be addressed.

1.1 Introduction

The use of fluoride (F), in several vehicles and different modes of application, is regarded as one of the main strategies for dental caries control worldwide. Its main mechanism of action is related to its presence in the oral environment (e.g., biofilm, saliva, mucosa, tooth surfaces), even at low concentrations, where it is able reduce

doi:10.1088/978-0-7503-3671-0ch1

demineralization (by protecting and/or incorporating into the apatite crystallites) and to catalyze remineralization [1, 2]. Furthermore, the application of high fluoride concentrations promotes the formation of calcium fluoride (CaF_2) globules on tooth surfaces, which are dissolved under acidic conditions, allowing the release of calcium (Ca) and F in the oral environment [2–4].

Robust evidence attests that F varnishes, gels, mouthwashes and toothpastes promote significant effects on the caries dynamic [5–8]. Nonetheless, the marked decrease in caries incidence and prevalence in most industrialized countries owing to the widespread availability of F, has been accompanied by increasing concerns on its side effects, such as acute or chronic (i.e., enamel fluorosis) toxicity [9, 10]. Furthermore, the polarization of this disease in low-income countries [11], as well as the lack of substantial action on dental erosion [12], has encouraged the search for novel strategies aiming to enhance the preventive effect of F-products, aiming to maximize the therapeutic effects, while minimizing unwanted side-effects.

Among the most studied compounds, cyclophosphates salts have been extensively assessed over recent years. It has been extensively documented that the addition of sodium trimetaphosphate (TMP) or sodium hexametaphosphate (HMP) in F-containing products substantially improves their protective effects on dental caries dynamics, under *in vitro*, *in situ*, and *in vivo* conditions [13–21]. Also, significant effects have been reported against dental erosion [22–29]. Taking into account the promising results described above, recent studies have further assessed the performance of these compounds when applied in the nano-size form, with more pronounced effects of nanoparticles over micrometric ones.

Special attention has also been given to silver nanoparticles, which have been shown to act synergistically with fluoride on microorganisms related to dental caries [30–33], besides demonstrating promising protective effects on enamel de- and re-mineralization processes [33]. Another agent comprehensively studied is nano-hydroxyapatite (nano-HAp). In spite of being often assessed alone, the incorporation of nano-HAp to F-vehicles has demonstrated promising beneficial effects on parameters related to dental caries, dental erosion and dentin hypersensitivity, due to its similar composition to the hard dental tissues, and to the increased surface area of nanoparticles, which may facilitate the availability of calcium phosphate ions to enamel [34, 35].

In addition, nano-sized CaF_2 has presented promising effects regarding parameters related to biofilms, mechanical behavior, and F release of restorative materials such as nanocomposites [36, 37] and resin-modified glass ionomer (RMGI) [38], resulting in the improvement of the preventive properties without compromising the mechanical capacity of the materials. Bioactive glass (nBAG) is another nanoparticle that has gained interest in recent years. It is often used on glass ionomer cement (GIC) and RMGI [39], as it is able to maintain a high alkaline pH that leads to a supersaturated ion environment, resulting in antibacterial effects [40]. Also, nBAG acts directly on the enamel and dentine remineralization, acting by dentine tubules occlusion and promoting rapid apatite formation [39–41].

Considering the context above, this chapter addresses the main findings on the association of F and nano-sized cyclophosphates (TMP and HMP), nano-HAp, silver nanoparticles, nano-CaF_2, and nano bioglass, from studies with *in vitro*, *in situ* and *in vivo* protocols assessing the effects on enamel de-/re-mineralization, erosive wear, and on dental biofilms. Insights on the mechanism of action, future steps for research and the current level of evidence for each strategy will also be addressed. It is noteworthy that some of the technologies above have been also studied alone (i.e., without fluoride), but will not be addressed in the present chapter due to its scope.

1.2 Nano-sized sodium trimetaphosphate

Sodium trimetaphosphate (TMP) is an inorganic phosphate widely employed in the food and cosmetic industry for several decades. This polyphosphate has been shown to increase the effects of fluoridated products on de- and re-mineralizaion, tooth erosion, dentine tubule obliteration, and biofilm formation [20, 22, 23, 42]. When co-administered with fluoride in dentifrices, gels, varnishes or mouthwashes, the resulting therapeutic/protective effects are similar to or greater than those attained by using twice as much fluoride [22–24, 26, 27], which is highly desirable from a clinical standpoint. Regarding biofilm formation, TMP alone is known to promote a reduction in the metabolism and components of the cellular matrix, in addition to forming a less compact biofilm, the latter favoring acid neutralization by saliva through the biofilm's layers, thus preventing tooth demineralization and/or enhancing remineralization [20]. The synergism between fluoride and TMP observed in laboratory studies was confirmed in a randomized clinical trial, given that children who brushed with a toothpaste containing 500 ppm F containing TMP developed significantly fewer caries lesions than those who used a conventional (1100 ppm F) toothpaste [31].

In light of the promising results of conventional TMP (microparticulate) and considering the advantages of nanoparticles over micrometric ones, intensive research has focused on studies with nano-sized TMP (nTMP) over recent years. An *in vitro* study assessed the effects of fluoridated toothpastes (1100 ppm F) containing micrometric TMP or nTMP on enamel remineralization, showing that the enamel treated with nTMP-containing toothpaste became 62% harder compared with 1100 ppm F, and 32% harder than its counterpart containing micrometric TMP [42]. This study also showed that the nTMP-containing toothpaste was more effective in increasing enamel uptake by ~100% and ~150% compared with 1100 ppm F and TMP toothpastes, respectively. Similar effects have been reported on the protection of enamel against demineralization and erosive tooth wear. The use of a nTMP-containing toothpaste was 44% more effective in reducing demineralization and promoted enamel F uptake 73% higher than its counterpart with micrometric TMP [43]. In addition, the toothpaste supplemented with nTMP was 30% more effective against erosive tooth wear compared with micrometric TMP, in an *in vitro/ex vivo* protocol. Interestingly, the effects of the nTMP toothpaste were similar to those attained by the use of a 5000 ppm F toothpaste, reinforcing its potential use in erosion-prone patients, including children and adolescents. The main studies assessing the effects of nano-sized sodium trimetaphosphate on variables related to dental caries, erosion and biofilms are summarized in table 1.1.

Table 1.1. Summary of the main studies assessing the effects of nano-sized sodium trimetaphosphate related to different variables assessed

Author (year)	Technology	Protocol	Vehicle	Main findings
Silva *et al* (2019) [101]	nTMP	*In vitro*	GIC	1.25% CHX and 14% nTMP added to a GIC increased its antimicrobial/antibiofilm action and its ability to reduce enamel demineralization, with minimal effects on the mechanical properties of RMGIC.
Emerenciano *et al* (2018) [102]	nTMP	*In situ*	Toothpaste	1100 ppm NaF/nTMP promoted a greater protective effect against enamel demineralization and significantly affected the composition of biofilm formed *in situ* when compared to 1100 ppm NaF toothpaste.
Danelon *et al* (2018) [28]	nTMP	*In vitro*	Toothpaste	The addition of 3% nTMP to 1100 ppm NaF toothpastes significantly increased the protective effect against enamel erosion compared with its counterparts with mTMP or without TMP. This effect was not influenced by the presence of acquired enamel pellicle and saliva.
Danelon *et al* (2017) [43]	nTMP	*In vitro*	Toothpaste	Fluoride toothpaste containing nTMP at 3% significantly decreased enamel demineralization compared to its counterparts without TMP or supplemented with mTMP.
Souza *et al* (2016) [103]	nTMP	*In situ*	Toothpaste	The protective effect of 250 ppm NaF-nTMP dentifrice was similar to a conventional dentifrice (1100 ppm F) for most of the variables studied, having a more pronounced effect on the subsurface lesion.
Danelon *et al* (2015) [42]	nTMP	*In situ*	Toothpaste	Addition of 3% nTMP to a conventional toothpaste (1100 ppm F) promoted an additional remineralizing effect of artificial caries lesions compared with counterparts with mTMP or without TMP.

TMP: sodium trimetaphosphate; **nTMP:** nano-sized sodium trimetaphosphate; **mTMP:** micro-sized sodium trimetaphosphate; **NaF:** sodium fluoride; **RMGIC:** resin-modified glass ionomer cement; **CHX:** chlorhexidine.

These results taken together support the hypothesis that the effects of nTMP are similar to those reported for micrometric particles, but potentiated by the properties of nano-sized materials, such as their high ratio of surface area to volume, as well as a high percentage of atoms on the surface compared to larger particles, which makes them more reactive compared to micrometric particles [43]. In brief, TMP has been suggested to retain charged ions of CaF^+ and Ca^{++} by replacing Na^+ from a cyclic structure [23]. At acidic pH, these linkages are broken, releasing Ca^{++} and CaF^+, which can further take part in a series of events that ultimately would lead to the formation of neutrally charged species ($CaHPO_4^\circ$ and HF_0) that have a higher diffusion coefficient into the enamel than charged species [44]. Furthermore, the association of F and TMP has been shown to enhance the formation of CaF_2 deposits on the exposed tooth surfaces [22, 28, 42–45], which are known to be key players on the dynamics of both caries and erosion [4, 46–48].

1.3 Nano-sized sodium hexametaphosphate

HMP is an inorganic cyclophosphate that has demonstrated extensive effects on enamel de- and re-mineralization processes, on dental erosion, as well as on oral biofilms. Due to its great affinity to metallic ions (Mg^{2+}, Ca^{2+}, K^+, Al^+, Fe^{3+}), HMP interacts with the dental surface, which substantially reduces mineral solubility [14, 15, 18]. Due to these features, when co-administered with F, this phosphate enhances the effects of dentifrices or gels on enamel de- and re-mineralization processes, in comparison to conventional products not supplemented with HMP [13–15, 17]. A synergistic effect between F and HMP administered in the form of a gel was also observed on dental erosion in an *in situ/ex vivo* model. It was shown that a low-F gel (4500 ppm F) promoted significantly lower enamel wear and mineral loss compared with a conventional formulation containing twice as much F (9000 ppm F) [49]. In addition to HMP's effects on the dental hard tissues, promising results have been demonstrated on biofilms *in vitro*. HMP, associated or not with other actives such as silver or F, was shown to have antimicrobial potential, besides interfering with the composition of the biofilm's extracellular matrix [50–52]. Although the mechanism by which HMP acts on biofilms is still not completely clear, it is possible that the great affinity that this phosphate presents to metallic ions (forming ionic complexes) helps HMP to bind to Ca^{2+} and $Mg^{2+}$from the bacterial cell walls, thus increasing cell permeability [53, 54].

Similarly to TMP, the great effect of HMP on the above-mentioned caries and erosion parameters has also encouraged the evaluation of this phosphate as nanoparticles (nHMP), given that reducing the size of the particles could enhance its effects in comparison to the commercial version of the product (i.e., microparticles) [55]. In that sense, although no evidence is available on the effects of nHMP over conventional HMP on dental erosion, *in vitro* and *in situ* studies attest to the additional benefits of nHMP over microparticles on enamel de- and re-mineralization processes, and on biofilms. The most studied vehicle for coadministration of F and nHMP is dentifrice, which has been shown to promote superior protective effects against enamel demineralization [55, 56], besides promoting an additional remineralizing effect of artificial caries lesions [57] compared with a F-toothpaste containing micrometric HMP. In addition to the effects on tooth enamel, the association between F and nHMP was shown to significantly reduce the cariogenicity of the dental biofilm formed *in situ*, as it significantly increased F and Ca concentrations and significantly reduced the amount of insoluble extracellular

Table 1.2. Summary of the main studies assessing the effects of nano-sized sodium hexametaphosphate related to different variables assessed

Author (year)	Technology	Protocol	Vehicle	Main findings
Danelon *et al* (2019) [57]	nHMP	*In situ*	Toothpaste	The addition of nHMP to a F-toothpaste was able to promote an additional remineralizing effect of artificial caries lesions compared with its counterparts with mHMP or without HMP.
Garcia *et al* (2018) [56]	nHMP	*In situ*	Toothpaste	F-dentifrice supplemented with nHMP demonstrated a greater protective effect against enamel demineralization and on the composition of biofilm when compared with a conventional F-toothpaste or a F-toothpaste containing mHMP.
Hosida *et al* (2018) [104]	nHMP	*In vitro*	GIC	HMP incorporation, especially in the nano-sized form, in restorative GIC increased antimicrobial activity and F release, and decreased enamel demineralization, despite reduced its physico-mechanical properties, in comparison to a conventional GIC without nHM.
Dalpasquale *et al* (2017) [55]	nHMP	*In vitro*	Toothpaste	The addition of nHMP to a F-toothpaste enhanced its protective effects against enamel demineralization when compared to its counterpart containing mHMP or a conventional F-toothpaste.

HMP: sodium hexametaphosphate; **nHMP:** nano-sized sodium hexametaphosphate; **mHMP:** micro-sized sodium hexametaphosphate; **GIC:** glass ionomer cement.

polysaccharides in comparison to its counterparts containing micrometric HMP or containing F alone [57]. Furthermore, the addition of nHMP to a GIC was shown to improve F release and its antimicrobial activity, in spite of reducing the physico-mechanical properties of the product [58]. Table 1.2 summarizes the main studies assessing the effects of nano-sized sodium hexametaphosphate on tooth structures and on biofilms.

1.4 Nano-hydroxyapatite

Hydroxyapatite (HAp) is one of the most studied compounds for the control of dental caries, erosion and/or hypersensitivity, since it is a mineral essentially composed by calcium and phosphate [$Ca_{10}(PO_4)_6(OH)_2$] found in dental enamel, dentine and bone [59]. Knowledge on the composition of the hard dental structures can help to better understand how nano-HAp might be an interesting alternative for the enhancement of F-vehicles.

Briefly, permanent dental enamel is an acellular and avascular structure composed mainly of inorganic material, essentially hydroxyapatite (accounting for ~85% of the composition), arranged as apatite crystals, which then form the enamel prisms. The remaining volume—disposed among the crystals—is composed of water (~12%) and organic material (~3%, mainly collagen) [2, 60, 61]. As for dentine, although its structure is also composed of mineral content in the form of hydroxyapatite, the volume is lower. Dentine tissue is constituted of ~47% of minerals, ~33% organic material and ~20% water, which makes dentine more prone to acid dissolution in comparison to enamel. Unlike dental enamel, dentine is a vascular and cellular tissue, which means that it is susceptible to response to outer stimuli [2, 62].

In turn, nano-HAp consists of a bioactive and biocompatible material, which acts by facilitating the availability of calcium phosphate ions to enamel, due to its similar composition to the hard dental tissues and to the increased surface area of the nanoparticles compared with micrometric ones [34, 35]. Although most of the studies have proposed the incorporation of nano-HAp in oral care products (e.g., toothpastes, gels and varnishes) as an alternative to F, this compound has also been studied in association with F, in products for the control of dental caries, erosion and dentine hypersensitivity. In brief, the effects of nano-HAp on variables related to dental caries have been assessed in a larger number of studies compared to those for dental erosion or dentin hypersensitivity. The data, however, is somehow conflicting. Although some studies reported that nano-HAp enhanced the protective and therapeutic effects of F-toothpastes on enamel de- and re-mineralization processes [63], most of them either did not include a suitable positive control group (i.e., a conventional F-dentifrice) or adopted a protocol insufficient to attest such promising trends, as detailed below.

The main studies assessing the effects of nano-HAp on several variables are described in table 1.3. A toothpaste containing nano-HAp combined with F led to a lower demineralization depth in comparison to a placebo toothpaste [35]; however, this study did not include a positive control group containing the active compounds alone, which hinders any conclusion on the actual effects of this association. Another work also observed that a toothpaste containing 7% nano-HAp and 1000 ppm NaF promoted a higher remineralizing effect compared with a F-only toothpaste. Nonetheless, the only parameter analyzed was surface hardness, and the experimental groups had to be adjusted prior to statistical analysis due to protocol limitations [64]. Besides F-toothpastes, the addition of nano-HAp to a fluoridated varnish was also shown to promote a higher remineralizing effect of enamel surface in comparison to a conventional F-varnish or to a paste containing casein phosphopeptide-amorphous calcium phosphate (CPP-ACP) [64]. Despite the interesting effects reported above, some methodological issues were identified, which suggest that further *in vitro* studies are still demanded prior their study under clinical conditions.

In addition to the above-mentioned findings, other studies have reported less promising trends on the effects of nano-HAp combined with F on caries-related variables. An experimental toothpaste containing nano-HAp, regardless of the

Table 1.3. Summary of the main studies assessing the effects of nano-hydroxyapatite related to different variables assessed

Author (year)	Technology	Protocol	Vehicle	Main findings
Ionescu *et al* (2020) [105]	Nano-HAp	*In vitro*	Toothpaste	F, Mg, Sr-carbonate substituted n-HAp toothpaste reduced the microbial colonization on resin-based composite.
Leal *et al* (2020) [66]	Nano-HAp	*In vitro*	Toothpaste	Nano-HAp did not improve the effect of dentifrices containing 5000 μg F/g on root dentin demineralization.
Demito *et al* (2019) [35]	Nano-HAp	*In situ*	Undefined product	Groups treated with nano-HAp combined with 9000 ppm NaF presented lower demineralization depth in relation to the negative control group.
Kooshki *et al* (2019) [64]	Nano-HAp	*In vitro*	Varnish	F-varnish containing nano-HAp promoted a higher remineralizing effect of enamel surface in comparison to a conventional F-varnish (not supplemented with nano-HAp) or to a paste containing casein phosphopeptide-amorphous calcium phosphate (CPP-ACP).
Soares *et al* (2018) [71]	Nano-HAp	*In vitro*	Gel	1.23% F gel containing nano-HAp modified the composition and morphology of the enamel surface.
Yu *et al* (2017) [78]	Nano-FHAp and nano-HAp	*In vitro*	Powder	nano-FHAp formulation showed higher plugging rate and penetrating depth into the tubules in comparison to F or nano-HAp.
Ebadifar *et al* (2017) [63]	Nano-HAp	*In vitro*	Toothpaste	The F-toothpaste containing nano-HAp was more effective than the F-toothpaste without nano-HAp for the purpose of remineralization.
Souza *et al* (2015) [68]	Nano-HAp	*In situ*	Toothpaste	Nano-HAp/F toothpaste was the only product able to significantly reduce dentine demineralization in comparison to toothpastes containing CPP-ACP + NaF, NaF alone or a placebo formulation. The test toothpaste also enhanced enamel remineralization compared to the placebo.

Taha *et al* (2015) [77]	Nano-FHAp	*In vitro*	Paste	Nano-FHAp pastes may offer immediate short-term relief of dentin hypersensitivity because of their ability to occlude tubules and to adhere to wet dentin surfaces; its long-term effect is due to ion release.
Comar *et al* (2013) [65]	Nano-HAp	*In vitro*	Paste	The experimental nano-HAp pastes, regardless of the addition of F, were unable to reduce dental demineralization.
Wang *et al* (2012) [70]	Nano-HAp	*In vitro*	Paste	nano-FAp paste with neutral pH value (7.5) for enamel erosion repair significantly enhanced the enamel surface hardness, reduced enamel mass loss and increased surface roughness.
Lin *et al* (2011) [69]	Nano-FHAp and nano-FAp	*In vitro*	GIC	The F release properties of the GIC was improved by incorporating nano-FAp or nano-FHAp, simultaneously maintaining a clinically sufficient bond strength when nano-FAp was added.
Jeong *et al* (2007) [67]	Nano-Carbonated apatite	*In vitro*	Toothpaste	The dentifrice containing 5% nano carbonated apatites and 25% silica was the most effective in remineralizing early caries lesion.

Nano-HAp: Nano-hydroxyapatite; **F:** fluoride; **Mg:** magnesium; **Sr:** strontium; **Nano-FHAp:** nano-fluorhydroxyapatite; **Nano-FAp:** nano-fluorapatite; **CPP-ACP:** amorphous casein-phosphate stabilized by casein phosphopeptides; **GIC:** glass ionomer cement.

addition of F, was unable to reduce dental demineralization compared with a placebo toothpaste [65]. Moreover, despite a toothpaste containing nano-HAp promoted protective effects against dentin demineralization similar to those observed for a 1100 ppm F toothpaste, no additional benefit was observed when nano-HAp was coadministered with F in formulations containing 1100 or 5000 μg F g^{-1} [66]. Furthermore, a F-dentifrice containing 30% nano-HAp decreased the enamel surface hardness, while the incorporation of 5 or 15% nano-HAp into a F-dentifrice did not significantlly promote an additional effect on the enamel surface hardness when compared to their counterpart without nano-HAp [67].

On the other hand, a well-designed *in situ* study compared the effects of a toothpaste containing 10% HAp, 0.2% NaF, nano-HAp/F with toothpastes containing amorphous casein-phosphate and NaF or NaF only and observed discrete differences among the treatments on dental de- and re-mineralization processes, although the nano-HAp containing toothpaste was the only one able to reduce dentine demineralization (ΔZ) and to improve enamel remineralization (ΔΔZ) compared to the negative control group (placebo) [68]. Furthermore, nano-HAp or nano-fluorhydroxyapatite (nano-FHAp), incorporated to a GIC led to a higher F release, without compromising its mechanical properties, in comparison to a conventional GIC not supplemented with nano-HAp or nano-FHAp [69].

Unlike for dental caries, little is available in the literature regarding the effect of nano-HAp associated with F on dental erosion. It was verified *in vitro* that a neutral pH (7.5) paste containing nano-HAp and F significantly enhanced the enamel surface hardness, avoided the enamel mass loss and increased surface roughness [70], which could be interesting for erosion repair. In addition, an acidulated phosphate F-gel (1.23% F) containing nano-HAp modified the composition and morphology of the enamel surface, in comparison to its counterpart without nano-HAp [71]. Due to the scarce data available on the effects of nano-HAp combined with F on dental erosion, this field still needs to be further explored in order to verify the actual effects of this combination for the control and/or treatment of dental erosion in humans.

Despite the robust body of evidence on the effect of nano-HAp alone on dentin hypersensitivity [72–77], the literature is scarce on the effect of nano-HAp associated with F for that purpose. It has been demonstrated *in vitro* that pastes containing nano-fluorhydroxyapatite (n-FHAp) may offer immediate short-term relief of dentin hypersensitivity because of their ability to occlude the tubules and to adhere to wet dentin surfaces [80]. Another study found that a new toothpaste containing nano-FHAp showed higher plugging rate and penetrating depth into the tubules in comparison to another product containing only F or nano-HAp [78].

Given the context above, it is safe to consider that the combination of nano-HAp and F is still an open topic for all the variables described in this section (i.e., dental caries, erosion and dentin hypersensitivity) due to the lack of more robust evidence on the action of these two compounds when co-administered. Thus, further studies should be performed to clarify the above-mentioned trends, preferably associating *in vitro* with *in situ* protocols, which are useful in providing different insights into mechanisms of action.

1.5 Silver nanoparticles

Fluoridated silver nanoparticles (or silver nanoparticles fluoride—SNF) have been proposed as an alternative to silver diamine fluoride (SDF), and are reported to have antimicrobial activity, along with the advantage of not staining dental and mucosa tissues [30, 79]. *In vitro* studies attest such effects against *Streptococcus mutans* [83, 84], and *Lactobacillus* [80], showing significant reductions on acidogenicity and adherence of these microorganisms [32]. The antimicrobial mechanism of SNF is related to their ability to penetrate the bacteria cell wall and cause denaturation of the plasma membrane, in addition to interruption on DNA replication, leading to damage to the bacterial membrane [81, 82]. SNF can also release Ag ions to the cell's interior and produce reactive oxygen species, disturbing protein synthesis and leading to cell lysis [83]. Due to their smaller particle size, this compound acts against both planktonic and colonized bacteria [84].

SNF also has the ability to prevent demineralization of sound enamel and to reduce the incidence of enamel caries lesions, with effects on both mineral loss and lesion depth, assessed by transverse microradiography [85]. The effects of solutions containing fluoride and SNF particles were shown to be dose-dependent, which were also able to promote increases in surface mineral density owing to mineral deposition [85]. Also, an *in vitro* study demonstrated that treatment of dentin subsurface lesions with a solution containing 5% NaF and 4000 ppm Ag nanoparticles promoted a similar remineralizing effect compared with 5% NaF alone, without promoting significant color changes compared with the respective controls (placebo or 5% NaF) [86]. Similar findings were observed in another *in vitro* study assessing enamel remineralization using optical coherence tomography, demonstrating that silver nanoparticles do not interfere with fluoride owing to the ionic stability of the latter, which do not react when associated with other biomaterials [87].

Another effect is related to the ability to not stain tooth and mucosa tissues [33, 86]. This is a clear advantage over SDF, as tooth staining has been widely documented as a side effect when applied to carious lesions, leading to a tooth black stain and minor injuries to the oral mucosa tissue [88]. An *in vivo* study showed the use of SNF had a success rate of 66.7% when compared to the control group (water) in arresting tooth decay and staining dental tissue [89]; this success rate is similar to that attained by SDF use (66.9%), as reported in a previous study [90]. Interestingly, a randomized clinical trial evaluated once a year the application of 5% SNF and 38% SDF in primary carious molar teeth, showing that both groups inhibited dentine caries progression and presented equivalent effects; SNF had a success rate of 77% and SDF 71%, the latter not leading to tooth staining [89]. As aforementioned about the staining, this black stain in carious tissue occurs due to the oxidation process of the ionic silver chloride contained in SDF formulation [86, 91], and with nanoparticle size reduction, leads to the non-teeth-staining due to the contact surface increase [92]. The main studies assessing the effects of SNF are summarized in table 1.4.

1.6 Nano-CaF_2

Unlike the above-mentioned agents, which have been added mostly to F-containing toothpastes or solutions, nano-CaF_2 has been extensively studied combined with restorative materials or with products for orthodontic cementation, such as composites and RMGIC, leading to promising effects on parameters related to biofilms and mechanical properties, besides enhancing their F release. The main results from studies assessing the effects of nano-CaF_2 are displayed in table 1.5.

It was shown *in vitro* that experimental composites containing nano-CaF_2 produced high F release at low filler levels. Furthermore, these composites presented higher flexural strength and elastic modulus in comparison to a control composite without nano-CaF_2 or a RMGIC [93]. Another study verified that a novel CaF_2 nanocomposite combined with chlorhexidine (CHX) reduced *S. mutans* biofilm formation, acid production, CFU-counting and metabolic activity, in comparison to commercial composites (containing or not F) or to an RMGIC. Although such

Table 1.4. Summary of the main studies assessing the effects of fluoride and silver nanoparticles related to different variables assessed

Author (year)	Technology	Protocol	Vehicle	Main findings
Aldhaian *et al* (2021) [85]	SNF	*In vitro*	Solution	Fluoride and silver nanoparticles reduced mineral loss and lesion depth in a dose-response manner.
Yin *et al* (2020) [86]	SNF	*In vitro*	Solution	NaF solution with AgNPs remineralized artificial dentine caries without significant tooth stain compared to its counterparts.
Yin *et al* (2020) [33]	SNF	*In vitro*	Solution	NaF solution with AgNPs showed antibacterial activity against SM and also no tooth staining effect .
Zhao *et al* (2020) [91]	SNF	*In vitro*	Solution	NaF + PEG-AgNPs and SDF had a similar remineralizing effect in artificial dentine caries. Moreover, NaF with PEG-AgNPs does not stain dentine caries.
Tirupathi *et al* (2019) [79]	SNF	RCT	Varnish and solution	Annual application of SNF has the same clinical efficacy as SDF in preventing the progression of dentinal caries and does not stain tooth.
Costa e Silva *et al* (2019) [87]	SNF	*In vitro*	Solution	SNF has an efficient effect in remineralizing dental enamel.
Teixeira *et al* (2018) [32]	SNF	*In vitro*	Toothpaste	SNF formulation had a better antibacterial effect compared to NaF dentifrices and similar action on the demineralization of enamel indicating their potential effectiveness to prevent caries.
Freire *et al* (2017) [31]	SNF	RCT	Solution	SNF decreased number of CFUs of SM, proving to be an effective dental biofilm inhibitor.
Santos Jr *et al* (2014) [89]	SNF	RCT	Solution	The use of SNF decreased risk of caries at seven days (81%) and five (72.2) and 12 months (66.7) when compared to control group (H_2O).
Targino *et al* (2014) [30]	SNF	*In vitro*	Solution	SNF proved to be antimicrobial agent, with low toxicity to living cells, and potential advantage of not staining teeth black.

RCT: Randomized clinical trial; **PEG-AgNPs:** Polyethylene glycol-coated silver nanoparticles; **SDF:** Sodium diamine fluoride; **SNF:** silver nanoparticles fluoride; **NaF:** Sodium Fluoride; **CFU:** colony-forming units; **SM:** *Streptococcus mutans.*

reductions were more evident for the CHX-containing CaF_2 composite, a substantial trend of decrease in metabolism and acid production was also observed for a nano-CaF_2 composite without CHX, in comparison to a F-free composite [36]. It was also verified from another nanocomposite containing CaF_2 that F release values

Table 1.5. Summary of the main studies assessing the effects of nano-sized calcium fluoride related to different variables assessed

Author (year)	Technology	Protocol	Vehicle	Main findings
Fei *et al* (2020) [95]	Nano-CaF_2	*In vitro*	Pit and fissure sealant	Experimental sealant containing nano-CaF_2 and dimethylaminohexadecyl methacrylate had increased hardness and increased F ion release compared with the commercial control. The pH in biofilm medium in contact with the experimental sealant was higher (pH 6.8) than that of commercial sealant (pH 4.7). No changes in paste flow and enamel bond strength were observed for the experimental sealant compared with the control.
Shin *et al* (2019) [96]	Nano-CaF_2	*In vitro* and *in vivo* (animal)	Nano composite hydrogel	F released from nano-CaF_2 in the composite hydrogel not only functioned as a stimulating agent for cell proliferation and migration, but also inhibited the growth of bacteria *in vitro*.
Yi *et al* (2019) [38]	Nano-CaF_2	*In vitro*	Orthodontic cement	The novel F ion-rechargeable orthodontic cement containing nano-CaF_2 presented acceptable enamel shear bond strength, good biocompatibility, and sustained F ion recharge and re-release (1.8-fold higher than a commercial formulation without nano-CaF_2).
Yi *et al* (2019) [94]	Nano-CaF_2	*In vitro*	Resin-modified glass ionomer cement (RMGIC)	The addition of Nano-CaF_2 and dimethylaminohexadecyl methacrylate to a RMGIC led to stronger antibacterial and remineralization capabilities, and greater enamel hardness than the commercial control (GC Ortho LC), without compromising the orthodontic bracket-enamel shear bond strength and biocompatibility.
Cheng *et al* (2012) [36]	Nano-CaF_2	*In vitro*	Nanocomposite	The novel CaF_2 nanocomposite with chlorhexidine was shown to reduce biofilm formation, acid production, CFU and metabolic activity of biofilms.
Xu *et al* (2010) [37]	Nano-CaF_2	*In vitro*	Nanocomposite	CaF_2 nanoparticles produced high F release at low filler levels, thereby making room in resin for reinforcement glass.

Nano-CaF_2: nano-sized calcium fluoride; **F:** fluoride; **RMGIC:** resin-modified glass ionomer cement; **CFU:** colony-forming units.

exceeded releases of traditional and RMGIC materials. In addition, the experimental CaF_2-nanocomposite presented relatively high strength and sustained release of F ions in comparison to traditional composites or RMGIC [93].

Promising data have also been reported for the incorporation of nano-CaF_2 to RMGIC. The addition of nano-CaF_2 to a conventional RMGIC (GC Ortho LC, Fuji, Aichi-ken, Japan) reduced several parameters related to biofilms (i.e., metabolic activity, production of extracellular polysaccharides and lactic acid, total microbial counts, and the counts of total streptococci and *S. mutans*), in comparison to a conventional RMGIC or to an adhesive system for orthodontic cementation, especially when combined with dimethylaminohexadecyl methacrylate (DMAHDM). Furthermore, the RMGIC containing nano-CaF_2 promoted an increase in enamel hardness and a decrease in lesion depth of artificial enamel lesions [94]. In addition, another nano-CaF_2-containing orthodontic cement presented higher F release in comparison to a conventional RMGIC (GC Ortho LC, Fuji, Aichi-ken, Japan) [38].

The incorporation of nano-CaF_2 in a pit and fissure sealant was also shown to increase F release and hardness compared with a counterpart without nano-CaF_2 (Helioseal F, Ivoclar, Mississauga, ON, Canada) [95]. Moreover, a novel alginate nanocomposite hydrogel containing nano-CaF_2 showed that F released from the nano-CaF_2 in the composite hydrogel functioned as a stimulating agent for cell proliferation and migration, in addition to inhibiting bacteria growth *in vitro*. This combination has also been shown to promote a decrease in the bacterial load, in comparison to its counterpart without nano-CaF_2 [96].

1.7 Bioactive glass

In addition to the above-mentioned compounds, which have been studied more extensively in vehicles and modes of administration, other lesser studied compounds deserve attention, whose results are summarized in table 1.6.

Bioactive glass or bioglass (BAG) is another material with marked remineralizing effects on dental structures, and has been used as an strategy to enhance the effects of GICs [39, 40]. *In vitro* studies have shown that nanosized BAG particles create a supersaturate ionic environment on dentine tubules (leading to dentine tubule occlusion), and a fast apatite formation, which prevents dentine hypersensitivity compared to micrometric particles [40, 41]. Moreover, a study assessing enamel demineralization and antibacterial effects of orthodontic bonding resins showed that the addition of fluoride to BAG resulted in higher antibacterial activity in a concentration-dependent trend, enhanced protective effects against demineralization, and the ability to prevent white spot lesions in comparison with a commercially orthodontic bonding adhesive [41].

Furthermore, it has been shown that teeth sonochemically coated with magnesium fluoride (MgF_2) nanoparticles significantly inhibited *S. mutans* biofilm formation *in vitro* in comparison to uncoated teeth [97]. In addition, the incorporation of nano-chitosan (nano-CH) in a conventional GIC led to higher compressive strength, wear resistance, as well as F release for 7 days *in vitro*, in comparison to its counterpart without nano-CH [98].

Table 1.6. Summary of the main studies assessing the effects of nanoparticles of bioactive glass, magnesium fluoride, amorphous calcium phosphate and β-tricalcium phosphate, related to different variables assessed

Author (year)	Technology	Protocol	Vehicle	Main findings
Nam *et al* (2019) [41]	BAG	*In vitro*	BAG and OBA	BAG presented a higher antibacterial activity in a concentration-dependent, better anti-demineralization effect, and ability to prevent white spot lesion when compared to its counterpart.
Yli-Urpo *et al* (2004) [39]	BAG	*In vitro*	GIC and RMGIC	RMGIC with BAG had the ability to remineralize dentine *in vitro*.
Eshed *et al* (2013) [97]	Nano-MgF_2	*In vitro*	Artificial teeth coated with MgF_2 nanoparticles	The artificial tooth surface covered with an nano-MgF_2 layer successfully inhibited *S. mutans* biofilm formation by over 60%.
Karimi *et al* (2019) [100]	Nano-ACP	*In vitro*	GIC	The working time of the formulated resin-modified glass ionomer decreased significantly upon increase of nano-ACP content from 5 to 20%. Nano-ACP-incorporated RMGICs showed improved photopolymerization and setting. The addition of nano-ACP to of RMGICs did not affect compressive strength or F release. The biocompatibility of the test RMGICs improved at ~20%. Alkaline phosphatase activity and osteogenic differentiation of mesenchymal stem cells noticeably increased after exposure to ACP-incorporated RMGIC.
Seyedlar *et al* (2019) [106]	Nano-TCP	*In vitro*	Solution	Nano-TCP solution was not more effective than the NaF 0.05 wt% in preventing demineralization.
Iafisco *et al* (2018) [99]	Nano-ACP	*In vitro*	Biomimetic amorphous calcium phosphate doped with F ions	The experimental material showed good ability to partially occlude the tubules of acid-etched dentin and to restore demineralized enamel into its native structure.

(*Continued*)

Table 1.6. (*Continued*)

Author (year)	Technology	Protocol	Vehicle	Main findings
Kumar *et al* (2017) [98]	Nano-CH	*In vitro*	GIC	Nano-CH GIC had significantly higher compressive strength values, flexural strength, wear resistance and higher F release for 7 days, compared with a conventional GIC.

Nano-ACP: nano-sized amorphous casein-phosphate; **Nano-TCP:** nano-sized β-tricalcium phosphate; **Nano-CH:** nano-chitosan; **Nano-MgF$_2$**: nano-sized magnesium fluoride; **NaF:** sodium fluoride; **GIC:** glass ionomer cement; **F:** fluoride; wt%: weight percent; **BAG:** bioactive glass; **RMGIC:** resin modified glass ionomer cement; **OBA:** orthodontic bonding adhesive.

Furthermore, the effects of a nano-biomimetic amorphous calcium phosphate doped with F ions were assessed on dental remineralization and dentin tubules occlusion. This compound was shown to promote partial occlusion of the tubules of acid-etched dentin and to restore demineralized enamel into the original structure, with no significant influence on the chemical-physical features of ACP, apart from the quicker conversion in the crystalline phase when immersed in aqueous solution [99]. Nano-ACP was also evaluated when incorporated to an RMGIC, resulting in enhanced alkaline phosphatase activity, without compromising the material's compressive strength or F release [100]. Such effects noticeably increased osteogenic differentiation of mesenchymal stem cells after exposure to ACP-incorporated RMGIC [100].

1.8 Conclusions

Considering the above-mentioned context on the nanoparticles presented, most of these compounds can be considered as promising alternatives for the enhancement of the F-vehicles on dental caries (effects on dental hard tissues and biofilms), on dental erosion and on dentine hypersensitivity. Despite relevant data on the mechanisms of action of these technologies achieved by *in vitro* protocols, it is important to emphasize that the literature is still scarce on data from *in vivo* studies. Thus, future research should focus on *in situ* and/or *in vivo* methodologies, as they are paramount in the determination of the actual effects of these nanoparticles.

References

[1] Featherstone J D 1999 Prevention and reversal of dental caries: role of low level fluoride *Community Dent. Oral Epidemiol.* **27** 31–40

[2] Buzalaf M A R, Pessan J P, Honorio H M and ten Cate J M 2011 Mechanisms of action of fluoride for caries control *Monogr. Oral Sci.* **22** 97–114

[3] RØlla G 1988 On the role of calcium fluoride in the cariostatic mechanism of fluoride *Acta Odontol. Scand.* **46** 341–5

[4] Ten Cate J M and Buzalaf M A R 2019 Fluoride mode of action: once there was an observant dentist *J. Dent. Res.* **98** 725–30

[5] Marinho V C, Chong L Y, Worthington H V and Walsh T 2016 Fluoride mouthrinses for preventing dental caries in children and adolescents *Cochrane Database Syst. Rev.* **29** CD002284

[6] Marinho V C, Worthington H V, Walsh T and Chong L Y 2015 Fluoride gels for preventing dental caries in children and adolescents *Cochrane Database Syst. Rev.* **15** CD002280

[7] Marinho V C, Worthington H V, Walsh T and Clarkson J E 2013 Fluoride varnishes for preventing dental caries in children and adolescents *Cochrane Database Syst. Rev.* CD002279

[8] Walsh T, Worthington H V, Glenny A M, Marinho V C and Jeroncic A 2019 Fluoride toothpastes of different concentrations for preventing dental caries *Cochrane Database Syst. Rev.* **3** CD007868

[9] Sampaio C, Delbem A C B, Paiva M F, Zen I, Danelon M, Cunha R F and Pessan J P 2020 Amount of dentifrice and fluoride concentration influence salivary fluoride concentrations and fluoride intake by toddlers *Caries Res.* **54** 234–41

[10] Whitford G M 2011 Acute toxicity of ingested fluoride *Monogr. Oral Sci.* **22** 66–80

[11] Frencken J E, Sharma P, Stenhouse L, Green D, Laverty D and Dietrich T 2017 Global epidemiology of dental caries and severe periodontitis—a comprehensive review *J. Clin. Periodontol.* **44** S94–105

[12] Huysmans M C, Young A and Ganss C 2014 The role of fluoride in erosion therapy *Monogr. Oral Sci.* **25** 230–43

[13] Wefel J S, Stanford C M, Ament D K, Hogan M M, Harless J D, Pfarrer A M, Ramsey L L, Leusch M S and Biesbrock A R 2002 *In situ* evaluation of sodium hexametaphosphate-containing dentifrices *Caries Res.* **36** 122–8

[14] da Camara D M, Pessan J P, Francati T M, Santos Souza J A, Danelon M and Delbem A C 2015 Synergistic effect of fluoride and sodium hexametaphosphate in toothpaste on enamel demineralization *in situ J Dent.* **43** 1249–54

[15] da Camara D M, Pessan J P, Francati T M, Souza J A, Danelon M and Delbem A C 2016 Fluoride toothpaste supplemented with sodium hexametaphosphate reduces enamel demineralization *in vitro Clin. Oral Investig.* **20** 1981–5

[16] Wierichs R J, Zelck H, Doerfer C E, Appel P, Paris S, Esteves-Oliveira M and Meyer-Lueckel H 2017 Effects of dentifrices differing in fluoride compounds on artificial enamel caries lesions *in vitro Odontology* **105** 36–45

[17] Gonçalves F M C, Delbem A C B, Pessan J P, Nunes G P, Emerenciano N G, Garcia L S G, Báez Quintero L C, Neves J G and Danelon M 2018 Remineralizing effect of a fluoridated gel containing sodium hexametaphosphate: an *in vitro* study *Arch. Oral Biol.* **90** 40–4

[18] Neves J G, Danelon M, Pessan J P, Figueiredo L R, Camargo E R and Delbem A C B 2018 Surface free energy of enamel treated with sodium hexametaphosphate, calcium and phosphate *Arch. Oral Biol.* **90** 108–112

[19] Freire I R, Pessan J P, Amaral J G, Martinhon C C R, Cunha R F and Delbem A C B 2016 Anticaries effect of low-fluoride dentifrices with phosphates in children: a randomized, controlled trial *J. Dent.* **50** 37–42

[20] Cavazana T P, Hosida T Y, Pessan J P, Sampaio C, Monteiro D R and Delbem A C B 2019 Activity of sodium trimetaphosphate, associated or not with fluoride, on dual-species biofilms *Biofouling* **35** 710–18

[21] Cavazana T P, Pessan J P, Hosida T Y, Sampaio C, Amarante V O Z, Monteiro D R and Delbem A C B 2020 Effects of sodium trimetaphosphate, associated or not with fluoride, on the composition and pH of mixed biofilms, before and after exposure to sucrose *Caries Res.* **54** 358–68

[22] Manarelli M M, Delbem A C, Lima T M, Castilho F C and Pessan J P 2014 *In vitro* remineralizing effect of fluoride varnishes containing sodium trimetaphosphate *Caries Res.* **48** 299–305

[23] Manarelli M M, Vieira A E, Matheus A A, Sassaki K T and Delbem A C 2011 Effect of mouth rinses with fluoride and trimetaphosphate on enamel erosion: an *in vitro* study *Caries Res.* **45** 506–9

[24] Moretto M J, Delbem A C, Manarelli M M, Pessan J P and Martinhon C C 2013 Effect of fluoride varnish supplemented with sodium trimetaphosphate on enamel erosion and abrasion: an *in situ*/*ex vivo* study *J. Dent.* **41** 1302–6

[25] Moretto M J, Magalhães A C, Sassaki K T, Delbem A C and Martinhon C C 2010 Effect of different fluoride concentrations of experimental dentifrices on enamel erosion and abrasion *Caries Res.* **44** 135–40

[26] Pancote L P, Manarelli M M, Danelon M and Delbem A C 2014 Effect of fluoride gels supplemented with sodium trimetaphosphate on enamel erosion and abrasion: *in vitro* study *Arch. Oral Biol.* **59** 336–40

[27] Cruz N V, Pessan J P, Manarelli M M, Souza M D and Delbem A C 2015 *In vitro* effect of low-fluoride toothpastes containing sodium trimetaphosphate on enamel erosion *Arch Oral Biol.* **60** 1231–6

[28] Danelon M, Pessan J P, Prado K M, Ramos J P, Emerenciano N G, Moretto M J, Martinhon C C R and Delbem A C B 2020 Protective effect of fluoride varnish containing trimetaphosphate against dentin erosion and erosion/abrasion: an *in vitro* study *Caries Res.* **22** 1–4

[29] Danelon M, Pessan J P, Santos V R D, Chiba E K, Garcia L S G, de Camargo E R and Delbem A C B 2018 Fluoride toothpastes containing micrometricor nano-sizedsodium trimetaphosphate reduceenamelerosion *in vitro Acta Odontol. Scand.* **76** 119–24

[30] Targino A G, Flores M A, dos Santos Junior V E, de Godoy Bené Bezerra F, de Luna Freire H, Galembeck A and Rosenblatt A 2014 Aninnovative approach totreating dental decay in children. A new anti-caries agent *J. Mater. Sci. Mater. Med.* **25** 2041–7

[31] Freire P L L, Albuquerque A J R, Sampaio F C, Galembeck A, Flores M A P, Stamford T C M and Rosenblatt A 2017 AgNPs: the new allies against *S. mutans* biofilm a pilot clinical trial and microbiological assay *Braz. Dent. J.* **28** 417–22

[32] Teixeira J A *et al* 2018 Effects of a new nano-silver fluoride-containing dentifrice on demineralization of enamel and *Streptococcus mutans* adhesion and acidogenicity *Int. J. Dent.* **8** 2018 1351925

[33] Yin I X, Zhao I S, Mei M L, Lo E C M, Tang J and Li Q 2020 Synthesis and characterization of fluoridated silver nanoparticles and their potential as a non-staining anti-caries agent *Int. J. Nanomed.* **15** 3207–15 A

[34] Tschoppe P, Zandim D L, Martus P and Kielbassa A M 2011 Enamel and dentine remineralization by nano-hydroxyapatite toothpastes *J. Dent.* **39** 430–7

[35] Demito C F, Costa J V D, Fracasso M L C and Ramos A L 2019 Efficacy of fluoride associated with nano-hydroxyapatite in reducing enamel demineralization adjacent to orthodontic brackets: *in situ* study *Dental Press. J. Orthod.* **24** 48–55

[36] Cheng L, Weir M D, Xu H H, Kraigsley A M, Lin N J, Lin-Gibson S and Zhou X 2012 Antibacterial and physical properties of calcium-phosphate and calcium-fluoride nanocomposites with chlorhexidine *Dent. Mater.* **28** 573–83
[37] Xu H H, Moreau J L, Sun L and Chow L C 2010 Novel CaF_2 nanocomposite with high strength and fluoride ion release *J. Dent. Res.* **89** 739–45
[38] Yi J, Weir M D, Melo M A S, Li T, Lynch C D, Oates T W, Dai Q, Zhao Z and Xu H H K 2019 Novel rechargeable nano-CaF_2 orthodontic cement with high levels of long-term fluoride release *J. Dent.* **90** 103214
[39] Yli-Urpo H, Vallittu P K, Narhi T O, Forsback A-P and Variparta M 2004 Release of silica,calcium, phosphorous and fluoride from glass ionomer cement containingbioactive glass *J. Biomater. Appl.* **19** 5–20
[40] Erol-Taygun M, Zheng K and Boccaccini A R 2013 Nanoscale bioactive glasses in medical applications *Int. J.App.l Glass Sci.* **4** 136–48
[41] Nam H J, Kim Y M and Kwon Y H *et al* 2019 Fluorinated bioactive glass nanoparticles: enamel demineralization prevention and antibacterial effect of orthodontic bonding resin *Materials (Basel).* **12** 1813
[42] Danelon M, Pessan J P, Neto F N, Camargo E R and Delbem A C B 2015 Effect of toothpaste with nano-sized trimetaphosphate on dental caries: *in situ* study *J. Dent.* **2459** 1–8
[43] Danelon M, Pessan J P, Souza-Neto F N, de Camargo E R and Delbem A C B 2017 Effect of fluoride toothpaste with nano-sized trimetaphosphate on enamel demineralization: an *in vitro* study *Arch. Oral Biol.* **78** 82–7
[44] Cochrane N J, Saranathan S, Cai F, Cross K J and Reynolds E C 2008 Enamel subsurface lesion remineralisation with casein phosphopeptidestabilised solutions of calcium, phosphate and fluoride *Caries Res.* **42** 88–97
[45] Delbem A C, Souza J A, Zaze A C, Takeshita E M, Sassaki K T and Moraes J C S 2014 Effect of trimetaphosphate and fluoride association on hydroxyapatite dissolution and precipitation *in vitro Braz. Dent. J.* **25** 479–48
[46] Rios D, Honório H M, Magalhães A C, Delbem A C B, Machado M A A M, Silva S M B and Buzalaf M A R 2007 Effect of salivary stimulation on erosion of human and bovine enamel subjected or not to subsequent abrasion: an *in situ/ex vivo* study *Caries Res.* **40** 218–23
[47] Magalhães A C, Rios D, Delbem A C, Buzalaf M A and Machado M A 2007 Influence of fluoride dentifrice on brushing abrasion of eroded human enamel: an *in situ/ex vivo* study *Caries Res.* **41** 77–9
[48] Ganss C, Lussi A, Grunau O, Klimek J and Schlueter N 2011 Conventional and anti-erosion fluoride toothpastes: effect on enamel erosion and erosion-abrasion *Caries Res.* **45** 581–9
[49] Conceição J M, Delbem A C, Danelon M, da Camara D M, Wiegand A and Pessan J P 2015 Fluoride gel supplemented with sodium hexametaphosphate reduces enamel erosive wear *in situ J. Dent.* **43** 1255–60
[50] Humphreys G, Lee G L, Percival S L and McBain A J 2011 Combinatorial activities of ionic silver and sodium hexametaphosphate against microorganisms associated with chronic wounds *J. Antimicrob. Chemother.* **66** 2556–61
[51] Mendes-Gouvêa C C, do Amaral J G, Fernandes R A, Fernandes G L, Gorup L F, Camargo E R, Delbem A C B and Barbosa D B 2018 Sodium trimetaphosphate and hexametaphosphate impregnated with silver nanoparticles: characteristics and antimicrobial efficacy *Biofouling* **34** 299–308

[52] Hosida T Y, Pessan J P, Cavazana T P, Sampaio C, Monteiro D R and Delbem A C B 2021 Effect of sodium hexametaphosphate and fluoride on dual-species biofilms of *Candida albicans* and *Streptococcus mutans Biofouling* (Accepted)

[53] Shibata H and Marioka T 1982 Antibacterial action of condensed phosphates on the bacterium *Streptococcus mutans* and experimental caries in the hamster *Arch. Oral Biol.* **27** 809–16

[54] Choi I K, Wen W W and Smith R W 1993 The effect of a long chain phosphate on the adsorption of collectors on kaolinite *Miner. Eng.* **6** 1191–7

[55] Dalpasquale G, Delbem A C B, Pessan J P, Nunes G P, Gorup L F, Neto F N S, de Camargo E R and Danelon M 2017 Effect of the addition of nano-sized sodium hexametaphosphate to fluoride toothpastes on tooth demineralization: an *in vitro* study *Clin. Oral Investig.* **21** 1821–7

[56] Garcia L S G, Delbem A C B, Pessan J P, Dos Passos Silva M, Neto F N S, Gorup L F, de Camargo E R and Danelon M 2018 Anticaries effec to F toothpaste with nano-sized sodium hexametaphosphate *Clin. Oral Investig.* **23** 3535–42

[57] Danelon M, Garcia L G, Pessan J P, Passarinho A, Camargo E R and Delbem A C B 2019 Effect of fluoride toothpaste containing nano-sized sodium hexametaphosphate on enamel remineralization: an *in situ* study *Caries Res.* **53** 260–7

[58] Hosida T Y, Delbem A C B, Morais L A, Moraes J C S, Duque C, Souza J A S and Pedrini D 2019 Ion release, antimicrobial and physio-mechanical properties of glass ionomer cement containing micro or nano-sized hexametaphosphate, and their effect on enamel demineralization *Clin. Oral Investig.* **23** 2345–54

[59] Pepla E, Besharat L K, Palaia G, Tenore G and Migliau G 2014 Nano-hydroxyapatite and its applications in preventive, restorative and regenerative dentistry: a review of literature *Ann. Stomatol. (Roma).* **5** 108–14

[60] Featherstone J D B and Adrian L 2006 Understanding the chemistry of dental erosion *Dental Erosion.* vol 20 (Basel: Karger Publishers) 66–76

[61] ten Cate J M 2008 Remineralization of deep enamel dentine caries lesions *Aust. Dent. J.* **53** 281–5

[62] Frank R M and Voegel J C 1980 Ultrastructure of the human odontoblast process and its mineralisation during dental caries *Caries Res.* **14** 367–80

[63] Ebadifar A, Nomani M and Fatemi S A 2017 Effect of nano-hydroxyapatite toothpaste on microhardnessofartificial carious lesions created on extracted teeth *J. Dent. Res. Dent. Clin. Dent. Prospects. Winter* **11** 1–14

[64] Kooshki F, Pajoohan S and Kamareh S 2019 Effects of treatment with three types of varnish remineralizing agents on the microhardness of demineralized enamel surface *J Clin. Exp. Dent.* **11** e630–5

[65] Comar L P, Souza B M, Gracindo L F, Buzalaf M A R and Magalhães A C 2013 The impact of experimental pastes containing hydroxyapatite nanoparticles, with or without fluoride, on the dental demineralization *in vitro Braz. Dent. J.* **24** 273–8

[66] Leal A M C, Beserra dos Santos M V, da Silva Filho E C, Menezes de Carvalho A L, Tabchoury C P M and Vale G C 2020 Development of an experimental dentifrice with hydroxyapatite nanoparticles and high fluoride concentration to manage root dentin demineralization *Int. J. Nanomedicine.* **15** 7469–79

[67] Jeong S H, Hong S J, Choi C H and Kim B I 2007 Effect of new dentifrice containing nano-sized carbonated apatite on enamel remineralization *Key Eng. Mater.* **330–332** 291–4

[68] Souza B M, Comar L P, Vertuan M, FernandesNeto C, Buzalaf M A and Magalhães A C 2015 Effect of an experimental paste with hydroxyapatite nanoparticles and fluoride on dental demineralisation and remineralisation *in situ Caries Res.* **49** 499–507

[69] Lin J, Zhu J, Gu X, Wen W, Li Q, Fischer-Brandies H, Wang H and Mehl C 2011 Effects of incorporation of nano-fluorapatite or nano-fluorohydroxyapatite on a resin-modified glass ionomer cement *Acta Biomater.* **7** 1346–53

[70] Wang C, Zhang Y, Wei J and Wei S 2012 Effects of different pH conditions on enamel erosion repair by nano fluorapatite pastes *J. Nanosci. Nanotechnol.* **12** 7346–53

[71] Soares L E S, da Silva Magalhães J, Marciano F R and Lobo A O 2018 Surface characteristics of a modified acidulated phosphate fluoride gel with nano-hydroxyapatite coating applied on bovine enamel subjected to an erosive environment *Microsc. Res. Tech.* **81** 1456–66

[72] Browning W D, Cho S D and Deschepper E J 2012 Effect of a nano-hydroxyapatite paste on bleaching-related tooth sensitivity *J Esthet. Restor. Dent.* **24** 268–76

[73] Vano M, Derchi G, Barone A and Covani U 2014 Effectiveness of nano-hydroxyapatite toothpaste in reducing dentin hypersensitivity: a double-blind randomized controlled trial *Quintessence Int.* **45** 703–11

[74] Amaechi B T, Mathews S M, Ramalingam K and Mensinkai P K 2015 Evaluation of nanohydroxyapatite-containing toothpaste for occluding dentin tubules *Am. J. Dent.* **28** 33–9

[75] Wang L, Magalhães A C, Francisconi-Dos-Rios L F, Calabria M P, Araújo D, Buzalaf M, Lauris J and Pereira J C 2016 Treatment of dentin hypersensitivity using nano-hydroxyapatite pastes: a randomized three-month clinical trial *Oper. Dent.* **41** E93–101

[76] Vano M, Derchi G, Barone A, Pinna R, Usai P and Covani U 2018 Reducing dentine hypersensitivity with nano-hydroxyapatite toothpaste: a double-blind randomized controlled trial *Clin. Oral Investig.* **22** 313–20

[77] Taha S T, Han H, Chang S R, Sovadinova I, Kuroda K, Langford R M and Clarkson B H 2015 Nano/micro fluorhydroxyapatite crystal pastes in the treatment of dentin hypersensitivity: an *in vitro* study *Clin. Oral Investig.* **19** 1921–30

[78] Yu Q, Liu H, Liu Z, Peng Y, Cheng X, Ma K and Ji Y 2017 Comparison of nanofluoridated hydroxyapatite of varying fluoride content for dentin tubule occlusion *Am. J. Dent.* **30** 109–15

[79] Tirupathi S, Nirmala S V S G, Rajasekhar S and Nuvvula S 2019 Comparative cariostatic efficacy of a novel Nano-silver fluoride varnish with 38% silver diamine fluoride varnish a double-blind randomized clinical trial *J. Clin. Exp. Dent.* **11** e105–12

[80] Kasraei S, Sami L, Hendi S, Alikhani M Y, Rezaei-Soufi L and Khamverdi Z 2014 Antibacterial properties of composite resins incorporating silver and zinc oxide nanoparticles on *Streptococcus mutans* and *Lactobacillus Restor. Dent. Endod.* **39** 109

[81] Morones J R, Elechiguerra J L, Camacho A, Holt K, Kouri J B, Ramirez J T and Yacaman M J 2005 The bactericidal effect of silver nanoparticles *Nanotechnology* **16** 2346–53

[82] Peng J J-Y, Botelho M G and Matinlinna J P 2012 Silver compounds used in dentistry for caries management: a review *J. Dent.* **40** 531–41

[83] Jacob S J, Finub J S and Narayanan A 2012 Synthesis of silver nanoparticles using *Piper longum* leaf extracts and its cytotoxic activity against Hep-2 cell line *Colloids Surf. B. Biointerfaces* **1** 212–4

[84] Pérez-Díaz M A, Boegli L, James G, Velasquillo C, Sánchez-Sánchez R, Martínez-Martínez R E, Martínez-Castañón G A and Martinez-Gutierrez F 2015 Silver nanoparticles

with antimicrobial activities against Streptococcus mutans and their cytotoxic effect *Mater. Sci. Eng.* C **55** 360–6

[85] Aldhaian B A, Balhaddad A A, Alfaifi A A, Levon J A, Eckert G J, Hara A T and Lippert F 2021 *In vitro* demineralization prevention by fluoride and silver nanoparticles when applied to sound enamel and enamel caries-like lesions of varying severities *J. Dent.* **104** 103536

[86] Yin I X, Yu O Y, Zhao I S, Mei M L, Li Q L, Tang J and Chu C H 2020 Inhibition of dentine caries using fluoride solution with silver nanoparticles: an *in vitro* study *J. Dent.* **103** 103512 B

[87] Costa e Silva A V, Teixeira J A and Melo Júnior P C *et al* 2019 Remineralizing potential of nano-silver-fluoride for tooth enamel: an optical coherence tomography analysis *Pesqui. Bras. Odontopediatria Clín. Integr.* **19** e4002

[88] Rosenblatt A, Stamford T C M and Niederman R 2009 Silver diamine fluoride: a caries 'silver-fluoride bullet' *J. Dent. Res.* **88** 116–25

[89] dos Santos V E Jr, Vasconcelos Filho A, Targino A G R, Flores M A P, Galembeck A, Caldas A F Jr and Rosenblatt A 2014 A new 'Silver-Bullet' to treat caries in children–nano silver fluoride: a randomised clinical trial *J. Dent.* **42** 945–51

[90] dos Santos V E Jr, Vasconcelos F M, Ribeiro A G and Rosenblatt A 2012 Paradigm shift in the effective treatment of caries in schoolchildren at risk *Int. J. Dent.* **62** 47–51

[91] Zhao I S, Yin I X, Mei M L, Lo E C M, Tang J, Li Q and Chu C H 2020 Remineralising dentine caries using sodium fluoride with silver nanoparticles: an *in vitro* study *Int. J. Nanomed.* **15** 2829–39

[92] Hernández-Sierra J F, Ruiz F, Cruz Pena D C, Martínez-Gutiérrez F, Martínez A E, de Jesús Pozos Guillén A, Tapia-Pérez H and Martínez C 2008 The antimicrobial sensitivity of Streptococcus mutans to nanoparticles of silver, zinc oxide, and gold *Nanomedicine* **4** 237–40

[93] Xu H H, Moreau J L, Sun L and Chow L C 2008 Strength and fluoride release characteristics of a calcium fluoride based dental nanocomposite *Biomaterials* **29** 4261–7

[94] Yi J, Dai Q, Weir M D, Melo M A S, Lynch C D, Oates T W, Zhang K, Zhao Z and Xu H H K A 2019 Nano-CaF_2-containing orthodontic cement with antibacterial and remineralization capabilities to combat enamel white spot lesions *J. Dent.* **89** 103172

[95] Fei X, Li Y, Weir M D, Baras B H, Wang H, Wang S, Sun J, Melo M A S, Ruan J and Xu H H K 2020 Novel pit and fissure sealant containing nano-CaF_2 and dimethylaminohexadecyl methacrylate with double benefits of fluoride release and antibacterial function *Dent. Mater.* **36** 1241–53

[96] Shin D Y, Cheon K H, Song E H, Seong Y J, Park J U, Kim H E and Jeong S H 2019 Fluorine-ion-releasing injectable alginate nanocomposite hydrogel for enhanced bioactivity and antibacterial property *Int. J. Biol. Macromol.* **123** 866–77

[97] Eshed M, Lellouche J, Banin E and Gedanken A 2013 MgF_2 nanoparticle-coated teeth inhibit *Streptococcus mutans* biofilm formation on a tooth model *J. Mater. Chem.* B **1** 3985–91

[98] Kumar R S, Ravikumar N, Kavitha S, Mahalaxmi S, Jayasree R, Sampath Kumar T S and Haneesh M 2017 Nanochitosan modified glass ionomer cement with enhanced mechanical properties and fluoride release *Int. J. Biol. Macromol.* **104** 1860–5

[99] Iafisco M, DegliEsposti L, Ramírez-Rodríguez G B, Carella F, Gómez-Morales J, Ionescu A C, Brambilla E, Tampieri A and Delgado-López J M 2018 Fluoride-dopedamorphous

calcium phosphate nanoparticles as a promising biomimetic material for dental remineralization *Sci. Rep. Nov.* 19 **8** 17016

[100] Karimi M, Hesaraki S, Alizadeh M and Kazemzadeh A 2019 Effect of synthetic amorphous calcium phosphate nanoparticles on the physicochemical and biological properties of resin-modified glass ionomer cements *Mater.Sci. Eng. C. Mater. Biol. Appl.* **98** 227–40

[101] Silva M E R, Danelon M, Santos Souza J A, Silva D F, Pereira J A, Pedrini D, de Camargo E R, Botazzo Delbem A C and Duque C 2019 Incorporation of chlorhexidine and nano-sized sodium trimetaphosphate into a glass-ionomer cement: Effect on mechanical and microbiological properties and inhibition of enamel demineralization *J. Dent.* **84** 81–8

[102] Emerenciano N G, Botazzo Delbem A C, Pessan J P, Nunes G P, Souza Neto F N, de Camargo E R and Danelon M 2018 *In situ* effect of fluoride toothpaste supplemented with nano-sized sodium trimetaphosphate on enamel demineralization prevention and biofilm composition *Arch. Oral Biol.* **96** 223–9

[103] Souza M D B, Pessan J P, Lodi C S, Souza J A S, Camargo E R, Souza Neto F N and Delbem A C B 2017 Toothpaste with nanosized trimetaphosphate reduces enamel demineralization *JDR Clin. Trans. Res.* **2** 233–40

[104] Hosida T Y, Delbem A C B, Morais L A, Moraes J C S, Duque C, Souza J A S and Pedrini D 2019 Ion release, antimicrobial and physio-mechanical properties of glass ionomer cement containing micro or nanosized hexametaphosphate, and their effect on enamel demineralization. *Clin. Oral Investig.* **23** 2345–54

[105] Ionescu A C, Cazzaniga G, Ottobelli M, Garcia-Godoy F and Brambilla E 2020 Substituted nano-hydroxyapatite toothpastes reduce biofilm formation on enamel and resin-based composite surfaces *J. Funct. Biomater.* **11** 36

[106] Seyedlar R M, Rezvani M, Barari S, Imani M, Nodehi A and Atai M 2019 Synthesis of plate-like β-tricalcium phosphate nanoparticles and their efficiency in remineralization of incipient enamel caries *Prog. Biomater.* **8** 261–76

IOP Publishing

Nanotechnology for Dentistry Applications

Ashutosh Kumar Shukla

Chapter 2

Role of angiogenic inorganic nanomaterials in dentistry applications

Papia Basuthakur and Chitta Ranjan Patra

Biomaterials offer a wide range of applications like restorative, preventive and regenerative treatments. Extensive research on dental therapeutic strategies over the past few decades have led to the discovery of a plethora of biomaterials (such as metals, alloys, polymers, dental cements, amalgams, ceramic materials and many more). Among them, biomaterials having angiogenic properties are employed for the treatment of oral diseases. Angiogenic properties play an important role for regeneration, restoration and revascularization of injured or traumatic dental tissues, dentine, dental pulp, root etc. Conventional biomaterials such as drugs and growth factors are often used for enhancement of angiogenesis in oral diseases. Several drawbacks (such as high cost, low sustainability of growth factors, drugs and many more) have compelled researchers to integrate nanotechnology into dentistry. Various nanomaterials (inorganic, polymer, carbon based, metal based, drug/growth factor conjugated) have been explored for development of effective therapeutic strategy that facilitate angiogenesis. Recent reports suggest that inorganic nanomaterials with angiogenic properties might have promising dental applications. However, no comprehensive book chapter or review article illustrates the useful dental applications of pro-angiogenic inorganic nanomaterials. In this book chapter, we discuss how nanotechnology is exploited for solving dental problems by using therapeutic angiogenesis. We further discuss about the possible routes to overcome the challenges to the path of successful clinical outcomes.

2.1 Introduction to dentistry

The dental pulp contains a network of connective tissue comprising blood vessels, nerve endings, collagen fibers, extracellular matrix etc. The maintenance of oxygen and nutrient supply and evacuation of toxic waste are regulated by blood vessels in the oral cavity. Thus, angiogenesis becomes a pre-requisite for recovery from several

doi:10.1088/978-0-7503-3671-0ch2

pathological oral and dental problems (such as repair of wounded oral cavity, regeneration of dental implants, recovery from occlusal trauma, pulp calcification etc). Conventional therapies such as use of pro-angiogenic growth factors, drugs and regeneration of dental pulp using engineered devices have only brought little success. There is an urgent need for more reliable and comfortable alternative treatment options for precise maintenance of the oral cavity. To this end, nanotechnology can solve unmet medical challenges. Nano-dentistry is a nascent but rapidly evolving concept for dental therapies. However, not much has been explored on the potential role of angiogenic inorganic nanoparticles in dentistry. Some inorganic nanoparticles like zinc, magnesium, cerium, titanium and many others have been reported for applications in dental angiogenesis. Additionally, based on reports of promising applications of pro-angiogenic inorganic nanoparticles in treatment of vascular pathophysiology (ischemia, wound healing, cardiovascular diseases etc) we can predict the meritorious prospects of these pro-angiogenic nanoparticles for their application in vascular dentistry. In this book chapter, we discuss how scientists have exploited nanotechnology's role in providing long-term success in treatment of dental problems. Apart from that, we also discuss the future challenges that nanotechnology faces in realizing its full potential.

2.1.1 Anatomy and development of the tooth

Teeth are three-dimensional architecture found in the human body [1]. They have three parts crown, neck, and root. Teeth originate from oral ectodermal cells and neural crest–derived mesenchyme. Dental tissue is comprised of a soft dental pulp at the bottom surrounded on the top by mineralized tissues like enamel and dentin [2]. Enamel and dentin are differentiated from the ameloblasts and odontoblast, respectively [3]. Enamel, which is one of the toughest structures in the human body [4], is formed of inorganic hydroxyapatite and non-collagenous proteins (namely enamelin, amelogenin, and ameloblastin) [5]. The tooth crown is composed of a mineralized outer enamel layer consisting of organic matter and proteins. The organic proteins of the enamel layer aid in biomineralization and structural organization of the enamel surface [6]. The dentin layer and the inner pulp dental tissue exist underneath the enamel layer. Dental pulp comprises the dentin, cementum, and periodontal ligament layers, that help in binding the tooth to the underlying alveolar bone [7]. On the other hand, the root of teeth forms when the cementoblasts of dental follicular mesenchyme differentiate into cementum [8].

Teeth formation begins from the time of embryogenesis and continues through the prenatal period, childhood and adolescence and ultimately leads to the development of permanent teeth. The dental tissues in the oral cavity form altogether in a regulated manner, being supported by different types of cells (osteoclasts, osteoblasts, cementoblasts, ameloblasts, odontoblast and mesenchymal tissue) and communicated by a set of intercellular signals that control and co-ordinate the morphogenesis, regeneration, and differentiation of different dental tissues [2, 9, 10, 11]. Since human teeth cannot regenerate after eruption, restoration of dental functions depends on the external applications of biomedical engineered non-dental cells [12]. In this book chapter we will discuss in detail the different types of non-dental biomaterials particularly nanomaterials that hold promising applications in dental tissue regeneration.

2.1.2 Angiogenesis mechanisms in human dental pulp

The process of angiogenesis involves the generation of new blood vessels from the pre-existing blood vessels [13]. Physiological processes like wound healing or pathological processes like stroke [14, 15], tumor development [16, 17], ischemia [18, 19] trigger the release of pro-angiogenic factors (such as vascular endothelial growth factor, platelet derived growth factor, epidermal growth factor, fibroblast growth factor, angiopoietin, $\alpha_v\beta_5$ integrins etc) in the local environment inside the blood vessels. This in turn stimulates the endothelial cells (EC) to proliferate and migrate by secreting the matrix metalloproteinases (MMPs) that disrupt the basement membrane. These processes lead to anastomosis of blood vessels, being supported by other cells like pericytes in the surroundings [20, 21, 22].

Within the oral cavity, the highly vascularized dental pulp possesses regenerative potential [23]. This dental pulp contains progenitor or postnatal stem cells that can differentiate into odontoblast-like cells and secrete reparative dentine [24]. Dental mechanical processes like orthodontic tooth movement, occlusal trauma etc can induce stress on the human dental pulp. These mechanical stresses can result in reduced supply of oxygen, blood and nutrition to the dental tissues. During such injury or damage in the dental cavity, the blood vessels get damaged. These kind of stress are comparable to the conditions, in other parts of the body, where lack of blood perfusion is a cause of concern, such as ischemia, hypoxia, wound and several others. Therefore, healing of such ischemic tissues is associated with release of angiogenic growth factors that can aid in development of blood vessels to compensate for the adequate supply of blood to the affected tissues [25, 26]. A rate determining step for healing of traumatized tissue is angiogenesis. This process is complemented by a number of growth factors like VEGF, FGF, PDGF, EGF. In response to the wear and tear of dental tissues due to shear stress, the dental pulp also attempts to increase the supply of blood vessels, in a process called angiogenesis, in the affected regions for rapid healing. This process of angiogenesis can be enhanced by stimulating the local availability of stem cells, oxygen levels and nutrient supply [27, 28]. The dental pulp, which is a loose connective tissue is composed of various types of cells (fibroblasts, endothelial and inflammatory cells), extracellular matrix, blood vessels, collagen fibers, and also nerve endings. All these components of human dentin take part in self recovery by forming new blood vessels that can in turn augment the formation of mineralizaton of tissue [29]. The endothelial cells that form the blood vessels in and around the dental pulp, recruit the odontoblast cells at the site of injury to initiate the healing process and also stimulate inflammatory responses [30]. During injury, the pulp progenitor cells, under the influence of a growth factor like FGF-2 [31], stimulate the matured dental pulp cells to proliferate [32] and also simultaneously induce mineralized calcium production [33, 34]. In this chapter, we will discuss in detail the different types of non-dental biomaterials, particularly nanomaterials, that hold promising applications in dental tissue regeneration.

2.2 Conventional biomaterials used for dentistry

Dental tissue regeneration has become a significant aspect for increased life expectancy. Although regenerative medicine has shown great progress, the field of regenerative dentistry is faced with a lot of challenges [35]. Hence, modern dentistry is incorporating the benefits of biotechnology in oral care. Integration of material sciences with biomechanical sciences helps in conceptualization of design and material for surgical implants. Today, dental implants have gained a lot of acceptance and are being considered as a treatment option for dental tissue regeneration. Table 2.1 lists many more conventional biomaterials that are currently used for dental applications.

2.2.1 Benefits of use of biomaterial implants

The term biomaterial implants generally refers to a device used to support or replace a biological structure [50]. Although dental tissues have the potential to self repair, it is only limited. Under such circumstances doctors often rely on the use of external biocompatible synthetic materials [51]. Due to the unfavorable microenvironment of the oral cavity, the majority of the biomaterials in use are limited by short lifespan and lack of versatility in functions. Hence, there is an urgent need for development of bioactive dental biomaterials which are not only biocompatible but also provide a broad range of dental restoration potentials such as improved cellular environment, long life, physical stability, non-toxicity, enhanced mechanical properties, greater abrasion resistance, controlled surface structures and improved aesthetics [52]. Additionally, newly developed dental biomaterials also possess the property to ward off microbial infections from the oral cavities as well as exhibit migration and proliferation of cells for osteointegration of the implant [53]. An ordinary surgical intervention like replacement of missing teeth involves an optimal interaction between the body and the implanted material. Advancements in science and technology have made the improvement of materials in dental implants possible [54].

The dental tubule is protected from outside, by the enamel (crown) or cementum. Damage to the enamel or cementum gives access to solutes, irritants, bacteria etc to

Table 2.1. Conventional biomaterials used for dental applications.

Sr. No.	Biomaterials	Applications	Reference
1.	Commercial pure titanium	Dental implant	[36]
2.	Mercury amalgam	Dental fillings	[37]
3.	Silver compounds	Treatment of caries	[38, 39]
4.	Silver nanopaticles	Endodontic refilment	[40]
5.	Silver nanopaticles	Dental implants	[41]
6.	Zirconia	Dental prosthetics	[42, 43]
8.	Dental pulp stem cells (DPSCs)	Regenerative dentistry	[44, 45]
9.	PDGF, BMP, FGF	Periodontal regeneration	[46, 47, 48]
10.	Carbon based	Regeneration of peridontal tissue	[49]

the dental tubule, that may lead to diseases in the oral cavity [55]. To overcome such conditions, commercial desensitizing products are often used to occlude the dentinal tubule. But these products are associated with a lot of disadvantages, as they do not provide a proper barrier to the dentine tubules and fail to withstand the mechanical forces in the hostile oral microenvironment. Hence, scientists are keen to develop durable biomimetic biomaterials that can penetrate deep into the dentinal tubules thereby providing complete occlusion of the exposed dentin.

In the next section, we discuss some of the conventional biomaterials made of inorganic metals and their utility as implants for dental tissue regeneration.

2.2.1.1 Mesoporous silica based biomaterials

Chiang *et al* developed an effective, reliable and biocompatible biomaterial containing mesoposrous silica decorated on gelatin [56]. The biocomposite also contains calcium carbonate nanoparticles (CCMS), which are blended with phosphoric acid (CCMS-HP). The biomaterial can effectively release Ca^{2+} and PO_4^{3-}/ HPO_4^{2-} ions that penetrate deep (40 μm) into the dentinal tubules and form dicalcium phosphate dihydrate (DCPD), tricalcium phosphate (TCP) or hydroxyapatite (HAp) crystals, that can occlude the exposed surface of the dentinal tubules. The CCMS-HP successfully covered the exposed dentinal tubules with 5 min of treatment, as was observed by Sub-Micron X-ray Computed Tomography (Sub-μ-CT). The highly dense crystals were capable of preventing microbial infection and withstanding mechanical forces more than commercial products. CCMS-HP exhibited higher viability of 3T3 fibroblast cells in WST-1 and lactic acid dehydrogenase (LDH) assays. Additionally, quantitative alkaline phospohatase (ALP) assay indicated that the biomaterial stimulates ALPase activity through the dentin barrier. Also, the release of Ca^{2+} ions into the dental pulp cells helps in their remineralization, probably via enhanced expression of bone morphogenetic protein (BMP-2) and osteopontin. Further, *in vivo* inflammatory studies showed that CCMS-HP exhibits no inflammation when treated on dental pulp. In this manner, the group showed that biomimetic CCMS-HP can form crystals within dentinal tubules and provide long-lasting protection to the dentinal tissues

In another study, Zhang *et al* demonstrated nanotechnological application of MSNs in dental materials [57]. Usefulness of most of the dental implants for restoration of a decayed tooth are limited by short life span. In this regard, secondary caries is a leading cause of failure of dental composite restorations. Hence, dental composites that can inhibit caries but also have good mechanical properties are highly desirable. In their studies, MSN was used for encapsulating CHX to form a dental composite. The implant showed good antibacterial properties and prevented oral biofilm formation by *Streptococcus mutans* and *Lactobacillus casei.* The dental composite also showed high mechanical strength and surface integrity.

2.2.1.2 Chitosan and calcium phosphate based biomaterials

Recently, three dimensional biomaterial scaffolds composed of natural polymers have gained attention from researchers from the field of bone tissue engineering.

These natural polymers come with a lot of advantages like supporting cell adherence and growth with the help of their highly organized structures that mimic the extra cellular components, the polymers also stimulate immune response in the body [58]. To this end, chitosan (Cs), a linear polysaccharide, composed of units of glucosamine and N-acetyl glucosamine linked by β (1–4) glycosidic bonds is one of the most common biopolymers used in tissue engineering [59]. Although chitosan itself can offer beneficial properties like biocompatibility, bioresorbability, antibacterial, activity, haemostatic characteristics [60] and osteoconductivity [61], scientists have frequently tried to improvise its efficiency by conjugating it with bone-like elements such as calcium phosphate ceramics (Cp) [62] that provide superior stability and *in vivo* osteogenicity [63].

In this context, Tanase *et al* reported the effects of chitosan–calcium phosphate (Cs–CP) scaffolds, synthesized using biomimetic co-precipitation method on *in vitro* cytotoxicity and proliferation of cells associated with bone regeneration (MG63—osteoblast-like human cell line, HOS—primary human osteoblasts and HDMEC—human dermal microvascular endothelial cells) [64]. After thorough assessment of mechanical and structural properties using characterization techniques, the group evaluated their biological properties. Biological assays indicate that the cells show viability as comparable to the untreated controls. Immunogenic assays like ELISA revealed increased secretion of VEGF, which indicated its role in vascularization and bone remodeling. Further, immunofluorescent assays to determine the morphology of the cells clearly showed that cells attached to the biomaterials, resulting in widespread colonization over a period of 14 days. According to the authors, the study demonstrates that Cs–CP composite scaffolds can act as a promising biomaterial for bone-tissue engineering in the near future.

Linear polysaccharides like CS have been applied in a wide range of fields including oral implants due to their excellent biodegradable and biocompatible characteristics. Cao *et al* synthesized a thermosensitive hydrogel using chitosan (CS) and β-glycerophosphate pentahydrate (β-GP) [65]. In the reaction, β-GP aided the phase transformation of CS solution at physiological temperature. The hydrogel plays a dual role in the implant. On one hand, it acts as a sealing and lubricating agent in dental implant. On the other hand, it is also capable of sustainable delivery of drug antimicrobial agents, such as povidone-iodine (PVP-I), around the implant. This enabled the prevention of oral biofilm formation by infectious bacteria like *Escherichia coli*, *Staphylococcus aureus* and *Porphyromonas gingivalis.*

2.2.1.3 Titanium based biomaterials

Of late, titanium and titanium alloys have become the gold standard for dental implants, especially for replacement of teeth [66]. Besides having good mechanical properties, titanium has excellent biocompatibility due to the formation of a stable oxide layer upon exposure with air. This stable oxide layer enables osteo-integration of titanium implants. Reports suggest that many bioactive metal ions enhance intracellular functions of osteoblasts and osteogenic differentiation. Zinc, for instance, helps in bone metabolism [67, 68]. Magnesium also plays an important role in osteogenesis, adhesion of osteoblasts to implants and aids in blood vessel formation [69]. Blood vessels are indispensable for osteoblast function, bone tissue generation and bone integration. Both these elements have been found to enhance adhesion, proliferation and ALP activity of BMSCs or osteoblast [70, 71].

In one study, Yu *et al* explored the potential of titanium implants co-implanted with zinc and magnesium ions on osteo-inductivity and angiogenesis [72]. In this study, dental implants are developed by plasma immersion ion implantation (PIII). This Zn/Mg-PIII promoted initial adhesion and proliferation of rat bone marrow mesenchymal stem cells (rBMSCs) by increasing concentrations of Mg^2+, Zn^2+ ions and upregulating ALP, OCN and integrins (α1 and β1). Mg^2+ ions from Zn/Mg-PIII caused upregulation of MagT1 transporter, VEGF, KDR and HIF-1αin endothelial cells (human umbilical vein endothelial cells) and thus induced angiogenesis. Zn^2+ ions on the other hand, prevented the growth of anaerobic bacteria (*Porphyromonas gingival, Fusobacterium nucleatum* and *Streptococcusmutans).* Furthermore, *in vivo* studies (Micro-CT evaluation, fluorescent labeling, histological analysis and push-out test) revealed that the implants enhanced bonding strength at bone-implant interfaces, bone formation, angiogenesis and osteointegration. The group demonstrated a promising approach using Zn/Mg and Ti co-implantation technique for manufacturing dental implant with long-term stability.

Huang *et al* developed a titanium based dental implant with silver coatings [73]. The composite was found to possess excellent biocompatibility to human gingival fibroblast (HGF) cells. Besides, the presence of silver coatings prevented infections by *S. aureus*, a major pathogen causing implant-associated infections. The group demonstrated that TaN-Ag coatings on implants can offer antibacterial performance with compatible biological response.

2.2.1.4 Zirconia based biomaterials

Scientists have also attempted to incorporate different types of metal-free ceramics for dental implants. Among them, zirconia-based ceramics have been used most widely in dental prosthetics, due to their biocompatibility and mechanical properties, low plaque affinity and tooth-like color [74]. Material science engineers employ modern technologies to produce good dental biomaterials with perfection, precision and quality, that can offer aesthetics, functionality and biocompatibility while maintaining integrity by resisting against masticatory forces [75].

Recently, ceramic implants which are chemically stable, mechanically hard, and highly resistant to degradation are considered as an alternative to titanium implants. To this end, zirconia ceramic with a combination of mechanical properties, tooth-like colour, excellent biocompatibility and low plaque affinity have become one of the best biomaterials for implants [76, 77, 78].

Scarano *et al* demonstrated biocompatible and osteo-conductive effects of zirconia in this study [79]. The group inserted zirconia implants inside the tibia of mature male rabbits. After four weeks of implantation, the group observed that there was considerable bone formation around the area of the implant. Concomitantly, the number of osteoblasts also increased near the implant.

2.3 Contribution of nanotechnology in dentistry

The advent of nanotechnology has induced a lot of innovations in diagnosis and therapy of dentistry. Nanotechnology is now used for manufacturing of oral disease preventive drugs, teeth implantation, maintenance of oral health and many other pathological conditions. Now, we focus on various applications of nanoparticles in the field of dentistry.

2.3.1 Inorganic nanoparticles

Inorganic nanoparticles especially metallic nanoparticles have large surface area-to-volume ratios and they are capable of releasing ions into the local environment to get the desired effect. Researchers and pharmaceutical companies are using many inorganic nanoparticles that have good antibacterial effects for the manufacturing of antibiotics [80, 81]. A plethora of microorganisms incessantly try to invade the oral cavity from food and water and cause diseases like oral plaque, caries, periodontitis etc [82]. Medical practitioners usually recommend antibiotics for managing such dental infections [83]. But irrational use of antibiotics may lead to various side effects like gastrointestinal problems, nausea, vomiting, oral lessons, fatal anaphylactic shock, resistance from the invading microbes and many more [84]. Besides, the commercial antimicrobial agents often fail to achieve desired applications for a longer time due to rapid degradation, unsustainable release of antimicrobial factors and side effects [82]. Therefore, considerable efforts are urgently needed to develop better antimicrobial biomaterials that can serve as an alternative to currently used antibiotics [85]. In this regard, nanomaterials may offer a new treatment strategy for preventing dental infections. Nanomaterials, due to their small size, large surface area-to-mass ratio and high charge density are better equipped for interacting with negatively charged bacteria and induce effective antimicrobial activity [86]. Antimicrobial dental materials rely on release of an antimicrobial agent, and contact-killing to inhibit bacterial attachment and subsequent biofilm formation [87]. For instance, silver nanoparticles (AgNPs) possess versatile properties such as antibacterial, antifungal and antiviral properties without any adverse effects. The materials are being applied in various fields of dentistry like endodontics [88], dental prostheses [89], restorative dentistry [90]. AgNPs incorporation resist microbial colonization over dental materials and increase oral health and quality of life.

In this regard, Rodrigues *et al* employed a simple, cost-effective and biogenic route for the synthesis of AgNPs using extract of green tea leaves (*Camelia sinensis*) [91]. In this synthetic approach, catechin, forms the main component of tea extract. Its most important medicinal property is its anti-oxidant activity. It acts as a reducing agent for the metallic silver (Ag^+ to Ag°). and also stabilizes the formed nanoparticle. *In vitro* antimicrobial studies (such as minimum inhibitory concentration, biofilm formation and cytotoxicity assay) showed that AgNPs exhibit antibacterial effects against *S. mutans* (causative agent of oral caries). Additionally, AgNPs were surface coated with biocompatible silicon dioxide (SiO2) that not only imparts optical, stabilizing and catalytic properties, but also makes the surface more porous thus broadening its biomedical applications. Finally, the group revealed a promising approach for development of antimicrobial biomaterials that have the potential for preventing dental infections by bacteria.

Implant dentistry is a significant facet of restorative practices in dental tissues. It is now a conventional treatment option for replacing missing teeth and it involves osteointegration. The main function of dental implants is to augment the alveolar bone for integration of the implanted material after tooth extraction [92, 93]. Additionally, the implant material must strive to maintain facial esthetics and functions of the implants and prevent infections [94].

Nanotechnology aids in conquering the hurdles associated with bone regeneration in dentistry. In the ever expanding field of biotechnology, nanomedicine holds a strong frontier [95]. Gold nanoparticles (AuNPs) are widely used for disease therapy, diagnostics, drug delivery, owing to ease of synthesis and versatile surface functionalization. They are biocompatible, non-toxic and can be synthesized using a simple green-chemistry approach.

In one study, Jadhav *et al* explored the bone regenerative potential of AuNPs synthesized from phyto-chemicals of plant extract of *Salacia Chinensis* [96]. The group showed the as-sythesized AuNPs are considerably stable in various blood elements. The AuNPs exhibited biocompatibility against erythrocytes and periodontal fibroblasts. Further, osteoinductive assay revealed that AuNPs increase the proliferation of MG-63 cells. This signifies that AuNPs have the potential to cause bone regeneration. Therefore, the group demonstrated that AuNPs can be used as bone-inductive adjuvants that can form an osteous interface and maintain the emerging peri-implant bone during treatment of implants.

2.3.2 Polymer based nanomaterials

Polymeric materials (PMs) and polymeric films (PMFs) are increasingly being used in biology and medicine owing to their excellent surfaces, low production cost, good mechanical properties and appealing biological properties [97]. In dentistry, they are being employed in preventive and regenerative therapies because of their antimicrobial, drug delivering properties and prevention of corrosion and friction.

Many PMs prevent tooth decay by protecting the enamel layer against adverse oral environments. For instance, acidic dietary components of certain food and drinks may cause tooth erosion and dental caries [98]. Under such circumstances, the surface of tooth may become softened and roughened due to demineralization. Bayern *et al* prepared polymer modified citric acid solutions (PMCAS, Ph 3.3) using propylene glycol alginate (PGA), highly esterified pectin (HP) and gum arabic (GA) [99]. When absorbed on the surfaces of the enamel layer, such PMCAS protected the enamel layer from erosion by acid.

In another report, Carvalho and Lussi prepared a toothpaste containing fluoride-, stannous- and chitosan (F/Sn/chitosan-), that can effectively prevent enamel erosion and abrasion [100]. Chitosan is a deacetylated form of chitin and is a natural biopolymer. Since chitosan contains positive amino group, it can bind strongly to the negatively charged enamel via electrostatic bond. Chitosan can form multiple layers on the surface of enamel and provide protection against erosion. Polymer NPs aid in targeted delivery of drugs by entrapment or immobilization method. NPs inhibit microbial colonization by penetrating the biofilms and releasing the metal ions inside the biofilms [97]. Chitosan shells are often used to make hybrid NPs that have desirable biomedical applications.

In one study, Covarrubias *et al* developed hybrid NPs composed of copper and chitosan shell (CuCh NP), that can effectively prevent invasion of *Streptococcus mutans* in the oral cavity [101]. The copper present in the NPs disrupts bacterial cell walls and kills the bacteria formed in the biofilm. This anti-biofilm activity of copper NPs can be attributed to the adherence of copper NPs on the surface of teeth, facilitated by interaction between chitosan, hydroxyapatite of tooth and bacterial cell wall. Further, the group also found that such antimicrobial properties of CuChNP was more effective than conventional therapies to control dental plaque such as use of oral antimicrobial agents like cetyl-pyridinium chloride or chlorhexidine.

2.3.3 Carbon based nanomaterials

There are various forms of carbon based nanomaterials. Among them, carbon nanotubes, find applications in various nano-devices and composite materials due to their unique physical properties such as high stiffness, tensile strength and elastic modulus [102, 103].

In one study, Khan *et al* utilized electrospinning technique to form a nano-composite of multiwalled carbon nanotubes (MWCNTs) and nano-hydroxyapatite (nHA). Electrospining is a versatile, cost-effective technique used to make nano-fibers [104, 105], that has been used in a number of biomedical applications [106, 107]. Hydroxyapatite [HA, $Ca_{10}(PO_4)_6(OH)_2$], on the other hand, is a type of synthetic calcium phosphate (CP) that is similar to hard biological structures and hence, can be used for regeneration of biological structures. Since dentine, enamel as well as bone are hard structures HA has found application in dental applications. For this study [49], HA functionalized CNT fibers were silanized with γ-methacry-loxypropyl-trimethoxysilane (MPTS), and electrospun using polyvinyl alcohol (PVA/HA/CNT). The fibers were found to possess excellent biocompatibility, as was observed by CAM assay. Such highly biocompatible HA/CNT can aid in regeneration of pulp-dentine and periodontal tissues [49]. Readers who are interested to read more about the applications of nanomaterials in dentistry, can consult with the published review or book chapters [108, 109, 110].

2.4 Angiogenic biomaterials used in dentistry

Growth factors are naturally occurring proteins existing in living tissues and they regulate cellular growth, proliferation, and differentiation. Their activities are important for regeneration and healing of damaged or injured tissues (e.g., bone, cartilage) or cells (e.g., nerves). Hence, they can be used as therapeutic tool. Among many growth factors, including VEGF, PDGF, FGF, BMP, ILGBP-1 etc [111], two main growth factors, IGFBPs, IGFBP-3, are abundantly found in osteoblasts [112] and can bind to circulating IGF-I. Evidence suggests the involvement of IGFBP-3 in the later stages of human tooth development [113, 114].

Considering the constructive effects of IGFBP-3 in bone formation and regeneration [115], Bhattari *et al* carried out research to evaluate the utilization of IGFBP-3 on dental implants [111]. For this purpose, the group focused on delivery of growth factor IGFBP-3 to the dental tissue via implant surfaces. NPs on the other hand, can act as efficient delivery vehicles (both *in vitro* and *in vivo*) [116]. As a result, NP coated implants are often used for local delivery of drugs or genes. For their studies, the group conjugated IGFBP-3 plasmid DNA with chitosan AuNPs (Ch-AuNPs) on titanium (Ti) coated dental implants and evaluated their effect on regulation of osteogenesis. *In vitro* studies showed increased proliferation of MC-3T3 E1 osteoblast-like cells with intact cellular morphology on Ch-GNP/IGFBP-3 coated Ti surfaces of implants. Protein expression studies indicated overexpression of extracellular signal related kinase (ERK) bone morphogenetic protein (BMP)–2 and 7 and down regulation of stress activated protein c-Jun N-terminal kinase (JNK). Additionally, *in vivo* protein -GNP/IGFBP-3 coated implants induced activation of receptor activator of nuclear factor kappa-B ligand (RANKL), osteogenic molecules including BMP2/7 and osteopontin (OPN) and inhibition of receptor activator of nuclear factor kappa-B ligand (RANKL). Further, μCT analysis demonstrated that IGFBP-3 in the implant enhanced bone formation surrounding it. Thus, the study gives an indication about the beneficial utilization of IGFBP-3 in implants for the purpose of bone formation and regeneration.

2.5 Role of angiogenic nanoparticles in dentistry

In dentistry, angiogenesis is required for regeneration of dentin and dental pulp tissues after injuries, and revascularization of traumatic premature teeth, that in turn results in radicular dental pulp survival and continuity of root formation [117]. Reports suggest that inorganic elements can induce the angiogenesis process. Many of these elements are used to manufacture dental biomaterials. The release of these elements from biomaterials into dental cavity can induce angiogenesis in the surrounding tissues. Considering their undeniable influence on angiogenesis events, pro-angiogenic inorganic NPs are often used for manufacturing of dental biomaterial and implants that are applied in different treatment modalities for better regeneration and healing [117].

Of late, a lot of research work has been carried out in targeting angiogenesis by using inorganic NPs. Medicinal chemistry also plays a major role in shaping the therapeutic applications of pro-angiogenic inorganic NPs. Right from simple adjustments in synthesis procedures, structural manipulations to chemical coupling with proteins, growth factors and drugs, all are key for producing biomaterials with desirable applications [118]. Currently, design and development of nanotechnology based therapies using inorganic NPs form an integral part of modern medicine. Considering the huge impact of inorganic nanomaterials on therapeutic angiogenesis, inorganic NPs have been integrated into therapeutic dentistry. Their use for regenerative dentistry is still in its infancy. Very few studies have been attempted to evaluate their application in dental tissue regeneration.

Reports suggest that magnetic biomaterials have attractive cell and tissue responses. Not much attention has been given to their potential applications on dental tissue regeneration. For this purpose, Yun *et al* explored dental tissue regeneration potential of magnetic nanofiber scaffolds [119]. For this study, the group incorporated magnetic NPs (MNPs) into nanofiber matrices and explored their biological activities employing various *in vitro* assays. HDPCs treated with MNPs, showed enhanced survivability in a dose dependent manner. Concomitantly, the ALP activity and ability of HDPCs to form mineralized nodule formation with MNPs treatment, also increased in a dose and time dependent manner. Additionally, gene expression studies revealed that the MNPs upregulated the expressions of ALP, OPN and DSPP mRNAs. Further, the author and his team also observed that MNPs promoted the two most important determining steps of angiogenesis, i.e., tube formation and migration in HUVECs. The group also performed deeper molecular biology studies to uncover the intracellular signaling pathways affected by the MNPs on HDPCs. In this regard, protein expression studies revealed that magnetic nanoscaffolds upregulated the expressions of Wnt3a, phosphorylated glycogen synthase kinase-3β (GSK-3 β), B-catenin proteins. The phosphorylation level of pERK, p38, and c-Jun N-terminal kinase (JNK) was also significantly higher in MNPs treated cells than the controls. Immunofluorescence studies revealed that NPs induced nuclear translocation of nuclear factor-β (NF-κβ), increased expression of p-Iκβα, p65 and also caused downregulation of Iκβα. The present data suggest a new approach to regeneration of dentin-pulp by using magnetic nanofiber scaffolds.

Further, iron oxide NPs (IONPs), γ-Fe_2O_3 nanoparticles (γ-IONPs), are biocompatible and are often used in biomedicine for the purpose of diagnosis (magnetic resonance imaging), drug or gene delivery agent, cell targeting agent, cancer therapy etc [120]. In one study, Xia *et al* demonstrated γ-IONPs scaffolded into calcium phosphate cement (IONP-CPC) can improve stem cell spreading, differentiation and bone mineral synthesis [121]. *In vitro* studies using human dental pulp stem cells (hDPSCs) revealed remarkable adhesion and spreading potential of the cells. Gene expression studies revealed that the NPs induced higher expression of ALP, COLIa, RUNX2 genes after 7 days and OCN genes after 14 days. TEM images showed that the NPs were uptaken by the cells via endocytosis. The group also indicated that enhancement of cell functions was mediated by the nanotopography of IONP-CPC, release of IONPs from IONP-CPC and intake of IONPs by cells. Thus, the group explained that newly developed IONP-CPC scaffolds hold promising applications of bone tissue engineering and re-generative medicine applications. Readers can find out more examples of applications of pro-angiogenic inorganic nanoparticles in dentistry in the following references [28, 122, 123].

Additionally, based on reports of promising applications of pro-angiogenic inorganic nanoparticles in treatment of vascular pathophysiology (ischemia, wound healing, cardiovascular diseases etc) we can predict the meritorious prospects of these pro-angiogenic nanoparticles for their application in vascular dentistry [124, 125, 126, 127, 128, 129].

The table 2.2 lists out some angiogenic nanoparticles that have important biomedical applications.

2.6 Challenges in application of nanotechnology in dentistry

Today, nanotechnology finds enormous applications in biology and medicine [138, 139]. However, despite their beneficial properties like smaller size, good optical, mechanical

Table 2.2. Angiogenic nanomaterials for various biological applications.

Sr. No.	Nanomaterials	Biological applications	Ref.
1.	Carbon nanotubes	Myocardial ischemia	[130]
2.	zinc-bioglass NPs	Dentistry	[131]
3.	Europium hydroxide nanorods	Recovery of myocardial infarction, tissue engineering, amelioration of heavy metal (Cd) toxicity,	[124, 132, 133]
4.	Terbium hydroxide nanorods	Wound healing	[129]
5.	Zinc oxide nanoflowers	Angiogenesis (Cerbral ischemia)	[15]
6.	Graphene oxide NPs	Angiogenesis	[134]
7.	Zinc oxide and zinc peroxide	Tissue regeneration	[135]
8.	Stem cell membrane coated poly (lactic-*co*-glycolic acid) (PLGA) NPs	Recovery of limb ischemia	[136]
9.	Cerium oxide NPs	Angiogenesis	[137]

and thermal properties and high reactivity, they are not fully accepted by biomedical sciences. Although the potential applications of NPs in modern dentistry is appreciable, health care professionals are still concerned for their clinical trial. This is because of the fact that there are reports on their potential toxic impact on living systems, compared to their bulk counterparts. The use of nanomedicines or nanomaterials in biological systems is still laden with numerous challenges at various steps including lack of biocompatibility, non-specificity, reactivity with biomolecules, poor stability, lack of co-relationship between *in vitro* and *in vivo* experimental outcomes, lack of information on effective dose range etc. Besides, NPs are often manufactured using toxic chemicals and following adverse synthesis conditions, which increase the toxicity potential of NPs.

Toxicity of NPs depends upon their size, shape, morphology, charge, surface functionalization as well as the ions released by them into the biological environment. Reports suggest that these nanomaterials, after entering into the body distribute and accumulate into different organs and tissues, for a long time and interfere with biological moieties and exert potential hazardous effects. However, a large number of studies need to be conducted, at both cellular and molecular level to explore their safety, efficacy, physicochemical properties and pharmacodynamic profiles.

Recently, scientists have identified that corroding NPs from implants release ions into the surrounding tissues as an important cause of chronic inflammatory responses [140, 141]. The rate of corrosion and the severity of toxic effect depend on implant composition. For instance, for filling of dental cavities, amalgams are often used as a restorative agent. These amalgams use 50% mercury (% w/w) as a component along with other elements. Since mercury is a highly toxic non-radioactive element, its use may impart toxicity within the body. As a result, adequate knowledge is necessary to maintain the use of mercury far below the permissible limit in order to minimize the risks of adverse health effects [142].

2.7 Conclusions

Conventional treatment options for common dental problems involve the use of antibiotics, filler materials, growth factors and dental stem cells imitating extracellular matrix and dental morphology [143]. Integration of nanotechnology and the above-mentioned approaches can be the way forward for regenerative dentistry in the near future [12]. Scientists' constant endeavors have led to the surge in the production and application of various forms of nano-biomaterials (composites, injectable materials, fibers etc) that mimic oral microenvironment and exhibit low toxicity, enhanced antimicrobial activities, improved protein–surface interactions for enhanced cell proliferation, improved surface topography and many other significant features. All these properties have rendered maximum responses for dental restoration.

Abbreviation

ALP	Alkaline phosphatase
ATCC	American Type Culture Collection
AuNPs	Gold nanoparticles
BMGs	Bulk metallic glasses
BMP-2	Bone morphogenetic protein
rBMSCs	Rat bone marrow mesenchymal stem cells
CCMS	Calcium carbonate nanoparticles
CAM	Chorio-allantoic membrane
Co-Cr	Cobalt-chrome
CuCh	NPs copper incorporated chitosan shell nanoparticles
DCPD	Dicalcium phosphate dihydrate
HDPCs	Human dental pulp cells
HDMEC	Human dermal microvascular endothelial cells
HP	Highly esterified pectin
HUVEC	Human umbilical vein endothelial cells
IONP-CPCs	Iron oxide incorporated calcium phosphate cement scaffolds
JNK	c-Jun N-terminal kinase
MG63	Osteoblast-like human cell line
MPTS	γ-methacryloxypropyl- trimethoxysilane
MIC	Minimum inhibitory concentration
MMPs	Matrix metalloproteinases
MWCNTs	Multi-walled carbon nanotubes
OPN	Osteopontin
PLGA	Poly(lactic-co-glycolic acid)
PGA	Propylene glycol alginate,
PMCAS	Polymer modified citric acid solutions
TCP	Tricalcium phosphate
ZnO	Zinc oxide

Acknowledgements

Financial support from Nano Mission-DST (SR/NM/NS-1252/2013; GAP 570), New Delhi, to C.R.P. is duly acknowledged. P.B.T. is thankful to CSIR, New Delhi for supporting her Senior Research Fellowship. We thank Director, CSIR-IICT

(Ms. No. IICT/Pubs./2021/017, dated January 20, 2021) for providing all the required facilities to carry out the work.

References

[1] Thesleff I and Tummers M 2009 Tooth organogenesis and regeneration *StemBook* (Cambridge, MA: Harvard Stem Cell Institute)

[2] Goldberg M, Kulkarni A B, Young M and Boskey A 2011 Dentin: structure, composition and mineralization: the role of dentin ECM in dentin formation and mineralization *Front. Biosci.* **3** 711

[3] Balic A 2018 Biology explaining tooth repair and regeneration: a mini-review *Gerontology* **64** 382–8

[4] Lynnerup N and Klaus H D 2019 Fundamentals of human bone and dental biology: structure, function, and development *Ortner's Identification of Pathological Conditions in Human Skeletal Remains* (Amsterdam: Elsevier) pp 35–58

[5] Yen A-H and Yelick P 2011 Dental tissue regeneration–a mini-review *Gerontology* **57** 85–94

[6] Giacaman R, Perez V and Carrera C 2016 Mineralization processes in hard tissues: teeth. *Biomineralization and Biomaterials* (Amsterdam: Elsevier) pp 147–85

[7] Chen H and Liu Y T 2014 *Advanced Ceramics for Dentistry* (Amsterdam: Elsevier) pp 5–21

[8] Zhou T, Pan J and Wu P *et al* 2019 Dental follicle cells: roles in development and beyond *Stem Cells Int.* **2019** 9159605

[9] Kim T-H, Bae C-H and Lee J-Y *et al* 2015 Temporo-spatial requirement of Smad4 in dentin formation *Biochem. Biophys. Res. Commun.* **459** 706–12

[10] Olley R, Xavier G M and Seppala M *et al* 2014 Expression analysis of candidate genes regulating successional tooth formation in the human embryo *Front. Physiol.* **5** 445

[11] Tompkins K 2006 Molecular mechanisms of cytodifferentiation in mammalian tooth development *Connect. Tissue Res.* **47** 111–8

[12] Xavier J R, Desai P, Varanasi V G, Al-Hashimi I and Gaharwar A K 2015 Advanced nanomaterials: promises for improved dental tissue regeneration *Nanotechnology in Endodontics* (Berlin: Springer) pp 5–22

[13] Adair T H and Montani J-P A 2010 *Colloquium Series on Integrated Systems Physiology: from Molecule to Function* 2010 (Williston, VT: Morgan & Claypool Life Sciences) pp 1–84

[14] Zhao M, Wang J, Xi X, Tan N and Zhang L 2018 SNHG12 promotes angiogenesis following ischemic stroke via regulating miR-150/VEGF pathway *Neuroscience* **390** 231–40

[15] Barui A K, Jhelum P and Nethi S K *et al* 2020 Potential therapeutic application of zinc oxide nanoflowers in the cerebral ischemia rat model through neuritogenic and neuroprotective properties *Bioconjugate Chem.* **31** 895–906

[16] Thurston G and Kitajewski J 2008 VEGF and Delta-Notch: interacting signalling pathways in tumour angiogenesis *Br. J. Cancer* **99** 1204–9

[17] Das S, Roy A and Barui A K *et al* 2020 Anti-angiogenic vanadium pentoxide nanoparticles for the treatment of melanoma and their *in vivo* toxicity study *Nanoscale* **12** 7604–21

[18] Jeong I-H, Bae W-Y, Choi J-S and Jeong J-W 2020 Ischemia induces autophagy of endothelial cells and stimulates angiogenic effects in a hindlimb ischemia mouse model *Cell Death Dis.* **11** 1–11

[19] Johnson T, Zhao L, Manuel G, Taylor H and Liu D 2019 Approaches to therapeutic angiogenesis for ischemic heart disease *J. Mol. Med.* **97** 141–51

[20] Ucuzian A A, Gassman A A, East A T and Greisler H P 2010 Molecular mediators of angiogenesis *J. Burn Care Res.* **31** 158–75
[21] Neve A, Cantatore F P, Maruotti N, Corrado A and Ribatti D 2014 Extracellular matrix modulates angiogenesis in physiological and pathological conditions *BioMed Res. Int.* **2014** 756078
[22] Krock B L, Skuli N and Simon M C 2011 Hypoxia-induced angiogenesis: good and evil *Genes Cancer* **2** 1117–33
[23] Mao J 2011 Tooth regeneration: current status *Preprosthetic and Maxillofacial Surgery* (Amsterdam: Elsevier) pp 363–73
[24] Tatullo M, Marrelli M, Shakesheff K M and White L J 2015 Dental pulp stem cells: function, isolation and applications in regenerative medicine *J. Tissue Eng. Regen. Med.* **9** 1205–16
[25] Aranha A M, Zhang Z, Neiva K G, Costa C A, Hebling J and Nör J E 2010 Hypoxia enhances the angiogenic potential of human dental pulp cells *J. Endodontics* **36** 1633–7
[26] Hilkens P, Fanton Y and Martens W *et al* 2014 Pro-angiogenic impact of dental stem cells *in vitro* and *in vivo Stem Cell Res.* **12** 778–90
[27] Roberts-Clark D and Smith A 2000 Angiogenic growth factors in human dentine matrix *Arch. Oral Biol.* **45** 1013–6
[28] Saghiri M A, Asatourian A, Sorenson C M and Sheibani N 2015 Role of angiogenesis in endodontics: contributions of stem cells and proangiogenic and antiangiogenic factors to dental pulp regeneration *J. Endodontics* **41** 797–803
[29] Caviedes-Bucheli J, Gomez-Sosa J, Azuero-Holguin M, Ormeño-Gomez M, Pinto-Pascual V and Munoz H 2017 Angiogenic mechanisms of human dental pulp and their relationship with substance P expression in response to occlusal trauma *Int. Endod. J.* **50** 339–51
[30] Kim Y-S, Min K-S, Jeong D-H, Jang J-H, Kim H-W and Kim E-C 2010 Effects of fibroblast growth factor-2 on the expression and regulation of chemokines in human dental pulp cells *J. Endodontics* **36** 1824–30
[31] Vaseenon S, Chattipakorn N and Chattipakorn S C 2020 The possible role of basic fibroblast growth factor in dental pulp *Arch. Oral Biol.* **109** 104574
[32] Takeuchi N, Hayashi Y and Murakami M *et al* 2015 Similar *in vitro* effects and pulp regeneration in ectopic tooth transplantation by basic fibroblast growth factor and granulocyte-colony stimulating factor *Oral Dis.* **21** 112–32
[33] Ishimatsu H, Kitamura C and Morotomi T *et al* 2009 Formation of dentinal bridge on surface of regenerated dental pulp in dentin defects by controlled release of fibroblast growth factor–2 from gelatin hydrogels *J. Endodontics* **35** 858–65
[34] Kim J, Park J C and Kim S H *et al* 2014 Treatment of FGF-2 on stem cells from inflamed dental pulp tissue from human deciduous teeth *Oral Dis.* **20** 191–204
[35] Baranova J, Büchner D, Götz W, Schulze M and Tobiasch E 2020 Tooth formation: are the hardest tissues of human body hard to regenerate? *Int. J. Mol. Sci.* **21** 4031
[36] Aparicio C, Gil F and Planell J 2002 Human-osteoblast proliferation and differentiation on grit-blasted and bioactive titanium for dental applications *J. Mater. Sci., Mater. Med.* **13** 1105–11
[37] Bengtsson U G and Hylander L D 2017 Increased mercury emissions from modern dental amalgams *Biometals* **30** 277–83
[38] Peng J-Y, Botelho M and Matinlinna J 2012 Silver compounds used in dentistry for caries management: a review *J. Dent.* **40** 531–41

[39] Swift E Jr 1989 *In vitro* caries-inhibitory properties of a silver cermet *J. Dent. Res.* **68** 1088–93
[40] Pissiotis E and Spngberg L 2000 Reaction of bony tissue to implanted silver glass ionomer and a reinforced zinc oxide-eugenol cement *Oral Surg. Oral Med. Oral Pathol. Oral Radiol. Endodontol.* **89** 623–9
[41] Sheikh F A, Barakat N A and Kanjwal M A *et al* 2010 Electrospun titanium dioxide nanofibers containing hydroxyapatite and silver nanoparticles as future implant materials *J. Mater. Sci., Mater. Med.* **21** 2551–9
[42] Denry I and Kelly J R 2008 State of the art of zirconia for dental applications *Dent. Mater.* **24** 299–307
[43] Malkondu Ö, Tinastepe N, Akan E and Kazazoğlu E 2016 An overview of monolithic zirconia in dentistry *Biotechnol. Biotechnol. Equip.* **30** 644–52
[44] Huang C-C, Narayanan R, Alapati S and Ravindran S 2016 Exosomes as biomimetic tools for stem cell differentiation: applications in dental pulp tissue regeneration *Biomaterials* **111** 103–15
[45] Rodríguez-Lozano F-J, Insausti C-L and Iniesta F *et al* 2012 Mesenchymal dental stem cells in regenerative dentistry *Med. Oral Patol. Oral Cir. Bucal* **17** e1062
[46] Nevins M, Camelo M, Nevins M L, Schenk R K and Lynch S E 2003 Periodontal regeneration in humans using recombinant human platelet-derived growth factor-BB (rhPDGF-BB) and allogenic bone *J. Periodontol.* **74** 1282–92
[47] Sigurdsson T J, Lee M B, Kubota K, Turek T J, Wozney J M and Wikesjö U M 1995 Periodontal repair in dogs: recombinant human bone morphogenetic protein-2 significantly enhances periodontal regeneration *J. Periodontol.* **66** 131–8
[48] Murakami S, Takayama S and Kitamura M *et al* 2003 Recombinant human basic fibroblast growth factor (bFGF) stimulates periodontal regeneration in class II furcation defects created in beagle dogs *J. Periodontal Res.* **38** 97–103
[49] Khan A, Hussain A and Sidra L *et al* 2017 Fabrication and *in vivo* evaluation of hydroxyapatite/carbon nanotube electrospun fibers for biomedical/dental application *Mater. Sci. Eng.* C **80** 387–96
[50] Khan W, Muntimadugu E, Jaffe M and Domb A J 2014 Implantable medical devices *Focal Controlled Drug Delivery* (Berlin: Springer) pp 33–59
[51] Ratner B D 2001 Replacing and renewing: synthetic materials, biomimetics, and tissue engineering in implant dentistry *J. Dent. Educ.* **65** 1340–7
[52] LaVenus S, Louarn G and Layrolle P 2010 Nanotechnology and dental implants *Int. J. Biomater.* **2010** 915327
[53] Inoue T and Matsuzaka K 2015 Surface modification of dental implant improves implant–tissue interface *Interface Oral Health Science 2014* (Berlin: Springer) pp 33–44
[54] Ananth H, Kundapur V, Mohammed H, Anand M, Amarnath G and Mankar S 2015 A review on biomaterials in dental implantology *Int. J. Biomed. Sci.: IJBS* **11** 113
[55] Love R and Jenkinson H 2002 Invasion of dentinal tubules by oral bacteria *Crit. Rev. Oral Biol. Med.* **13** 171–83
[56] Chiang Y-C, Lin H-P and Chang H-H *et al* 2014 A mesoporous silica biomaterial for dental biomimetic crystallization *ACS Nano* **8** 12502–13
[57] Zhang J, Wu R and Fan Y *et al* 2014 Antibacterial dental composites with chlorhexidine and mesoporous silica *J. Dent. Res.* **93** 1283–9
[58] Cheung H-Y, Lau K-T, Lu T-P and Hui D 2007 A critical review on polymer-based bio-engineered materials for scaffold development *Compos. Part B: Eng.* **38** 291–300

[59] Muzzarelli R A 2011 Chitosan composites with inorganics, morphogenetic proteins and stem cells, for bone regeneration *Carbohydr. Polym.* **83** 1433–45

[60] Muzzarelli R A 2009 Chitins and chitosans for the repair of wounded skin, nerve, cartilage and bone *Carbohydr. Polym.* **76** 167–82

[61] Dash M, Chiellini F, Ottenbrite R M and Chiellini E 2011 Chitosan—a versatile semi-synthetic polymer in biomedical applications *Prog. Polym. Sci.* **36** 981–1014

[62] Pallela R, Venkatesan J, Janapala V R and Kim S K 2012 Biophysicochemical evaluation of chitosan-hydroxyapatite-marine sponge collagen composite for bone tissue engineering *J. Biomed. Mater. Res.* A **100** 486–95

[63] Eniwumide J, Yuan H, Cartmell S, Meijer G and de Bruijn J D 2007 Ectopic bone formation in bone marrow stem cell seeded calcium phosphate scaffolds as compared to autograft and (cell seeded) allograft *Eur. Cells. Mater.* **14** 30–8

[64] Tanase C, Sartoris A, Popa M, Verestiuc L, Unger R and Kirkpatrick C 2013 *In vitro* evaluation of biomimetic chitosan–calcium phosphate scaffolds with potential application in bone tissue engineering *Biomed. Mater.* **8** 025002

[65] Cao X, Cai X, Chen R, Zhang H, Jiang T and Wang Y 2019 A thermosensitive chitosan-based hydrogel for sealing and lubricating purposes in dental implant system *Clin. Implant Dent. Relat. Res.* **21** 324–35

[66] Prasad S, Ehrensberger M, Gibson M P, Kim H and Monaco E A Jr 2015 Biomaterial properties of titanium in dentistry *J. Oral Biosci.* **57** 192–9

[67] Gaffney-Stomberg E 2019 The impact of trace minerals on bone metabolism *Biol. Trace Elem. Res.* **188** 26–34

[68] Yamaguchi M 1998 Role of zinc in bone formation and bone resorption *J. Trace Elem. Exp. Med.* **11** 119–35

[69] Zreiqat H, Howlett C and Zannettino A *et al* 2002 Mechanisms of magnesium-stimulated adhesion of osteoblastic cells to commonly used orthopaedic implants *J. Biomed. Mater. Res.* **62** 175–84

[70] Hu H, Zhang W, Qiao Y, Jiang X, Liu X and Ding C 2012 Antibacterial activity and increased bone marrow stem cell functions of Zn-incorporated TiO_2 coatings on titanium *Acta Biomater.* **8** 904–15

[71] Jin G, Cao H, Qiao Y, Meng F, Zhu H and Liu X 2014 Osteogenic activity and antibacterial effect of zinc ion implanted titanium *Colloids Surf.* B **117** 158–65

[72] Yu Y, Jin G, Xue Y, Wang D, Liu X and Sun J 2017 Multifunctions of dual Zn/Mg ion co-implanted titanium on osteogenesis, angiogenesis and bacteria inhibition for dental implants *Acta Biomater.* **49** 590–603

[73] Huang H-L, Chang Y-Y, Lai M-C, Lin C-R, Lai C-H and Shieh T-M 2010 Antibacterial TaN-Ag coatings on titanium dental implants *Surf. Coat. Technol.* **205** 1636–41

[74] Grech J and Antunes E 2019 Zirconia in dental prosthetics: a literature review *J. Mater. Res. Technol.* **8** 4956–64

[75] Nistor L, Grădinaru M and Rîcă R *et al* 2019 Zirconia use in dentistry-manufacturing and properties *Curr. Health Sci. J.* **45** 28

[76] Abd El-Ghany O S and Sherief A H 2016 Zirconia based ceramics, some clinical and biological aspects *Future Dent. J.* **2** 55–64

[77] Oliva J, Oliva X and Oliva J D 2008 Zirconia implants and all-ceramic restorations for the esthetic replacement of the maxillary central incisors *Eur. J. Esthetic Dent.* **3** 174–85

[78] Gautam C, Joyner J, Gautam A, Rao J and Vajtai R 2016 Zirconia based dental ceramics: structure, mechanical properties, biocompatibility and applications *Dalton Trans.* **45** 19194–215
[79] Scarano A, Di Carlo F, Quaranta M and Piattelli A 2003 Bone response to zirconia ceramic implants: an experimental study in rabbits *J. Oral Implantol.* **29** 8–12
[80] Priyadarsini S, Mukherjee S and Mishra M 2018 Nanoparticles used in dentistry: a review *J. Oral Biol. Craniofac. Res.* **8** 58–67
[81] Hamouda I M 2012 Current perspectives of nanoparticles in medical and dental biomaterials *J. Biomed. Res.* **26** 143–51
[82] Song W and Ge S 2019 Application of antimicrobial nanoparticles in dentistry *Molecules* **24** 1033
[83] Bennadi D 2014 Antimicrobial stewardship—an alarming call in dentistry *Int. J. Pharm. Sci.* **6** 46–9
[84] Heta S and Robo I 2018 The side effects of the most commonly used group of antibiotics in periodontal treatments *Med. Sci.* **6** 6
[85] Sánchez-López E, Gomes D and Esteruelas G *et al* 2020 Metal-based nanoparticles as antimicrobial agents: an overview *Nanomaterials* **10** 292
[86] Cao W, Zhang Y and Wang X *et al* 2018 Novel resin-based dental material with anti-biofilm activity and improved mechanical property by incorporating hydrophilic cationic copolymer functionalized nanodiamond *J. Mater. Sci., Mater. Med.* **29** 162
[87] Jiao Y, Tay F R, Niu L-n and Chen J-h 2019 Advancing antimicrobial strategies for managing oral biofilm infections *Int. J. Oral Sci.* **11** 1–11
[88] Rodrigues C T, De Andrade F and De Vasconcelos L *et al* 2018 Antibacterial properties of silver nanoparticles as a root canal irrigant against Enterococcus faecalis biofilm and infected dentinal tubules *Int. Endodontic J.* **51** 901–11
[89] Jadhav R D, Bhide S V, Prasad B and Shimpi J 2016 Applications of silver nanoparticles in prosthodontics: an overview *Eur. J. Prosthodont.* **4** 45
[90] Cheng L, Weir M D and Xu H H *et al* 2012 Antibacterial amorphous calcium phosphate nanocomposites with a quaternary ammonium dimethacrylate and silver nanoparticles *Dent. Mater.* **28** 561–72
[91] Rodrigues M C, Rolim W R and Viana M M *et al* 2020 Biogenic synthesis and antimicrobial activity of silica- coated silver nanoparticles for esthetic dental applications *J. Dent.* **96** 103327
[92] Heo D N, Ko W-K and Lee H R *et al* 2016 Titanium dental implants surface-immobilized with gold nanoparticles as osteoinductive agents for rapid osseointegration *J. Colloid Interface Sci.* **469** 129–37
[93] Salem D, Natto Z, Elangovan S and Karimbux N 2016 Usage of bone replacement grafts in periodontics and oral implantology and their current levels of clinical evidence—A systematic assessment *J. Periodontol.* **87** 872–9
[94] Misch C 2001 The importance of dental implants *Gen. Dent.* **49** 38–45
[95] Crisan L, Crisan B and Soritau O *et al* 2015 *In vitro* study of biocompatibility of a graphene composite with gold nanoparticles and hydroxyapatite on human osteoblasts *J. Appl. Toxicol.* **35** 1200–10
[96] Jadhav K, Rajeshwari H and Deshpande S *et al* 2018 Phytosynthesis of gold nanoparticles: characterization, biocompatibility, and evaluation of its osteoinductive potential for application in implant dentistry *Mater. Sci. Eng.* C **93** 664–70

[97] Rokaya D, Srimaneepong V, Sapkota J, Qin J, Siraleartmukul K and Siriwongrungson V 2018 Polymeric materials and films in dentistry: an overview *J. Adv. Res.* **14** 25–34

[98] Beyer M, Reichert J, Heurich E, Jandt K D and Sigusch B W 2010 Pectin, alginate and gum arabic polymers reduce citric acid erosion effects on human enamel *Dent. Mater.* **26** 831–9

[99] Beyer M, Reichert J, Sigusch B W, Watts D C and Jandt K D 2012 Morphology and structure of polymer layers protecting dental enamel against erosion *Dent. Mater.* **28** 1089–97

[100] Carvalho T and Lussi A 2014 Combined effect of a fluoride-, stannous-and chitosan-containing toothpaste and stannous-containing rinse on the prevention of initial enamel erosion–abrasion *J. Dent.* **42** 450–9

[101] Covarrubias C, Trepiana D and Corral C 2018 Synthesis of hybrid copper-chitosan nanoparticles with antibacterial activity against cariogenic Streptococcus mutans *Dental Mater. J.* **37** 379–84

[102] Liu Y J and Chen X 2003 Evaluations of the effective material properties of carbon nanotube-based composites using a nanoscale representative volume element *Mech. Mater.* **35** 69–81

[103] Guldi D M and Martín N 2010 *Carbon Nanotubes and Related Structures: Synthesis, Characterization, Functionalization, and Applications* (New York: Wiley)

[104] Zhang Y, Lim C T, Ramakrishna S and Huang Z-M 2005 Recent development of polymer nanofibers for biomedical and biotechnological applications *J. Mater. Sci., Mater. Med.* **16** 933–46

[105] Kai D, Tan M J and Prabhakaran M P *et al* 2016 Biocompatible electrically conductive nanofibers from inorganic-organic shape memory polymers *Colloids Surf.* B **148** 557–65

[106] Tetteh G, Khan A, Delaine-Smith R, Reilly G and Rehman I 2014 Electrospun polyurethane/hydroxyapatite bioactive Scaffolds for bone tissue engineering: the role of solvent and hydroxyapatite particles *J. Mech. Behav. Biomed. Mater.* **39** 95–110

[107] Tian M, Gao Y and Liu Y *et al* 2007 Bis-GMA/TEGDMA dental composites reinforced with electrospun nylon 6 nanocomposite nanofibers containing highly aligned fibrillar silicate single crystals *Polymer* **48** 2720–8

[108] Qasim S B and Rehman I U 2018 Application of nanomaterials in dentistry *Micro and Nanomanufacturing* vol. II (Berlin: Springer) pp 319–36

[109] Subramani K and Ahmed W 2012 *Nanobiomaterials in Clinical Dentistry* (Norwich, NY: William Andrew)

[110] Padovani G C, Feitosa V P and Sauro S *et al* 2015 Advances in dental materials through nanotechnology: facts, perspectives and toxicological aspects *Trends Biotechnol.* **33** 621–36

[111] Bhattarai G, Lee Y-H, Lee M-H, Park I-S and Yi H-K 2015 Insulin-like growth factor binding protein-3 affects osteogenic efficacy on dental implants in rat mandible *Mater. Sci. Eng.* C **55** 490–6

[112] Miyakoshi N, Richman C, Qin X, Baylink D J and Mohan S 1999 Effects of recombinant insulin-like growth factor-binding protein-4 on bone formation parameters in mice *Endocrinology* **140** 5719–28

[113] Kawai M and Rosen C J 2012 The insulin-like growth factor system in bone: basic and clinical implications *Endocrinol. Metab. Clin.* **41** 323–33

[114] Chard T 1994 Insulin-like growth factors and their binding proteins in normal and abnormal human fetal growth *Growth Regul.* **4** 91

[115] Raschke M, Wildemann B and Inden P *et al* 2002 Insulin-like growth factor-1 and transforming growth factor-β1 accelerates osteotomy healing using polylactide-coated implants as a delivery system: a biomechanical and histological study in minipigs *Bone* **30** 144–51

[116] Yilgor P, Hasirci N and Hasirci V 2010 Sequential BMP-2/BMP-7 delivery from polyester nanocapsules *J. Biomed. Mater. Res. Part* A **93** 528–36
[117] Saghiri M A, Asatourian A, Garcia-Godoy F and Sheibani N 2016 A new era considering inorganic trace elements and biological activity of dental biomaterials (angiogenic activity) *Acta Biomater. Odontol. Scand.* **2** 93–4
[118] Kargozar S, Ramakrishna S and Mozafari M 2019 Chemistry of biomaterials: future prospects *Curr. Opin. Biomed. Eng.* **10** 181–90
[119] Yun H-M, Kang S-K and Singh R K *et al* 2016 Magnetic nanofiber scaffold-induced stimulation of odontogenesis and pro-angiogenesis of human dental pulp cells through Wnt/MAPK/NF-κB pathways *Dent. Mater.* **32** 1301–11
[120] Shabestari Khiabani S, Farshbaf M, Akbarzadeh A and Davaran S 2017 Magnetic nanoparticles: preparation methods, applications in cancer diagnosis and cancer therapy *Artif. Cells Nanomed. Biotechnol.* **45** 6–17
[121] Xia Y, Chen H and Zhang F *et al* 2018 Injectable calcium phosphate scaffold with iron oxide nanoparticles to enhance osteogenesis via dental pulp stem cells *Artif. Cells Nanomed. Biotechnol.* **46** 423–33
[122] Liu W, Zhang G and Wu J *et al* 2020 Insights into the angiogenic effects of nanomaterials: mechanisms involved and potential applications *J. Nanobiotechnol.* **18** 1–22
[123] Alsuraifi A 2020 Metallic nanoparticles in dental biomaterials: a review *Int. J. Mol. Sci.* **3** 27–37
[124] Augustine R, Nethi S K, Kalarikkal N, Thomas S and Patra C R 2017 Electrospun polycaprolactone (PCL) scaffolds embedded with europium hydroxide nanorods (EHNs) with enhanced vascularization and cell proliferation for tissue engineering applications *J. Mater. Chem.* B **5** 4660–72
[125] Nethi S K, Veeriah V and Barui A K *et al* 2015 Investigation of molecular mechanisms and regulatory pathways of pro-angiogenic nanorods *Nanoscale* **7** 9760–70
[126] Barui A K, Veeriah V and Mukherjee S *et al* 2012 Zinc oxide nanoflowers make new blood vessels *Nanoscale* **4** 7861–9
[127] Patra C R and Barui A K 2013 Nanoflowers: a future therapy for cardiac and ischemic disease? *Nanomedicine* **8** 1735–8
[128] Barui A K, Nethi S K and Patra C R 2017 Investigation of the role of nitric oxide driven angiogenesis by zinc oxide nanoflowers *J. Mater. Chem.* B **5** 3391–403
[129] Nethi S K, Barui A K, Bollu V S, Rao B R and Patra C R 2017 Pro-angiogenic properties of terbium hydroxide nanorods: molecular mechanisms and therapeutic applications in wound healing *ACS Biomater. Sci. Eng.* **3** 3635–45
[130] Fan C, Joshi J and Li F *et al* 2020 Nanoparticle-mediated drug delivery for treatment of ischemic heart disease *Front. Bioeng. Biotechnol.* **8** 687
[131] Zhang J, Park Y-D and Bae W-J *et al* 2015 Effects of bioactive cements incorporating zinc-bioglass nanoparticles on odontogenic and angiogenic potential of human dental pulp cells *J. Biomater. Appl.* **29** 954–64
[132] Vemuri S K, Nethi S K, Banala R R, Goli P V S, Annapareddy V G R and Patra C R 2019 Europium hydroxide nanorods (EHNs) ameliorate isoproterenol-induced myocardial infarction: an *in vitro* and *in vivo* investigation *ACS Appl. Bio Mater.* **2** 1078–87
[133] Roy A, Nethi S K, Suganya N, Raval M, Chatterjee S and Patra C R 2020 Attenuation of cadmium-induced vascular toxicity by pro-angiogenic nanorods *Mater. Sci. Eng.* C **115** 111108
[134] Mukherjee S, Sriram P and Barui A K *et al* 2015 Graphene oxides show angiogenic properties *Adv. Healthcare Mater.* **4** 1722–32

[135] Ahtzaz S, Nasir M and Shahzadi L *et al* 2017 A study on the effect of zinc oxide and zinc peroxide nanoparticles to enhance angiogenesis-pro-angiogenic grafts for tissue regeneration applications *Mater. Des.* **132** 409–18
[136] Bose R J, Kim B J and Arai Y *et al* 2018 Bioengineered stem cell membrane functionalized nanocarriers for therapeutic targeting of severe hindlimb ischemia *Biomaterials* **185** 360–70
[137] Das S, Singh S and Dowding J M *et al* 2012 The induction of angiogenesis by cerium oxide nanoparticles through the modulation of oxygen in intracellular environments *Biomaterials* **33** 7746–55
[138] Kargozar S, Baino F, Hamzehlou S, Hamblin M R and Mozafari M 2020 Nanotechnology for angiogenesis: opportunities and challenges *Chem. Soc. Rev.* **49** 5008–57
[139] Kargozar S and Mozafari M 2018 Nanotechnology and nanomedicine: start small, think big *Mater. Today Proc.* **5** 15492–500
[140] Posada O M, Tate R and Grant M 2015 Toxicity of cobalt–chromium nanoparticles released from a resurfacing hip implant and cobalt ions on primary human lymphocytes *in vitro J. Appl. Toxicol.* **35** 614–22
[141] Lee Y J, Kim J and Oh J *et al* 2012 Ion-release kinetics and ecotoxicity effects of silver nanoparticles *Environ. Toxicol. Chem.* **31** 155–9
[142] Rathore M, Singh A and Pant V A 2012 The dental amalgam toxicity fear: a myth or actuality *Toxicol. Int.* **19** 81
[143] Bartold P M, Xiao Y, Lyngstaadas S P, Paine M L and Snead M L 2006 Principles and applications of cell delivery systems for periodontal regeneration *Periodontology 2000* **41** 123–35

Chapter 3

Nanotechnology in dental implants

Saptarshi Chatterjee, Shatarupa Biswas and Banashree Roy

Teeth are a vital sensory organ of any individual's life span. They give extra value to life by association with activities like chewing, speech and several facial activities, even smiling. However, this organ can be lost for several reasons like periodontal disease, carries, dental plague, dental trauma etc. To bring back smiling faces dental implantation is very much required. The word periodontal is actually related to periodontitis, which means the inflammation of tissues around teeth due to improper hygiene of teeth (broadly oral cavity). This condition finally leads to the loss of teeth. So implantation is highly required. Apart from sugar rich diets, poor oral hygiene triggers the formation of dental carries and disease by the action of bacteria (*S. mutans, Lactobacillus, Actinomycetes* etc) that also results in loss of teeth. In dental trauma the ligament that holds our teeth in the proper place feels some extra pressure that results in displacement of teeth and finally loss of teeth. So to gain better knowledge about this disease and its consequences it is very much necessary to take a tour through the implantation industry.

3.1 Dental implantation

It is a technique through which the function of this vital organ i.e. teeth is revived (partially/completely) in edentulous patients. The term 'osseointegration' is the base of dental implantation in which osteoblasts (cells actually employed to form new bone) grow and make contact with the surface of an artificial implant. Dental implant surgery involves the replacement of tooth roots with a metal, screw-like post that replaces the damaged portion of the teeth with an artificial one to restore functionality (figure 3.1).

3.2 History of dental implantology

The history of dental implants emerges at 3000 BC during the ancient Egyptian Civilization (first a copper stud was used in the mouth). As early as 2000 BC carved bamboo pegs were used as dental implants in china. The early era of dental surgery

doi:10.1088/978-0-7503-3671-0ch3

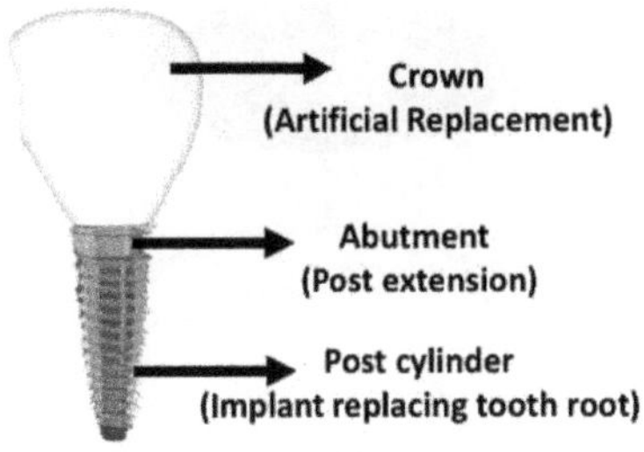

Figure 3.1. Dental implant technology.

was between 1600–1800 BC. A French dentist named Maggiolo J first described the technique of modern dental implants with 18-karat gold alloy in his book 'Le Manuel de I 'Art du Dentiste' [1]. The birth of recent dental implantology starts at the turn of 19th century. In 1891, Znamenski and Hillischer implanted teeth made of porcelain. The next 20 years was the time to innovate in the field of dental implantology. At this time the Strock brothers experimented with and introduced endodontic implants. The time between 1950–1970 saw the burst of new development and modifications of designs of earlier concepts. In 1951 the American Academy of Implant Dentistry was established [2]. Blade vents by Linkow and Cherchève and the ceramic bone screw by Sandhaus were the two important results of dental implantology at that time [3]. After 1970, the main focus of study was based on the factors that determine clinical success or failure of dental implants. In the end of the 1970s as per certain findings Brånemark *et al* introduced threaded screw implants made up of pure titanium [4]. He observed TiO_2/titanium oxide covering a metal surface could get direct bone–implant contact. In 1982 the first guidelines for implant dentistry were created at the Toronto Conference on Osseointegration. To date, there have been steady progress and improvements in the domain of dental implants.

3.3 Development and surface modifications

Among most recent developments of implants those that are associated with conical and cylindrical shapes are most cost effective and available. Surface structures as well as surface macroporosity play an important role in implantation that actually helps in the process of osseointegration. Recently, researchers worked on it and developed highly porous-surfaced implants with a trabecular bone-like surface topography (Trabecular Metal Zimmer®) based on tantalum at Dental Implant System (Parsippany, NJ, USA) [5, 6].

Strategies for surface modifications: Several strategies for surface modification have been developed:

(a) **Implant surface with micro roughness:** micro roughness on the implant surface increases the osseointegration capacity compared to smooth surfaces. Among multiple production methods of micro rough surface some important ones are grit blasting with silica, TiO_2 with subsequent acid etching, machining, sandblasting, micron sized metal bead coating etc [7].

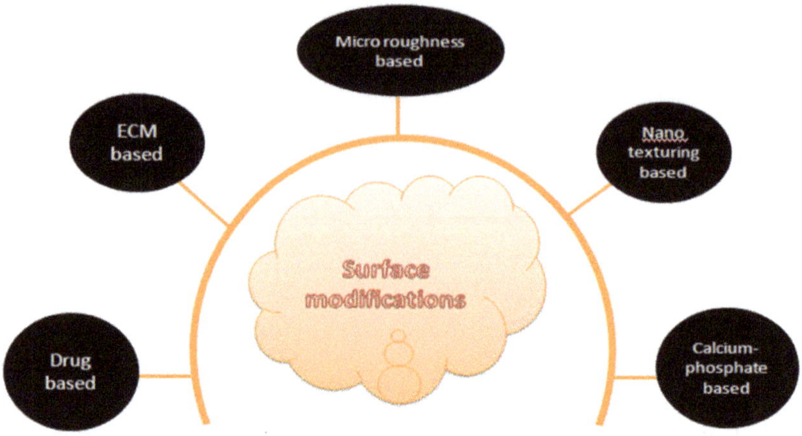

Figure 3.2. Several coating methods employed for surface modification.

(b) **Implant surface modification with nano-texturing:** This technique is based on mimicking the nano pattern of bone structure that may increase surface energy and helps in osseointegrtion. It deals with particles of nanometer scale (<100 nm). For example, TiO_2 nanoparticle can be attached to a titanium surface that ultimately leads to enhancement of osseointegration [8] (figure 3.2).

3.4 Complications during implantation

Apart from the usefulness of implantation, several complications exist for both short- and long- term implantations that include [9]:

1. **Implantation fracture**: The main culprit behind implant fracture is bone resorption (breaking down of bone tissue by osteoclasts). It actually decreases the amount of supporting bones around implants and thus the implant feels an immense stress. The detection is done by radiography or computer tomography (CT) as a radiolucent line.
2. **Improper osseointegration:** In this case the normal bone fails to grow properly and thus cannot integrate with an artificial implant due to erroneous osseointegration.
3. **Allergic reactions**: If someone is allergic to materials like titanium, this kind of implant can lead to major complications, including discomfort or swelling etc.
4. **Infection**: Sufficient and good dental hygiene is always recommended for dental implantation. Otherwise, it will lead to several major forms of infection where bacteria can adhere to the implant surface and form biofilm. This results in patients with hardening of tissue, redness and painful gingival inflammation surrounding the implants. These symptoms frequently lead to erosion and resorption of the bone. These problems are detected by radiography.
5. **Other complications**: Sinus problems, damage of implant due to immense force, failure in cementing, loss of marginal crestal bone, surgical trauma, failure in abutment, disease of gum etc.

Apart from the above-mentioned complications, implantation itself can cause several disturbances to natural vital structures surrounding the site of it. It can cause injury to these vital structures including the mandibular canal (neurovascular bundle), maxillary sinus, nasal cavity, and incisive foramen. Radiologists must analyze the images of these areas during pre and post-operative discussion. Dental implant failure is also associated with several factors: smoking, diagnosis with rheumatoid arthritis (an auto immune disease), diabetes etc.

3.5 Application of nanoparticles in dental implantation

3.5.1 Silver nanoparticles

In dentistry, silver (Ag) has been used since time immemorial in the form of metallic silver, silver nitrate and silver sulfadiazine for treatment of several bacterial infections [10]. Silver ion ($Ag+$) is a strong antibacterial agent with reasonable stability and broad spectrum antimicrobial effects on both Gram-positive and Gram-negative bacteria. The nanoparticles can continuously release silver ions, which may be considered the antibacterial activity to damage of the bacterial cell membrane [11]. The denaturation of cytoplasmic membrane can damage the organelles, generating reactive oxygen species (ROS) and the result is cell lysis. According to studies both *in vitro* and *in vivo*, silver-containing materials can inhibit bacterial attachment onto the dental implants. Silver nanoparticles are used as an antimicrobial agent in many fields of medicine and it is embedded in various film coatings which have been applied to implant surfaces to present antimicrobial activity. The Ag nanoparticles are known as biocompatible due to the low doses required in dental implants. Therefore, incorporating an appropriate amount of Ag not only enhances the antibacterial ability of the implant but also allows the surface to remain biocompatible [12]. Ag nanoparticles can be used in several treatments, like prosthetic treatment, endodontic treatment, dental implant treatment, orthodontic treatment, periodontal treatment etc (figure 3.3).

3.5.1.1 Prosthetic treatment

In Prosthetic treatment, acrylic resin is generally used to fabricate removable dentures. Stomatitis is a dental infection caused by opportunistic oral pathogens which can colonize acrylic materials. In the addition of acrylic resin, Ag nanoparticles can inhibit the growth of several bacteria like *Streptococcus mutans, Staphylococcus aureus* and *Escherichia coli* [13]. *Candida albicans* is one of the opportunistic pathogens on a denture base. Ag nanoparticles have antifungal properties against this pathogen [14]. Ag nanoparticles can increase the thermal conductivity and compressive strength of an acrylic base [15].

3.5.1.2 Endodontic treatment

Endodontic treatment is also known as Root Canal Treatment (RCT). Gutta-percha is commonly used in root canal obturation. In detailed studies, gutta-percha was coated with Ag nanoparticles and it had antibacterial and antifungal properties. Gutta-percha can prevent bacterial leakages [16].

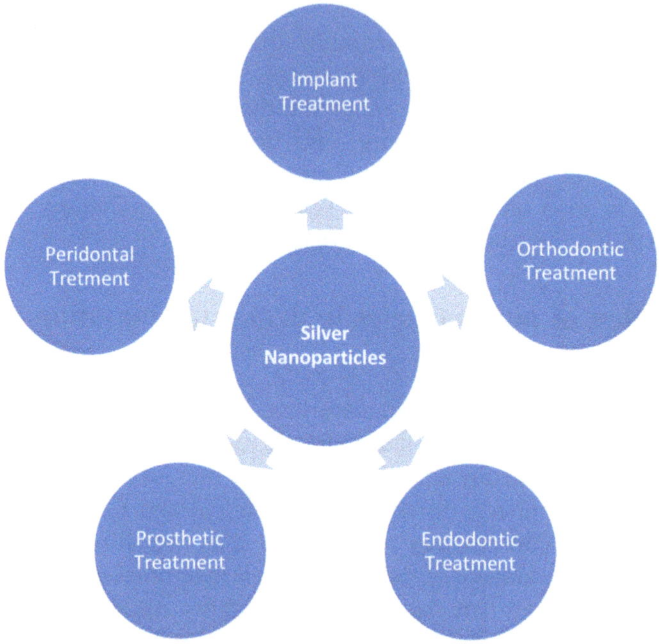

Figure 3.3. The antimicrobial application of silver nanoparticles in dentistry.

Sodium hypochlorite is considered as the gold standard for chemical disinfection of the root canal in endodontic treatment [17]. A study found that root canal irrigation with Ag nanoparticle solution could not affect the mechanical properties of dentine [18].

In dentistry, mineral trioxide aggregate has extensive applications, including pulp-capping, apexification and sealing perforations. To enhance its antimicrobial properties it is essential to add antimicrobial agents into mineral trioxide aggregate. Ag nanoparticles integrated with mineral trioxide aggregate can increase antimicrobial activity against anaerobic endodontic pathogens. Moreover, mineral trioxide aggregate combined with Ag nanoparticles, can increase the pH value and compressive strength with radiopacity and setting time [19]. In addition, Ag nanoparticles with mineral trioxide aggregate have good biocompatibility without inducing inflammatory reaction [20].

3.5.1.3 Dental implant treatment

Dental implant treatment is one of the most common and recognized ways to repair missing teeth. For their high strength, durability and biocompatibility, titanium implants are widely used in dentistry [21].

Peri-implant infection is a major impendence for implant treatment. Due to the certain risks of infection from peri-operation which continue in later life, the conventional antibacterial method is the finest method for this infection [22]. In detailed studies, the long-term antibacterial effect of titanium implants coated with Ag nanoparticles by steadily controlling Ag ion which is released with various coating methods [23]. On the implant surface the antibacterial effect can be continued for a long time by immobilizing Ag nanoparticles. At a very low concentration they can kill *Staphylococcus aureus* and *Pseudomonas aeruginosa*, which has no significant cytotoxic effects on osteoblastic cells [24]. In addition, Ag nanoparticles attached with titanium can improve the bone mineral density, bone formation and trabecular pattern with no harm to tissues adjoining the dental implant [25].

3.5.1.4 Orthodontic treatment

Orthodontic treatment is a way of straightening or moving teeth, to improve the appearance of the teeth and how they work. Orthodontics is a specialty of dentistry that deals with the diagnosis, prevention, and correction of mal-positioned teeth and jaws, and misaligned bite patterns. It can also focus on modifying facial growth, known as dentofacial orthopedics. Abnormal alignment of the teeth and jaws is common. In orthodontic treatment Ag nanoparticles can be used to prevent enamel caries, i.e., white spot lesion, which is a common issue for patients. Orthodontic elastomeric modules like ligatures can be united with Ag nanoparticles to prevent enamel caries [26], and they can prevent *Steptococcus mutans* in the materials caused by contact with bacteria, and can inhibit bacterial action in the region of brackets and wires by the release of silver ions [27]. In addition, the materials release silver ions for four months for long-term antibacterial effect.

3.5.1.5 Periodontal treatment

In periodontal treatment, periodontitis is a dominant chronic inflammatory disease caused by various microorganisms. Adequate infection control is very necessary for periodontal treatment with the disruption of the biofilm and suppression of inflammation [28]. Ag nanoparticles occupy antibacterial properties without the generation of bacterial resistance compared to the traditional antibiotics [29]. Ag nanoparticles can raise their bactericidal properties when they assemble with antibiotics, including cefotaxime, ceftazidime, meropenem and ciprofloxacin. Importantly, when nanoparticles combine with inactive antibiotics, they can restore strong antibacterial activity against multi-resistant bacterial strains [29].

In addition, Ag nanoparticles provide an anti-inflammatory effect with the help of modulation of the level of inflammatory cytokines and growth factors [30]. Periodontal dressing implicated with Ag nanoparticles can be applied to treat gingival wounds. However, in periodontal dressing Ag nanoparticles can give rise to inflammation and foreign body reaction [31].

3.5.2 Zinc oxide (ZnO) nanoparticles

ZnO nanoparticles have a large surface area relative to their size and high catalytic activity. The exact physical and chemical properties of ZnO nanoparticles depend on the different ways they are synthesized. In human skin ZnO nanoparticles become a suitable agent for addition to textiles and the surfaces that come in contact with the human body, such as sunscreens. ZnO nanoparticles have antibacterial activity in both Gram-positive and Gram-negative bacteria along with the spores which have high resistant capacity of temperature and pressure [32, 33]. Memarzadeh *et al* [34] tested a system containing a mixture of ZnO nanoparticles and nanohydroxyapatite that act as a coating material to reduce the bacterial fidelity and support the osteoblast growth. ZnO can be considered as an optimal coating for implants in both antimicrobial activity and biocompatibility. According to various studies, the substrate is entirely coated with ZnO nanoparticles which showed efficiency over that with composite surface coatings. However, all the tested cells maintained had normal morphology when attached onto the surface of nano ZnO coated substrates.

3.5.3 Copper oxide nanoparticles

Copper oxide nanoparticles appear as a brownish-black powder. They can be reduced to metallic copper when exposed to hydrogen or carbon monoxide under high temperature. CuO nanoparticles with unique biological, physical and chemical properties and antimicrobial effect, along with their low cost of preparation, attract the scientist's interest. CuO nanoparticles are a high potential antimicrobial agent. They are a proper alternative against biofilm formation within the oral cavity [35]. CuO nanoparticle's bactericidal activity depends on the size, stability and concentration when added to the growth medium [32].

3.5.4 Chlorhexidine (CHX) nanoparticles

CHX nanoparticles have a wide antimicrobial and antifungal spectrum and belong to the group of medicines known as antiseptic antibacterial agents. The application of CHX nanoparticles is widely distributed in medicine and dentistry [36]. They are mainly used in pre-surgical topical cleansing agents.

3.5.5 Al_2O_3 nanoparticles

Al_2O_3 is one of the main candidates in dental implant coatings because of its high hardness, sufficient stability and ability to improve osseointegration properties [37]. Using electrophoretic deposition method (EFD), Rsul *et al* [37] coated Ti dental implants with a nano-biocomposite of Al_2O_3 and $AgNO_3$. They implanted the screw type of dental implants in tibias of white male New Zealand rabbits. And the result is, the coated screws showed that bone trabeculae occupied a base of implant bed with osteoblast and osteocyte at 2 weeks. After 4 weeks, they saw progress in the healing process around the dental implant including new bone with haversian canals, i.e., osteoblast and osteocyte. They proposed that a multifunctional implant surface can be coated by alumina and silver nitrate.

3.6 Surface release of dental implant

To prevent bacterial colonization Ag nanoperticles are used in dentistry. In dentistry Ag nanoparticles have been applied in different areas, like endodontic treatment, dental implant treatment, orthodontic treatment [38–41]. However, resin-based composites holding Ag ion-implanted fillers that release the Ag ions have been found on oral cariogenic bacteria [42].

Therefore, from all these studies, Ag nanoparticles are released or Ag ions from those particles. Although, Ti based implants have been considered to be biologically inert, it has been found that the body can release non-dental implants as particulate debris over time [43, 44]. Ti particles released from implants have been found as regenerated bone and peri-implant tissues in animals [45, 46]. However, Ti ions also can be released from embedded implants in animals [47, 48].

3.7 Toxicity of nanoparticles

Currently, nanoparticles are widely used in the dental industry as implants. So we need to have complete idea about the toxicity of the materials along with physical, chemical properties and biocompatibilities. The nature as well as the entry route of the toxin has a greater impact on the toxicity. As the implants remain in direct touch with the oral cavity, it is much more necessary to evaluate the toxicity of every nanoparticle used in dental implantation. Here are some important examples of them.

Ag nanoparticles: Use of Ag nanoparticles is of past, present and future interest for the dental industry. Along with several advantages it also carries some disadvantages in the form of toxicity. Ag ions and Ag nanoparticles both have toxic effects. According to demonstration of different studies it can be concluded that the toxic effect of Ag nanoparticles is connected to the action of free Ag^+ ion in the medium [49]. Toxic effects associated with Ag nanoparticles includes: Argyria (high amount of exposure during dental filling leads to blue-gray discoloration of skin, internal organs, gums etc), oxidative stress, actin disruption, limits function of mitochondrion, obstruction of Na–K ATPase. According to an experiment it is known to induce ROS production as well as apoptosis and necrosis [50].

ZnO nanoparticles: Depending on the number of Zno nanoparticles and time, they can show toxic effects towards various kinds of mammalian cells like human erythrocytes, dermal fibroblast cell, gingival fibroblasts and ROS generation by the toxic effect of ZnO nanoparticles, which is an important consideration [51].

TiO_2 nanoparticles: Among all nanoparticles, TiO_2 is one of the extensively used ones in dentistry. Apart from inhalation, it can be acquired via implants. Long exposure to TiO_2 nanoparticles leads to skin risks. Other toxic effects include ROS production, inflammation, cell death; in severe cases cancer may occur (mainly in the TiO_2 production industry). In the case of titanium the fine particles are less toxic than nanoparticles. It also responsible for genotoxicity (actually it hampers DNA repair pathway) [52].

Apart from those above-mentioned, several other nanoparticles show severe toxic effects that include-zirconia (an agent that increases ROS production thus contributing to DNA damage), silica (causes silicosis, shows cytotoxicity, genotoxicity and sometimes immunotoxicity, even lung cancer can occur), copper oxide (causes oxidative stress and DNA damage that ultimately leads to cell death), etc. However, Al_2O_3 is an important material used in dentistry that is less toxic (except when high concentration is used).

Table 3.1. Personnel directly get exposed to nanoparticles.

From where?	When?	To whom?	How?
Dental Laboratories	During processing, polishing, packaging etc	Dental Lab personnel	Inhalation
Dental clinic	During implantation, surgeries and any medication with nanoparticles	Dentists and auxiliaries	Inhalation
Dental clinic	Application of unset material, oral medications, implantation, polishing, restoration, removal and finally the usage period	Patients	Inhalation and Ingestion

3.8 Personnel and environmental exposure

Nanoparticles can be exposed to the dental industry and environment by various ways. To build up any successful industry risk analysis it must be done such that it comprises risk assessment, risk management and proper risk communication. The prime component of risk assessment is the chance of exposure of hazard (here nanoparticles) and the quantification of exposure as well as the estimation of damage by it. Apart from the routine uptake of nanoparticles from several materials, a person can acquire them from the dental industry (table 3.1) Several processes of dentistry (from production to usage) involving nanomaterials lead to environmental exposure and most of the NPs are potentially hazardous. If the procedures of production are carried out under vacuum suction, water cooling etc, dust formation is not expected to happen, thus chances of exposure reduce in dental labs. One of the major sources is nano-waste generated after removal of dental restoration, which is responsible for greatly affecting the environment, so proper separation and an effluent system must be present in every dental clinic. For example, amalgam is widely used in dentistry but contains mercury that is harmful for our body as well as the aquatic environment (thus affecting the food chain) so a special separator is required.

3.9 Recent updates on nanoparticles in dental implants

Re-mineralisation treatment, endodontic treatment, periodontal treatment, prosthodontic treatment, orthodontic treatment are the basic focuses in nano-dentistry. Among all of these our main topic of interest is prosthodontic treatment that covers the area of replacement of teeth, dental restoration, complete denture and fixing, removal, implantation etc.

Recent updates of nanoparticles used in implantation have been aided to certain features including:

1. Better osseointegration;
2. Better mechanical anchorage;

3. Less porosity;
4. Non corrosive;
5. Less inflammatory action and less allergic;
6. Antimicrobial activity;
7. Easy to handle;
8. Reduction in cost etc.

Some advancements include:

- To improve the integration of implants to living bone and tissues, several coatings have been developed for example thin ceramic coatings, quercitrin nanocoating etc.
- Development of hydroxyapatite nanoparticles gives additional benefit to the implantation industry, like easy and smooth integration into dental tubules, dental hypersensitivity reduction, easy fitting to the narrow cavities on enamel generated by acidic erosion and easily absorbed by enamel of the teeth [53].
- The improvement of cobalt–chromium binary alloy to cobalt–chromium–molybdenum alloy via the intermediate cobalt–chromium–tungsten alloy has provided great impact on the implantation process [54].
- Titanium was used as implant in its pure form but the integration property was not satisfactory. Dorkhan *et al* [55] performed modification of titanium by anodic oxidation (pore range 50 nm) and observed the reduction in attachment of oral streptococci on the modified structure over the pure one. Nanostructure modification of titanium surface also increases the implant integration property.
- Emergence of zirconia ceramics has been found to provide better results in implantation, less bacterial adhesion, low cytotoxicity (negligible ion release), resistance against corrosion, tooth-like color, biocompatibility, making it more suitable to be used as an implant.
- Currently, several modified versions of titanium are the central choice in the dental implantation industry, but these modifications can have a severe impact on quality and rate of osseointegration. So to improve the quality of osseointegration, graphene is most promising as a coating because of its good antibacterial and osteogenic capacity.

3.9.1 Covid-19 and dental implants

According to recent research, the nature of dental problems has immensely changed and the proportion of dental related infection has increased during the pandemic. Non-urgent or non-emergency cases declined to 30% of pre-Covid-19 levels [56]. Dental droplets, salivary droplets, and several instruments of implantation are responsible for transmission of corona virus in the clinic and surrounding areas. Dental clinics should follow several general factors throughout this pandemic including use of PPE, maintaining physical distancing, use of hand sanitizer, maintaining thermal checking etc (figure 3.4).

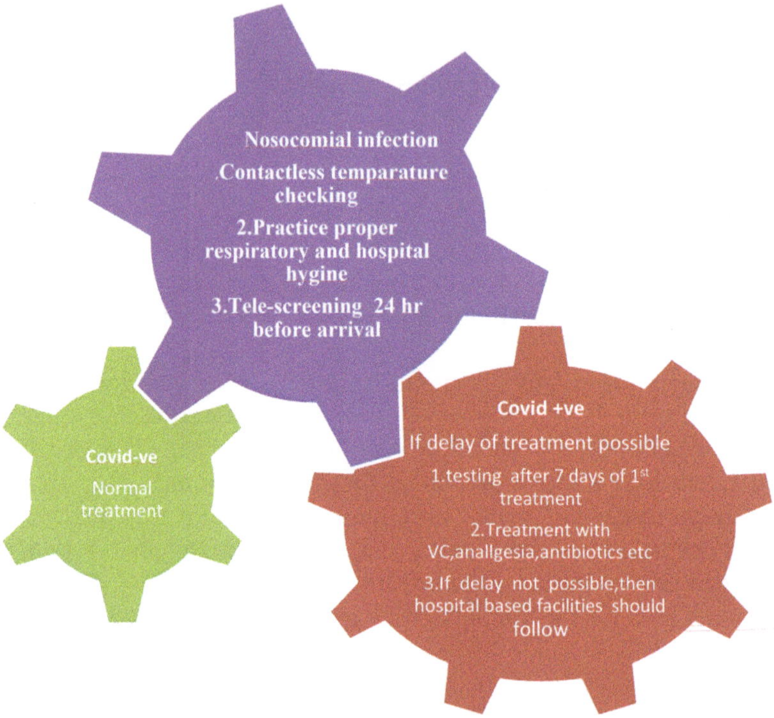

Figure 3.4. Dental patient management during COVID-19.

3.10 Conclusion

In this chapter we have discussed several important aspects of nanomaterials that are involved in the implantation industry. The continuous demand of implants in the dental industry actually triggers innovation as well as application of nanoparticles. Because of their antimicrobial activity, lower time consumption manner, cost effectiveness and high osseointegration capacity they have achieved high importance in the dental research area and dental industry. Apart from implantation, nanotechnology is spreading throughout several branches of dentistry with the use of nanomaterials.

References

[1] Ring M E 1995 A thousand years of dental implants: a definitive history—part 1 *Compend. Contin. Educ. Dent.* **16** 1060–9

[2] Sullivan R M 2001 Implant dentistry and the concept of osseointegration: a historical perspective *J. Calif. Dent. Assoc.* **29** 737–45 PMID: 11806052

[3] Linkow L I and Cherchève R 1970 *Theories and Techniques of Oral Implantology* ed M Jones vol 1 (St. Louis: CV Mosby)

[4] Brånemark P I, Hansson B O, Adell R, Breine U, Lindström J, Hallén O and Ohman A 1977 Osseointegrated implants in the treatment of the edentulous jaw. Experience from a 10-year period *Scand. J. Plast. Reconstr. Surg. Suppl.* **16** 1–132

[5] Bencharit S, Byrd W C, Altarawneh S, Hosseini B, Leong A, Reside G, Morelli T and Offenbacher S 2014 Development and applications of porous tantalum trabecular metal-enhanced titanium dental implants *Clin. Implant Dent. Relat. Res.* **16** 817–26

[6] Lee J W, Wen H B, Gubbi P and Romanos G E 2018 New bone formation and trabecular bone microarchitecture of highly porous tantalum compared to titanium implant threads: a pilot canine study *Clin. Oral Implant Res.* **29** 164–74

[7] Mendonça G, Mendonça D B, Aragão F J and Cooper L F 2008 Advancing dental implant surface technology–from micron- to nanotopography *Biomaterials* **29** 3822–35

[8] Prodanov L, Lamers E, Domanski M, Luttge R, Jansen J A and Walboomers X F 2013 The effect of nanometric surface texture on bone contact to titanium implants in rabbit tibia *Biomaterials* **34** 2920–7

[9] Liaw K, Delfini R H and Abrahams J J 2015 Dental implant complications *Semin. Ultrasound, CT, MR* **36** 427–33

[10] Mohamed Hamouda I 2012 Current perspectives of nanoparticles in medical and dental biomaterials *J. Biomed. Res.* **26** 143–51

[11] Bapat R A, Chaubal T V, Joshi C P, Bapat P R, Choudhury H, Pandey M, Gorain B and Kesharwani P 2018 An overview of application of silver nanoparticles for biomaterials in dentistry *Mater. Sci. Eng.* C **91** 881–98

[12] Wood N J, Jenkinson H F, Davis S A, Mann S, O'Sullivan D J and Barbour M E 2015 Chlorhexidine hexametaphosphate nanoparticles as a novel antimicrobial coating for dental implants *J. Mater. Sci. - Mater. Med.* **26** 201

[13] de Castro D T, do Nascimento C, Alves O L, de Souza Santos E, Agnelli J and Dos Reis A C 2018 Analysis of the oral microbiome on the surface of modified dental polymers *Arch. Oral Biol.* **93** 107–14

[14] Li Z, Sun J, Lan J and Qi Q 2016 Effect of a denture base acrylic resin containing silver nanoparticles on Candida albicans adhesion and biofilm formation *Gerodontology* **33** 209–16

[15] Bacali C *et al* 2020 Flexural strength, biocompatibility, and antimicrobial activity of a polymethyl methacrylate denture resin enhanced with graphene and silver nanoparticles *Clin. Oral Investig.* **24** 2713–25

[16] Mishra P and Tyagi S 2018 Surface analysis of gutta percha after disinfecting with sodium hypochlorite and silver nanoparticles by atomic force microscopy: an *in vitro* study *Dent. Res. J.* **15** 242–7

[17] Ioannidis K, Niazi S, Mylonas P, Mannocci F and Deb S 2019 The synthesis of nano silver-graphene oxide system and its efficacy against endodontic biofilms using a novel tooth model *Dent. Mater.* **35** 1614–29

[18] Suzuki T, Gallego J, Assunção W G, Briso A and Dos Santos P H 2019 Influence of silver nanoparticle solution on the mechanical properties of resin cements and intrarradicular dentin *PLoS One* **14** e0217750

[19] Vazquez-Garcia F, Tanomaru-Filho M, Chávez-Andrade G M, Bosso-Martelo R, Basso-Bernardi M I and Guerreiro-Tanomaru J M 2016 Effect of silver nanoparticles on physicochemical and antibacterial properties of calcium silicate cements *Braz. Dent. J.* **27** 508–14

[20] Samiei M, Ghasemi N, Asl-Aminabadi N, Divband B, Golparvar-Dashti Y and Shirazi S 2017 Zeolite-silver-zinc nanoparticles: biocompatibility and their effect on the compressive strength of mineral trioxide aggregate *J. Clin. Exp. Dent.* **9** e356–60

[21] Li P, Tong Z, Huo L, Yang F and Su W 2016 Antibacterial and biological properties of biofunctionalized nanocomposites on titanium for implant application *J. Biomater. Appl.* **31** 205–14

[22] Pokrowiecki R, Zaręba T, Szaraniec B, Pałka K, Mielczarek A, Menaszek E and Tyski S 2017 *In vitro* studies of nanosilver-doped titanium implants for oral and maxillofacial surgery *Int. J. Nanomed.* **12** 4285–97

[23] Lampé I *et al* 2019 Investigation of silver nanoparticles on titanium surface created by ion implantation technology *Int. J. Nanomed.* **14** 4709–21

[24] Salaie R N, Besinis A, Le H, Tredwin C and Handy R D 2020 The biocompatibility of silver and nanohydroxyapatite coatings on titanium dental implants with human primary osteoblast cells *Mater. Sci. Eng.* C **107** 110210

[25] Zhou W, Jia Z, Xiong P, Yan J, Li Y, Li M, Cheng Y and Zheng Y 2017 Bioinspired and biomimetic AgNPs/gentamicin-embedded silk fibroin coatings for robust antibacterial and osteogenetic applications *ACS Appl. Mater. Interfaces* **9** 25830–46

[26] Hernández-Gómora A E, Lara-Carrillo E, Robles-Navarro J B, Scougall-Vilchis R J, Hernández-López S, Medina-Solís C E and Morales-Luckie R A 2017 Biosynthesis of silver nanoparticles on orthodontic elastomeric modules: evaluation of mechanical and antibacterial properties *Molecules* **22** 1407

[27] Espinosa-Cristóbal L F, López-Ruiz N and Cabada-Tarín D *et al* 2018 Antiadherence and antimicrobial properties of silver nanoparticles against streptococcus mutants on brackets and wires used for orthodontic treatments *J. Nanomater.* **2018** 9248527

[28] Halkai K R, Mudda J A, Shivanna V, Rathod V and Halkai R S 2017 Biosynthesis, characterization and antibacterial efficacy of silver nanoparticles derived from endophytic fungi against *P. gingivalis J. Clin. Diag. Res.: JCDR* **11** ZC92–6

[29] Panáček A *et al* 2016 Silver nanoparticles strongly enhance and restore bactericidal activity of inactive antibiotics against multiresistant *Enterobacteriaceae Colloids Surf.* B **142** 392–9

[30] Singh P *et al* 2018 *In vitro* anti-inflammatory activity of spherical silver nanoparticles and monodisperse hexagonal gold nanoparticles by fruit extract of *Prunus serrulata*: a green synthetic approach *Artif. Cells Nanomed. Biotechnol.* **46** 2022–32

[31] Sugiharti R, Widyasar E, Rusminah N and Mustika I 2019 Evaluation of silver nanoparticles addition in periodontal dressing for wound tissue healing by 99mTc-ciprofloxacin *J. Young Pharm.* **11** 17–20

[32] Dizaj S M, Barzegar-Jalali M, Zarrintan M H, Adibkia K and Lotfipour F 2015 Calcium carbonate nanoparticles as cancer drug delivery system *Expert Opin. Drug Deliv.* **12** 1649–60

[33] Mehdipour M, TaghaviZenouz A, Bahramian A, Yazdani J, Pouralibaba F and Sadr K 2010 Comparison of the effect of mouthwashes with and without zinc and fluocinolone on the healing process of erosive oral *Lichen planus J. Dent. Res. Dent. Clin. Dent. Prosp.* **4** 25–8

[34] Memarzadeh K, Sharili A S, Huang J, Rawlinson S C and Allaker R P 2015 Nanoparticulate zinc oxide as a coating material for orthopedic and dental implants *J. Biomed. Mater. Res.* A **103** 981–9

[35] Anu K, Maleeka Begum S, Rajesh G and Renuka Devi K 2016 Wet biochemical synthesis of copper oxide nanoparticles coated on titanium dental implants *Int. J. Adv. Res. Sci. Eng. Technol.* **3** 1191–94

[36] Varoni E, Tarce M, Lodi G and Carrassi A 2012 Chlorhexidine (CHX) in dentistry: state of the art *Minerva Stomatol.* **61** 399–419

[37] Safarabadi M, Khansari N and Rezaei A 2014 An experimental investigation of HA/AL_2O_3 nanoparticles on mechanical properties of restoration materials *Eng. Solid Mech.* **2** 173–82

[38] Samiei M, Aghazadeh M, Lotfi M, Shakoei S, Aghazadeh Z and Vahid Pakdel S M 2013 Antimicrobial efficacy of mineral trioxide aggregate with and without silver nanoparticles *Iran. Endod. J.* **8** 166–70

[39] Lotfi M, Vosoughhosseini S, Ranjkesh B, Khani S, Saghiri M and Zand V 2011 Antimicrobial efficacy of nanosilver, sodium hypochlorite and chlorhexidine gluconate against *Enterococcus faecalis Afr. J. Biotechnol.* **10** 6799–803

[40] Flores C Y, Diaz C, Rubert A, Benítez G A, Moreno M S, Fernández Lorenzo de Mele M A, Salvarezza R C, Schilardi P L and Vericat C 2010 Spontaneous adsorption of silver nanoparticles on Ti/TiO_2 surfaces. Antibacterial effect on *Pseudomonas aeruginosa J. Colloid Interface Sci.* **350** 402–8

[41] Zhao L, Wang H, Huo K, Cui L, Zhang W, Ni H, Zhang Y, Wu Z and Chu P K 2011 Antibacterial nano-structured titania coating incorporated with silver nanoparticles *Biomaterials* **32** 5706–16

[42] Yamamoto K, Ohashi S, Aono M, Kokubo T, Yamada I and Yamauchi J 1996 Antibacterial activity of silver ions implanted in SiO_2 filler on oral streptococci *Dent. Mater.* **12** 227–9

[43] Meachim G and Williams D F 1973 Changes in nonosseous tissue adjacent to titanium implants *J. Biomed. Mater. Res.* **7** 555–72

[44] Galante J O, Lemons J, Spector M, Wilson P D Jr and Wright T M 1991 The biologic effects of implant materials *J. Orthop. Res.* **9** 760–75

[45] Martini D *et al* 2003 Detachment of titanium and fluorohydroxyapatite particles in unloaded endosseous implants *Biomaterials* **24** 1309–16

[46] Franchi M, Bacchelli B, Martini D, Pasquale V D, Orsini E, Ottani V, Fini M, Giavaresi G, Giardino R and Ruggeri A 2004 Early detachment of titanium particles from various different surfaces of endosseous dental implants *Biomaterials* **25** 2239–46

[47] Ducheyne P, Willems G, Martens M and Helsen J 1984 *In vivo* metal-ion release from porous titanium-fiber material *J. Biomed. Mater. Res.* **18** 293–308

[48] Ferguson A B J, Akahoshi Y, Laing P G and Hodge E S 1962 Characteristics of trace ions released from embedded metal implants in the rabbit *JBJS* **44** 323–36

[49] Noronha V T, Paula A J, Durán G, Galembeck A, Cogo-Müller K, Franz-Montan M and Durán N 2017 Silver nanoparticles in dentistry *Dent. Mater.* **33** 1110–26

[50] Beer C, Foldbjerg R, Hayashi Y, Sutherland D S and Autrup H 2011 Toxicity of silver nanoparticles—nanoparticle or silver ion? *Toxicol. Lett.* **208** 286–92

[51] Liao C, Jin Y, Li Y and Tjong S C 2020 Interactions of zinc oxide nanostructures with mammalian cells: cytotoxicity and photocatalytic toxicity *Int. J. Mol. Sci.* **21** 6305

[52] Biola-Clier M, Beal D, Caillat S, Libert S, Armand L, Herlin-Boime N, Sauvaigo S, Douki T and Carriere M 2017 Comparison of the DNA damage response in BEAS-2B and A549 cells exposed to titanium dioxide nanoparticles *Mutagenesis* **32** 161–72

[53] Li L, Pan H, Tao J, Xu X, Mao C, Gu X and Tang R 2008 Repair of enamel by using hydroxyapatite nanoparticles as the building blocks *J. Mater. Chem.* **18** 4079–84

[54] Tang N, Li Y P, Kurosu S, Matsumoto H, Koizumi Y and Chiba A 2011 Interfacial reactions between molten Al and a Co–Cr–Mo alloy with and without oxidation treatment *Corros. Sci.* **53** 4324–26

[55] Dorkhan M, Hall J, Uvdal P, Sandell A, Svensäter G and Davies J R 2014 Crystalline anatase-rich titanium can reduce adherence of oral streptococci *Biofouling* **30** 751–9

[56] Rutkowski J L, Camm D P and El Chaar E 2020 AAID white paper: management of the dental implant patient during the COVID-19 pandemic and beyond *J. Oral Implantol.* **46** 454–66

IOP Publishing

Nanotechnology for Dentistry Applications

Ashutosh Kumar Shukla

Chapter 4

Roles of nanomedicines in dental therapeutics

Nafiu Aminu and Muhammad Sohail Zafar

Nanotechnology has attracted enormous attention in medicine, as it addresses several issues connected with the conventional therapeutic systems that include lack of target ability, systemic toxicity, and poor pharmacokinetics. Since the emergence of nanomedicine from the concept of nanotechnology about three decades ago, there has been tremendous development in various areas of therapeutics, including dentistry, through innovations in drug and biologics delivery, diagnostics and imaging, implants, and tissue engineering. It has paved multiple ways in preventing and curing dental diseases, restoring dental tissues and bone, and diagnostics. Nanomedicines are envisioned to have potential for facilitating the achievement of perfect oral hygiene and are increasingly manifesting as the next generation of dental therapeutic platforms with high specificity and efficacy. This chapter focuses on the insights related to the roles of nanomedicines on the advancement of dental therapeutics. In addition, the chapter highlights the general aspects of nanomedicines and their prospects.

4.1 Introduction

Nanotechnology is increasingly becoming a more popular field of current research and has advanced in various disciplines, such as medicine, material sciences, engineering, pharmaceutical sciences, biological sciences, chemical sciences, toxicology, etc [1]. It is earning tremendous publicity in multiple fields and has already spawned several journals, book chapters, and books, and is becoming a hot source for a growing number of scientific papers, reviews, and consensus reports [2].

The word 'nanotechnology' is an amalgamation of 'nano', derived from the Greek word for dwarf, and 'technology', a noun that represents a discipline dealing with the art and science of applying scientific knowledge to practical problems. Nanotechnology is defined as the study, development, and application of ultra-small (1–100 nm) materials, systems, and devices that exhibits exceptional physicochemical and biological properties that are different from those found on larger particulates [3]. However, in some literatures, the size coverage of nanotechnology

doi:10.1088/978-0-7503-3671-0ch4

matter is extended to 1000 nm for most pharmaceutical applications [1, 4, 5]. The molecular level interaction exhibited by nanomaterials increases their overall affinity and efficacy in biological systems as compared with micro/macro sized particles. Moreover, the high surface area of nanoparticulate matter is a unique physical characteristic, which implies that there are more atoms available on the nano-systems' surface. This is particularly useful due to the potential for creating new and strong bonds, and thus, nanosystems are more reactive in comparison to their micro- and macro-particles counterparts which have less surface area [6].

Nanotechnology is a meaningful form of engineering in today's technological advancements, helping in making biomaterials, biomedical instruments, electronics, and many consumer products by controlling atoms and molecules [3]. It has revolutionized the medical and dental fields through enhancement of physical and mechanical properties of materials (e.g. restorative material) and improved the overall bonding (e.g. between dentin and biomaterials), which has in turn brought new diagnostic modalities and nanotherapeutic applications [6]. To sustain this development, there is significant increase in funding for nanotechnology-based research and development. For example, the United States (US) President's 2021 budget request showed that, more than \$1.7 billion is earmarked for national nanotechnology initiatives (NNI), as an increased investment compared with that of 2020 (\$1.4 billion) and previous years, to demonstrate strong support in the foundational research and innovation strategies for the betterment of the future [7]. With the 2021 request, the cumulative fund pumped to NNI reached a total of over \$31 billion since the establishment of the institution in 2001 [7]. This support reflects the continued research importance to manipulate matter at the nanoscale for the technological breakthroughs that benefit the people.

Nanotechnology manipulates matter at the nanoscale level and its application in medicine is called nanomedicine [8]. The concept of 'nanomedicine' was initially presented by Robert A Freitas Jr in 1993 and was defined as the science of preventing, diagnosing, controlling, and treating disease at the molecular level using nano-structures and nano-devices [8, 9]. Since then, researchers from different parts of the world have engaged in researches for various types of nanomedicine which resulted in tremendous exploration for diagnosis, prevention, and treatment of various local and systemic ailments [10]. This has opened a new paradigm for a wide range of biomedical applications (figure 4.1). Additionally, nanomedicines became particularly of interest for not only exploring new applications in the healthcare delivery, but also for adding flexibility to the existing systems, including the possibility of reformulating previously abandoned drug substances and products to be better, safer, and more efficacious medicines [11–13].

Nanomedicine under its branch called 'nanomaterials', has significantly developed in materials such as polymers, ceramics, metals, and composites which all have practical applications in healthcare [14]. Nanomaterials are described as materials that have components of less than 100 nm in at least one dimension [15]. These include clusters of atoms, grains, fibers, films, nanoholes, and composites that are within nanoscale size [8]. The unique features of nanomaterials make them a point of attraction from biomaterials scientists to drive benefits for various applications that

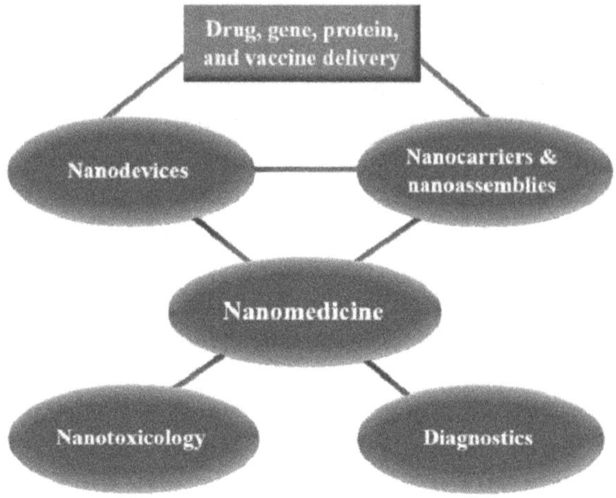

Figure 4.1. The different domains of nanomedicine.

can enhance the quality of life. Their properties can be altered and controlled by changing their synthetic conditions to suit certain desirable requirements.

Variety of nanomaterials have been developed or are in the developing stage for practical usage in a range of medical applications that include diagnosis, delivery of proteins or genes, tissue engineering etc. Nanotechnology applications extend to dentistry where it provides immense opportunities for the development of novel dental products that could be utilized for dental disease prevention, dental restoration, aesthetics purposes, periodontal disease and dental caries treatment, diagnostics and imaging, implantology, oral cancers, prosthetic dentistry, adhesive dentistry, and endodontics [16–19]. Nanoparticulate systems loaded with antimicrobial active agents are playing a vital role in tackling infection-based dental problems such as periodontal disease and dental caries, and this facilitates attainment of near-perfect oral hygiene [20]. Their immense antimicrobial potential can prevent biofilm build up, which avoids micro-leakage and secondary caries [21]. Examples of common nanoparticulate systems include: organic-based (biodegradable and non-toxic) like polymeric nanoparticles, polymeric nanofibers, sold lipid nanoparticles, liposomes, dendrimers, micelles, nucleic acid nanoassemblies, and protein constructs; inorganic-based such as silver nanoparticles, iron oxide nanoparticles, mesoporous silica nanomaterial, gold-based nanostructures; and carbon-based like carbon dots, carbon nanotubes, carbon nanofibers, fullerenes, and graphene. While the majority of such technologies are still in preclinical development, an increasing number of nanomedicine products are reaching the market or clinical trials, and have overcome many drug delivery challenges such as poor solubility and permeability, reduced harmful side effects, minimized drug degradation and loss, and increased drug bioavailability and accumulation at the target site [22]. These therapeutic benefits are achievable due to the attractive physicochemical properties that are associated with nanoparticulate systems' morphological attributes such as size, structure, and shape [23]. Therefore, their properties can be suitably altered by changing the morphological attributes to confer on them certain desired applications. This chapter provides an insight for the applications of nanodentistry and different roles of nanomedicines in dental therapeutics.

4.2 General aspect of nanomedicines

Today's scientists and engineers are experimenting with a range of nanoscale materials to take advantage of their superior features, such as stronger strength, lower weight, better light spectrum, and higher chemical reactivity, compared to their larger-scale ones [24]. In addition, the nanoscalic dimension of the nanomaterials gave them an ability to penetrate into extracellular and intracellular locations that other drug delivery systems may not be able to reach, including dental areas such as periodontal pockets that are located below the gum line [25]. This has given birth to the discipline called 'nanomedicine' which refers to a wide area of nanotechnology research that includes both diagnostic and therapeutic applications. It has the potential to revolutionize how the pharmaceutical and medical industries handle the pharmaceutical care concerns of the twenty-first century [26]. Nanotechnology has proven useful in a variety of scientific and technological breakthroughs in healthcare delivery, with a few notable examples including:

1. Nanomedicine, among its many emerging benefits, may enable the oral administration of drugs that are currently administered via injection route only. Such medicinal agents can be nanoencapsulated in a suitable polymer or lipid matrix, allowing them to comfortably pass through the gastrointestinal lining and permeate into the bloodstream, where their payload will be released and be bioavailable.
2. Nanoparticles, which can be as small as a virus, can easily enter sick cells, allowing for more accurate detection and treatment of diseases.
3. Nanomaterials have also been shown to stimulate the body system to regenerate damaged and lost cells.
4. Nanomedicines have been well utilized for targeted drug delivery, especially in cancer treatment to overcome the serious side effects challenges associated with many anticancer agents. They circumvent the healthy cells while ensuring effective guidance of the drug to the targeted site.

Basic nanodevices and nanomaterials, some of which are currently available, can be used in nanomedicine to solve a variety of essential health issues. Nanomedicines enhance existing procedures while also generating brand new ones [26–28]. Therefore, in whatever application of nanomedicine, it would continue to significantly improve a patient's quality of life by allowing for early diagnosis and/or more efficient treatment with fewer drug-related side effects. Overall, the advancement of these nanopharmaceuticals will continue to improve medical practice and clinical outcomes in the years ahead.

4.2.1 Forms of nanomedicines

As shown in figure 4.2, typical nanomedicines can be classified into different domains. They are also prepared through various techniques, multiple methods

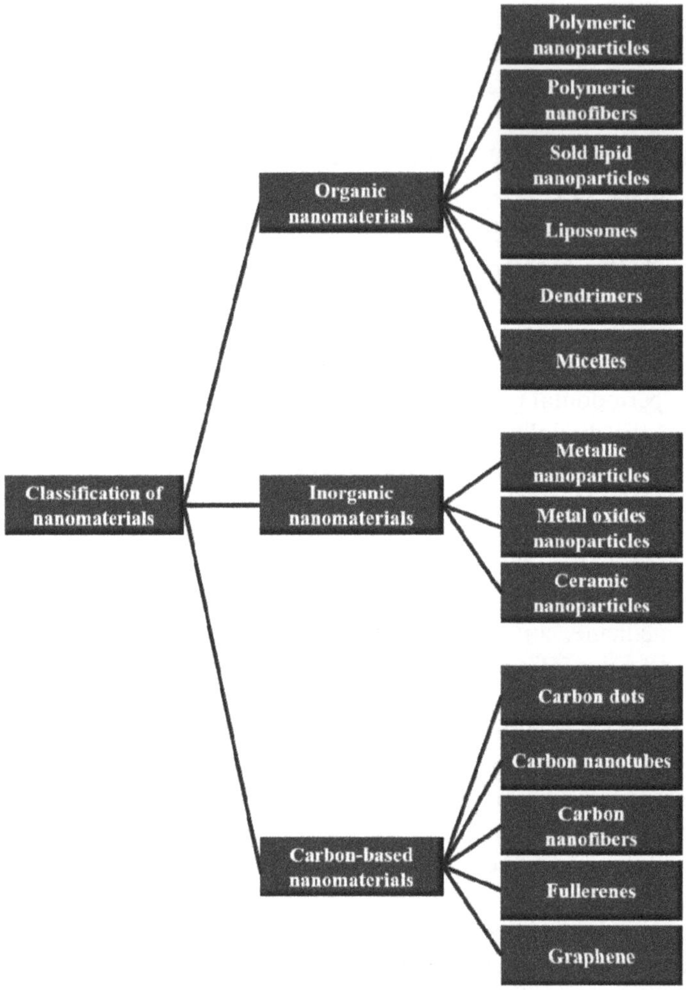

Figure 4.2. Various classes of nanomaterials.

for each nanomedicine. They can be characterized in a variety of ways, including physicochemical evaluations like crucial particle size and particle size distribution, drug loading (rate and capacity), chemical and physical stability of both drug and carrier, release rate, surface charge and character (hydrophilic, hydrophobic), and diffusion and rheology. Other aspects such as the absorption of the carrier with its encapsulated or attached drug, drug and carrier distribution in the body system, metabolism, and excretion are all covered by the biological evaluation of nanomedicines. Some of the classes of nanomedicines that are more relevant to dentistry have been discussed in the following section.

4.2.1.1 Nanoparticles/nanospheres/nanocapsules

Nanoparticles are colloidal solid particles that are in the range of 1 to 1000 nm in diameter [22]. They are generally fabricated by several manufacturing methods

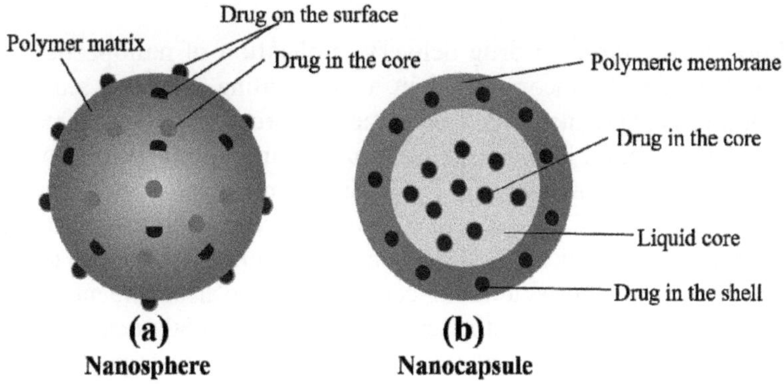

Figure 4.3. Categories of nanoparticles. (a) Nanosphere where drug is entrapped in the polymer matrix or adsorbed on the surface of the nanosphere or both. (b) Nanocapsule where drug is either entrapped inside the capsule core or adsorbed into its polymeric membrane or both.

which either involve size reduction of particles (e.g. by milling) to within nano-metre range, or molecular agglomeration (like precipitation methods) to produce the nanoparticles. The former method is applicable in the preparation of drug particles to produce their nanosized range, and it does not involve carrier material, whilst the latter is usually employed in the preparation of nanoparticles as carriers in which drug is loaded [5]. Drugs or other active compounds can be entrapped, adsorbed, or attached to the surface of the nanoparticles. Nanoparticles have emerged as one of the most promising vectors, with a broad range of applications including the crucial ones such as targeted delivery of drug, protein, and gene [29]. They allow efficient targeting and tissue penetration due to their sub-microscopic size. Nanoparticles are usually constructed from biodegradable or nonbiodegradable materials sourced from natural or synthetic polymers, lipids, phospholipids, or metals. Thus, there is existence of a variety of them such as polymeric nanoparticles, solid lipid nanoparticles, inorganic (metal, metal oxides, or metal sulfide) nanoparticles, magnetic nanoparticles, paramagnetic nanoparticles, quantum dots, gold nanoparticles, and many others, that are being studied for the diagnosis and treatment of various diseases. These nanoparticles can be subdivided into two categories:

1. **Nanospheres:** consisting of a dense polymeric matrix that are spherical in shape with drugs entrapped within them, adsorbed on their outer surface, or both (figure 4.3(a)).
2. **Nanocapsules:** comprising an inner core which is mostly liquid where drugs, genes, proteins, or other active substances are incorporated, and this is surrounded by a solid polymeric shell. The drugs or other active substances can also be adsorbed on the shell or be both entrapped in the core and adsorbed on the shell (figure 4.3(b)). Encapsulation of therapeutic agents into the nanocapsules is one of the strategies being employed in drug delivery for preserving them from degradation, destruction, and rapid metabolism throughout the desired delivery protocol, as well as shielding the body from hazardous effects of certain drugs (rapid release of some drugs that can be toxic).

In addition to the targeted drug delivery application of nanoparticles in dental therapeutics, they can be incorporated in an obturating material, sealer, irrigating solutions, toothpaste, and intracanal medicament to provide the desired results [30, 31]. Silver nanoparticles have been found very useful in prophylaxis, prevention, and disinfection of oral cavity infections [32, 33]. Silver and zinc oxide nanoparticles were found to improve the dental bond strength of the restoration and provided antibacterial activity when the nanoparticles were incorporated into conventional glass ionomer cement and resin-modified glass ionomer cement [34]. The arrangement of amelogen in protein into nanospheres provides the scaffolding for the initial nucleation and growth of enamel apatite crystals [35]. Furthermore, the incorporation of peptide nanospheres into composite restoratives and dental adhesives resulted in a significant reinforcement and improvement of mechanical properties [24].

The nanoparticles' benefits that are being utilized in dental applications can be showcased from the their characteristic aspects [36]:

1. They provide a large surface area;
2. They can be engineered to have antimicrobial, antifungal, and antiviral properties, and as a result, can prevent biofilm formation when antimicrobial nanoparticles are incorporated into resin composites;
3. They improve the mechanical properties of dental materials, which is particularly important in restorative dentistry;
4. They strengthen the bond between dentin and the biomaterial;
5. They stop cracks from spreading and white spot lesions from appearing;
6. They make porcelain restorations more fracture resistant.

Overall, nanoparticles, both nanospheres and nanocapsules, are promising alternatives for delivery systems, as they can be used not only for controlled delivery of drugs, but also for future dental regenerative procedures. Some of the applications of nanoparticles in dental therapeutics have been summarized in table 4.1, and the roles for selected ones have been discussed later in this chapter.

4.2.1.2 Nanofibers

Nanofibers are a subclass of biomaterials that have a wide range of biomedical applications [70]. Because of their similarities to the extracellular matrix, nanofibers have gained a lot of attention in nanomedicine. The production of nanofiber has piqued the interest of many researchers due to the unique qualities that are highly desirable for biomedical applications, such as the availability of more surface area for cellular interactions, binding sites to cell receptors, and protein absorption that make it a well-suited substrate for tissue engineering as well as other cell growth experiments [71]. Due to the unique properties of larger surface-to-volume ratio and the tunability of the polymers being used in their development, nanofibers have boosted drug encapsulation efficiencies and drug-loading capacities [72]. Additionally, nanofibers can improve material strength by facilitating maximum volume fraction packing by managing the alignment and orientation of the fibers [48]. For specific applications, material features like porosity, surface morphology,

Table 4.1. Some nanomedicines developed for various dental therapeutics.

Class of nanomedicine	Formulation components	Active compound(s)	Application(s)	Main features	References
Nanoparticles	PCL, Pluronic F-68	Triclosan	Periodontal infections	Higher drug entrapment efficiency. Absence of cytotoxicity.	[37, 38]
	2-hydroxyethyl methacrylate, ethylene glycol dimethacrylate, and methacrylic acid	PolymP-nCalciumZinc	Dental restorative/ mineralization	Improved mechanical performance of the resin–dentin interface.	[39]
	Polymerized acrylic resins	Zinc oxide	Restorative dentistry	Significantly improved flexural strength of auto-polymerized acrylic resins	[40]
	PCL	Zinc oxide	Periodontal tissue regeneration	Exerts both osteoconductivity and antibacterial properties. Demonstrates great potential forperiodontal tissue engineering.	[40]
	Mesoporous silica, PGA, N-cetyltri-methylammonium bromide	Chlorhexidine	Restorative dentistry	pH-sensitive drug release. Exhibits excellent bonding with the application of commercially available dentin adhesive.	[41]
	Conventional glass ionomer cement powder	Silver, amoxicillin	Restorative dentistry	Shows good characterization and stability. Shows synergistic antimicrobial and anti-biofilm activity.	[42]
	PLA, PEG, Collagen	Bovine serum albuminAcidic fibroblast growth factor	Tissue regeneration	Prolonged release of growth factor.	[43]
Nanomaterial	None	Calcium carbonate	Tooth remineralization	The dentifrice test shows significant mineral gain, indicating lesion remineralization.	[44]

	None	Graphene	Dental restorative in caries	Exhibits antimicrobial action against dental caries pathogens	[15, 45]
Nanogels	PCL and chitosan	Triclosan and flurbiprofen	Periodontitis	Improves penetration and retention time of triclosan and flurbiprofen into the periodontal pockets. pH responsiveness for targeted delivery into periodontal pocket. Increased efficacy through synergistic delivery of co-encapsulated agents.	[46]
Nanofibers	Scaffolds composed of PDS and halloysite aluminosilicate clay	Possible loading of bioactive molecules	Regenerative endodontics	Supports the attachment as well as proliferation of human-derived pulp fibroblast cells. Facilitates the delivery of growth factors and antimicrobial drugs. Exhibits a high level of biocompatibility.	[47, 48]
	PCL and PLA	Hydroxyapatite crystals	Tissue and bone regeneration	Enhances osteogenic potential and mechanical properties. Carriers of growth factors and drugs such as bone morphogenic protein-2 and antibiotics, respectively.	[48]
	PCL and PLA	nHAand zeolite	Dental tissue engineering	Significantly promotes the proliferation of human dental pulp-derived stem cells. Found to be a suitable scaffold for bone and tooth tissue engineering applications.	[49]

(*Continued*)

Table 4.1. (*Continued*)

Class of nanomedicine	Formulation components	Active compound(s)	Application(s)	Main features	References
	PLLA, mineraltrioxide aggregates		Dental tissue engineering	Exhibits nanofibrous structure, increased biomineralization, interconnected pores, and suitable mechanical properties. Supports the attachment and proliferation of dental pulp stem cells.	[50]
	PCL, chitosan, PEO		Cartilage tissue engineering	Supports cellular aggregation, chondrogenesis and matrix formation.	[51]
	Silk fibroin		Buccal mucosa repair	Shows significant repair performance. Shows better wound healing ability.	[52]
	PVA	Hyaluronic acid	Periodontal pocket drug delivery	Presented controlled release behaviour, anti-inflammatory effect, and good mucoadhesive strength.	[53]
	PLGA, gum tragacanth	Tetracycline hydrochloride	Periodontal disease treatment	Controlled drug release. Exhibits good biocompatibility, antibacterial and mechanical properties.	[54]
	Chitosan, PEO		Regeneration of periodontal tissues	Supports matrix deposition. Promotes proliferation of osteoblastic cells.	[55]
	Chitosan and thiolated chitosan	*Garcinia mangostana* extract	Dental caries prevention	Shows mucoadhesion and antibacterial activity against dental caries pathogens.	[56]
	PVA and gellan	Bioactive molecules	Tissue engineering application	Proven to be non-toxic and biocompatible by supporting the growth of murine embryonic stem cells.	[57]

Liposomes	PEG-2000CholesterolDMPC	Proteolytic enzyme (Collagenase)	Tissue regeneration	Enhanced periodontal remodeling. Supported soft and hard tissue recovery and reduced tooth relapse.	[58]
Nanocomposites	Composite resin	Condyloform II NFC	Restorative dentistry	Exhibits good tooth wear resistance.	[59]
	Nano-ceramic	Ceram-X Mono	Restorative dentistry	Demonstrates less mean shear bond strength when compared to traditional orthodontic composite Transbond XT.	[60]
	PEG 6000, polyacrylic acid, span 80	Chlorhexidine, amorphous calcium phosphate	Restorative dentistry	Effectively inhibits growth of *S. mutans* at a low concentration. Demonstrates mechanical and remineralization property.	[61]
	Bis-GMA, methacrylic acid, acrylic acid, TEGDMA	Hydroxyapatite, silicon dioxide, zirconium dioxide	Restorative dentistry	Proven to be highly effective in the improvement of mechanical characteristics. Exhibits better water resistance and lower volumetric shrinkage.	[62]
	HEC, CNF	Nepheline fluorapatite glass powder (for release of phosphate, calcium, and fluoride ions)	Preventive dentistry (prevention of dental caries)	Shows significant increase in mean enamel micro-hardness after 15 and 30 days remineralization as compared to the demineralized specimens.	[63]
	Bis-GMA, TEGDMA, camphorquinone, EDMAB	Silica, 1H-1,2,4-triazole	Restorative dentistry	Demonstrates significant enhancement of desirable properties compared to control samples.	[64, 65]

(Continued)

Table 4.1. (*Continued*)

Class of nanomedicine	Formulation components	Active compound(s)	Application(s)	Main features	References
	Dendritic methacrylate end-caped monomers, TEGDMA,	Modified silica	Restorative dentistry	Enhanced biological, physicochemical and mechanical properties.	[66]
	PMMA	Barium titanate		Shows better mechanical strength. Shows significant reduction in roughness values. Remarkably increases the tensile modulus values which leads to the promotion of the dental composites' longevity.	[67]
Nanocrystals	Polycrystalline material	Hydroxyapatite	Oral and maxillofacial surgery	Improves osteointegration for maxillofacial implants.	[68]
	Polycrystalline material	Hydroxyapatite	Periodontal tissue regeneration	Shows increased proliferation rate of periodontal ligament cells	[69]

Bis-GMA, 2,2 propyl bisphenylglycidyldimethacrylate; CNF, cellulose nanofibers; DMPC, dimyristoylphosphatidylcholine; EDMAB, ethyl-4-(dimethylamino) benzoate; HEC, hydroxyethyl-cellulose; nHA, nano-hydroxyapatite; PCL, poly-ε-caprolactone; PDS, polydioxanone; PEG, poly(ethylene glycol); PEO, polyethylene oxide; PGA, poly-L-glycolic acid; PLA, poly(lactic acid); PLGA, poly(lactic glycolic acid); PLLA, poly(L-lactide); PMMA, polymethyl methacrylate; PVA, poly(vinyl alcohol); TEGDMA, triethylene glycol dimethacrylate.

and geometry can be modified or functionalized by adjusting the versatile solution dope and spinning parameters [73]. Previously, nanofibrous scaffolds were solely known and utilized for drug and hormone delivery purposes, but it has now been discovered that they possess more advantageous features, allowing them to be used for preserving and delivering stem cells to human target organs [74]. Currently, such scaffolds are employed for local drug delivery to the target areas, and one of their key advantages is biocompatibility (as they can disintegrate without triggering any significant immune reaction) [75]. In addition to their drug delivery applications, nanofiber scaffolds have emerged as a vital contender for a variety of other biological applications, including wound dressings [76], tissue engineering scaffolds [77], regenerative medicine [78], and dental applications [55].

A vast array of nanofibers with high therapeutic applications in the dentistry area have been developed (table 4.1). Such applications include dental tissue regenerations in dentin, pulp, bone, neural tissues, ligament, and tooth [79–81]. Nanofibers were found to improve the mechanical characteristics of dental composite restoratives. The mechanism of the reinforcing effect can be attributed to the presence of a strong interfacial binding force between the distinct phases of the restorative material and a uniform distribution of the impregnated nanofillers, thereby preventing fracture progression [82]. A recent study showed that polyetherimide-based electrospun nanofibers have possible mechanical improvement of dental materials and hence, could be used as dental materials [83].

There are a variety of techniques for the production of nanofibers, such as template synthesis, phase separation, melt-blowing, self-assembly, nanofiber seeding, templating system, and electrospinning [48, 70, 84], etc. Amongst these methods, electrospinning turns out to be a highly resourceful and cheaper technique that can be utilized for continuous synthesis of nanofibers of various nanometer sizes [85]. Electrospinning favours incorporation of drugs and other biomolecules to obtain the desired properties in the final nanofibrous product. Nanofibers can be classified based on their origin type: polymeric nanofibers fabricated using electrospun poly-ε-caprolactone (PCL), poly(lactide-co-caprolactone), poly(ethylene glycol) (PEG), poly(lactic-co-glycolic acid) (PLGA), polyvinyl acetate, poly(ethylene oxide), Nylon 6, poly(vinyl alcohol) (PVA), Eudragit S100, poly (L-lactic acid) (PLLA), cellulose, chitosan, gelatin, alginate, and polyacrylonitrile/polymethyl methacrylate; inorganic nanofibers made of hydroxyapatite, glass, zirconium, quartz, and fibrillar silicate; and ceramic nanofibers [24, 72, 86, 87].

4.2.1.3 Nanocomposites

Tooth decay, also known as dental caries, is a common problem that affects people of all age categories. There are a number of options for tackling the problem, one of which is by using dental filling material. Recent advancements in the field of dentistry have resulted in the development of advanced filling materials known as resin nanocomposites [88]. Nanocomposites are heterogeneous/hybrid materials made by combining polymers with inorganic solids (clays to oxides), whereby at least one of their phases has nanoscale morphology (10–100 nm) such as nanoparticles, nanotubes, or lamellar nanostructure [89]. From a chemical standpoint, they typically consist of an organic phase (polymers) that includes 2,2-bis[p-(2′-hydroxy-3′-methacryloxypropoxy)phenylene]propane (Bis-GMA) or 1,6-bis(methacryloxy-2-ethoxycarbonylamino)-2,4,4-trimethylhexane (UDMA) as the basic monomers and triethylene glycol dimethacrylate (TEGDMA) as the co-monomer, and inorganic phase which are filler reinforcers (like silica, quartz, glass, oxide nanoparticles), as well as, organosilane coupling agents that bind the polymeric matrix to inorganic nanoparticles [90]. This makes their structure more complex than microcomposites, and it is highly influenced by compositions of component and interfacial interactions [89]. The rapid estimated demand for these striking potentially advanced materials makes them extremely useful in a wide range of biomedical fields including dentistry. Nanocomposite systems have the potential to improve the continuity of the tooth structure, and their nanosized filler particles provide a more stable and natural interface between the tooth's mineralized hard tissues, and this represents advancement of restorative biomaterials.

Polymer matrix nanocomposites have attracted a lot of attention in the research community over the last two decades because of their superior properties, such as stiffness, light weight, better aesthetic properties, and having higher specific strength than base polymers and conventional materials [91]. Hence, dental composite resins have largely displaced the much older and time-tested silver amalgam as the most popular restorative material available to dentistry. This is due to their metal-free tag, which allows them to mimic the shade of natural teeth, particularly in the highly aesthetic anterior regions, thus satisfying the patient's desire for a 'tooth-like' restoration. Adhesive bonded resin composites also have the benefit of preserving and reinforcing the tooth's remaining structure. Furthermore, unlike dental amalgams, these composites are mercury-free materials, hence posing no health risks [92].

Silica nanoparticles had been widely used as effective nanofillers that provide mechanical strength to the filling material, until recently when titania was found to be more efficient, and that led to the gradual replacement of the silica nanoparticles with titania nanoparticles as nanofillers [88]. Titania outperforms silica in a variety of ways. Another similar recent significant advancement is the development of new and more efficient monomers. In this context, a new strategy of using multi-functional or dendritic monomers had led to several advantages over bi-functional traditional dental monomers (e.g., Bis-GMA, TEGDMA), and others, which include low shrinkage stress and strain during photopolymerization, high mechanical characteristics, and high degree of conversion [66]. Furthermore, when compared to other molecules of similar molecular weight, these monomers have a high cross-link density (due to their greater percentage of reactive functional end groups), resulting in a three-dimensional network as well as a decrease in solubility and water sorption [66].

There are several techniques for fabrications of dental nanocomposites. The newer methods can effectively synthesize nanocomposites at the molecular level, followed by light curing. These materials have better mechanical, chemical, and biological properties than commercially available materials [88].

In clinical practice, due to the extreme oral environment conditions, restoration processes by using dental composite resins are usually faced with some requirement challenges such as good biocompatibility, excellent mechanical properties, water solubility or sorption, control for pH variations, caries-inhibition ability, low polymerization shrinkage and marginal leakage, colour matching, and stability, among others [93]. Through the nanotechnology innovations, most of these challenges have been addressed. Nanotechnology or molecular manufacturing may provide composite resin with filler particles that are significantly smaller, can be dissolved at higher concentrations, and are polymerized into the resin system with molecules designed to be compatible when coupled with a polymer and provide unique physical, mechanical, and optical properties. An example of this development is demonstrated by research findings from various researchers. Yushau *et al* (2019), prepared and characterized functional nanosilica incorporated dental resins as dental nanocomposites with better mechanical properties, low toxicity, and improved solubility and sorption [65]. The researchers developed the dental nanocomposites by embedding silica nanoparticles that were functionalized with 3-amino-1,2,4-triazole into bisphenol-A-glycidyl methacrylate/triethylene glycol dimethacrylate matrix by mechanical mixing of the monomer, filler, initiator combination, and then followed by LED light curing for 60 s. They characterized the developed dental nanocomposites by Fourier transform infrared (FTIR) spectroscopy, scanning electron microscopy (SEM), and thermogravimetric analysis techniques. Cell viability study was conducted to assess the cytotoxicity and stability of the nanocomposites. The results of this study showed the potentials of the nanocomposites on dental health improvement [65]. Another similar contribution was from Nikolaidis *et al* in 2021, who investigated the physicochemical behaviour as well as mechanical performance of dental nanocomposite resins that were developed from a variety of quaternary ammonium silane compounds (QASiC), methacrylate based dental resins, and silica nanofillers [90]. The synthesized dental nanocomposite resins were extensively characterized. The images of SEM revealed that most of the synthesized nanocomposites had a satisfactory dispersion of silica nanoclusters. Curing kinetics revealed an increase in both the autoacceleration effect and the degree of conversion, which was primarily caused by shorter QASiC. Polymerization shrinkage was discovered to be influenced by the type of silanized products used. The flexural modulus and strength of composites were increased by 74% and 19%, respectively, while compressive strength rose by up to 19% [90]. These findings could help to improve the design of multifunctional dental materials capable of meeting today's clinical challenges.

In another recent development, a team of researchers synthesized poly(2-hydroxyethylmethacrylate) (pHEMA) nanocomposites as a favourable dental material due their nontoxicity, tissue compatibility, and high water content [91]. The researchers fabricated the pHEMA nanocomposites by blending and an extrusion dispersion method with twin-screw extruder machine. Titanium dioxide (TiO_2) nanoparticles and nanoclay were dispersed in pHEMA which led to the formation of the nanocomposites. The nanocomposites were subsequently characterized by FTIR, X-ray diffraction (XRD), thermogravimetric analysis (TGA), differential scanning calorimetry (DSC), and field emission scanning electron microscopy (FESEM) techniques. The results of this study revealed that the developed pHEMA-based nanocomposites exhibited a significant change in microstructural behaviour, as well as improved thermo-mechanical properties [91]. A hybrid nanocomposites system was also developed from methyl methacrylate with methyl methacrylate monomer and gelatin-modified halloysite nanotubes, along with silane-coupled aluminum trihydrate for dental applications as an innovative nanofillers system [94]. The developed nanocomposites system was characterized with respect to some mechanical and utilitarian properties, such as hardness, abrasion resistance, and buffer solution absorption. The outcome of the study confirmed the synergistic effects of using a hybrid system for improvement of properties, hence, legitimized the usage of modified mineral nanofillers for the manufacture of dental nanocomposites. Another advantage of the developed hybrid nanocomposite materials is their impact on biofilm formation reduction, especially in the case of yeast, and biofilm formation limitation for specific bacterial strains [94].

4.3 Nanodentistry

When nanotechnology is applied to the field of dentistry it is called 'nanodentistry', hence it refers to the application of nanotechnology to the field of dentistry. Nanodentistry is the science and technology of using nanosized materials, dental nanorobotics, and tissue engineering to prevent, diagnose, and treat oral and dental diseases, as well as to reduce/eliminate pain and improve oral health [95]. Although many dental problems may be addressed with traditional ways, the new era of nanotechnology in dentistry will usher in novel approaches to dental disease diagnostics and therapy. Current research in nanodentistry focuses on prevention, diagnosis, reconstruction, regeneration, restoration, and rehabilitation fields [95]. These are critical for maintaining oral health and repairing oral lesions with nanomaterials, and the development of novel diagnostic and therapeutic modalities [96]. Nanodentistry has the required versatility to uniquely overcome many of the most challenging impediments in dental therapeutics which hinder effective dental treatment and perfect oral health.

Historically, R A Freitas Jr invented the term nanodentistry at the beginning of the 21st century. Freitas described how nanorobots could be utilized in dentition and orthodontics regeneration. He also outlined his plans to use nanobots and nanomaterials to create dental products like dentifrices and futuristic dentifrobots. According to this vision, therapeutic interventions in the field of nanodentistry would include nanosurgery, nanodrugs, and a plethora of other nanomaterials that would have a profound impact on clinical dentistry, significantly altering our current views and understanding [96]. The introduction of a new era of dental materials will include all parts of dentistry, including conservative and aesthetic dentistry, prosthodontics, orthodontic surgery, and more [97]. In the near future, nanodentistry will make it possible to maintain comprehensive oral health by utilizing nanomaterials, dental nanorobotics, and tissue engineering [98]. Local anaesthesia, permanent hypersensitivity cure, dentition renaturalization, covalently bonded diamondised enamel, complete orthodontic realignments could be achieved in a single medical visit, and continuous oral health maintenance using mechanical dentifrobots are some of the new potential treatments range in emerging painless dentistry [99, 100].

4.4 Various roles of nanomedicines in dental therapeutics

Due to the complexity of various oral diseases, tremendous research efforts have been made in recent years and new nanotechnological approaches for dentistry have emerged. These novel strategies may help effective delivery of antimicrobial molecules, tissue regeneration, and overall maintenance of oral health, by acting in a more specific site and then not being cytotoxic [101]. In this context, nanomedicines appear to be versatile platforms capable of acting both as carriers for the conveyance and release of therapeutic molecules and as a scaffold for the formation of new dental tissue and bone [102]. Nanomedicines play several therapeutic roles in different dental specialties (figure 4.4). Some of these roles are discussed in the following sections.

4.4.1 Preventive dentistry

The goal of modern dentistry is to prevent rather than treat biofilm-based oral disorders such as periodontal diseases, dental caries, and endodontic. Nanodentistry provides novel ways for preventing these oral illnesses. The most important preventive dentistry strategy is to halt or prevent biofilm build-up on teeth and gums in order to interrupt the production of dental plaque. For example, dental caries, which causes destruction of tooth structure as a result of acids released by *Streptococcus mutans, Lactobacillus spp, Streptococcus sobrinus* bacteria, can be controlled by inhibiting the bacterial action, reversing the process of demineralization and encouraging remineralization. In this regard, anti-caries vaccine development was attempted [104]. Several nanoparticles (e.g., silver, zinc oxide, titanium oxide, iron oxide, copper oxide, and polyethylenimine) that are incorporated into dental adhesives and composites were found to inhibit the growth of bacteria via different mechanisms [33, 40]. These mechanisms comprise the inhibition of the active transport along with sugars' metabolism, bacterial cell membrane disruption, generation of reactive oxygen species, causing disruption in the electron flow across the bacterial membrane, hindering DNA replication, and magnesium ions displacement which is required for oral biofilm enzymatic activity [104]. These nanoparticles were successful in an *in vitro* model lowering of *S. mutans* and *Lactobacillus acidophilus* biofilms [105, 106]. A study revealed that antimicrobial nanoparticulate quaternary ammonium poly (ethylene imine) nanoparticles incorporated into dental composite resins served as a viable tool for antibiofilm agents [107]. Antibacterial nanocoating on tooth surfaces was also found to be

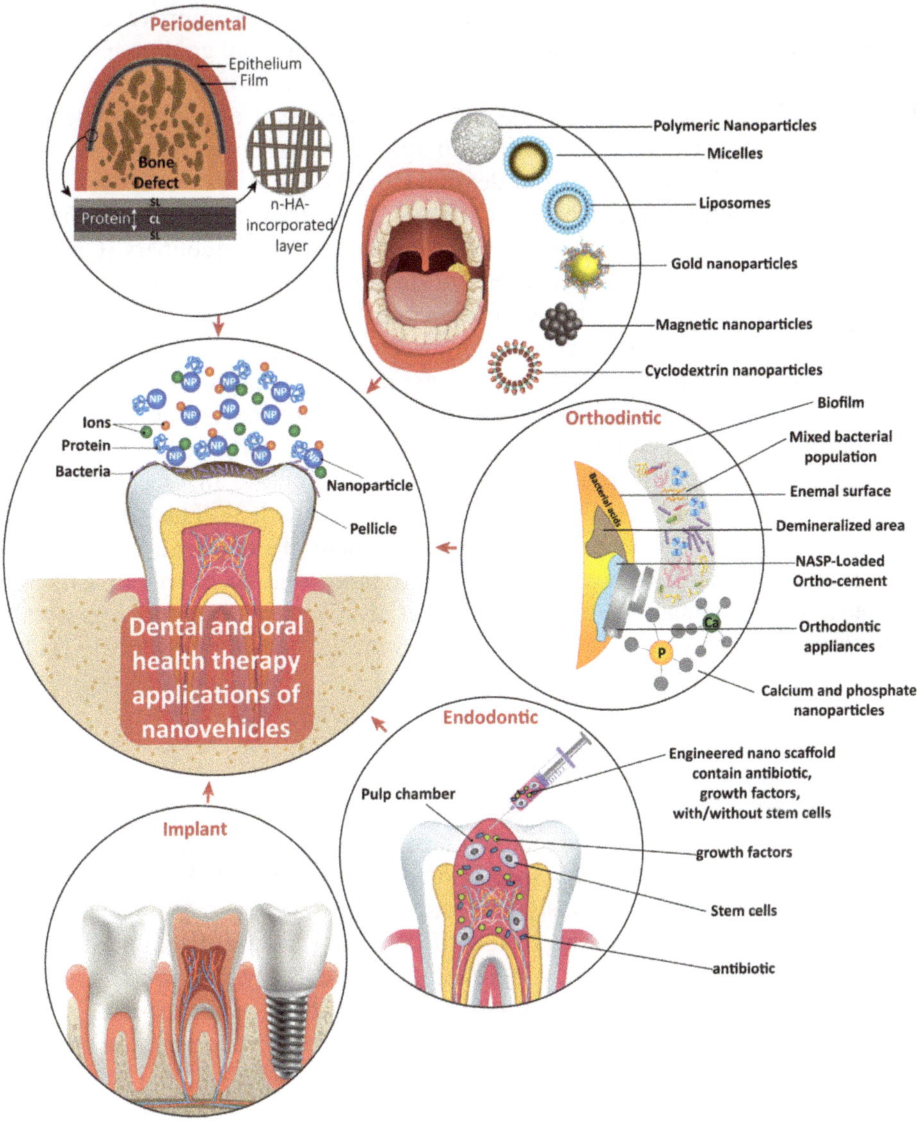

Figure 4.4. Applications of nanomedicines in some dental disciplines [103], copyright (2021) The Authors. *Advanced Science* published by Wiley-VCH GmbH. HA, hydroxyapatite; NACP, nano-sized amorphous calcium phosphate.

effective in killing bacteria, limiting bacterial adherence, and maintaining the integrity of the tooth, even in the presence of biological fluids (saliva) [108].

Studies have shown that, nanotechnology could give tools for early caries prevention and treatment through the use of customized toothpastes, mouth rinses, and varnishes, that can facilitate remineralization of early caries lesions [109, 110]. By integrating nanogold or nanosilver colloidal particles between toothbrush bristles, researchers created a nano-toothbrush that offers effective prevention against periodontal disease. In addition to its potential to improve mechanical plaque clearance, the researchers discovered that the additional gold or silver has an antimicrobial impact, which could lead to a large reduction in periodontal disease [6]. Nano-hydroxyapatite (n-HA) has sparked widespread attention in dentistry, owing to its crystal structural similarities to apatite crystals, which provides it with the ability to heal dental enamel and dentine [109, 110].

In a recently published study by Zaki *et al* (2021), an oral nanocomposite film was prepared using hydroxyethyl-cellulose (HEC) and cellulose nanofibers (CNFs) for the purpose of prevention and treatment of dental caries [63]. The nanocomposite film was suitably loaded with a blend of nepheline (75%) and fluorapatite (25%) glass powder that serve as bioactive glass powder. The developed nanocomposite exhibited uniformity in thickness and release of high concentrations of calcium and fluoride ions. The results also indicated a significant increase in mean enamel micro-hardness after 15 and 30 days remineralization as compared to the demineralized specimens, hence, the prepared nanocomposite films are an effective approach for remineralization of early demineralized tooth lesions [63]. This strategy could open up new avenues for the prevention and treatment of early dental caries. It was also found that the presence of silver, gold, and zinc oxide nanoparticles in toothpaste can significantly reduce the bacterial effects of *Streptococcus mutans*, the bacteria that causes dental caries, thus preventing dental caries disease [111].

4.4.2 Nanodiagnostics/imaging

Early identification of diseases such as cancer at the cellular and molecular levels, pharmacokinetics, and detection of biomarkers and pathogens are all potential uses of nanorobotics devices in medicine. For example, a unique form of biosensor surface modification for the highly specific detection of short RNAs has been used to accomplish early cancer diagnosis and disease progression tracking [112, 113]. Utilizing polymeric nanoparticles to identify cancer-related cell modifications and then recognizing cancer at an early stage using minimally invasive procedures vastly improves prognosis [114]. Sensor systems such as nanoelectromechanical systems, oral fluid nanosensor test, and optical nanobiosensors are examples of nanodiagnosis applications.

The nanoelectromechanical systems act by transforming the cantilever array sensor and biochemical signal into sensitive technology suited for the germs detection (bacteria, viruses and fungi), and also DNA at the molecular level. Therefore, nanoelectromechanical systems can be used to detect germs and diagnose oral cancers and diabetes mellitus [112]. This technology is useful in diagnosing oral cancer and diabetes mellitus, and also in detecting bacteria, viruses, and fungi.

Oral fluid nanosensor tests are micro-electromechanical systems comprised of sensor array chips that are being used to detect cancer's electrochemical sensors such as salivary protein and RNA biomarkers. The method is based on efficiently binding target RNA molecules or proteins to the sensor's surface and employs micro-fluid technology, which is described as the automation of analytical laboratory procedures into a single device or 'chip.' A combination of four salivary mRNA biomarkers

(ODZ, IL-1b, IL-8, and salivary transkriptome-SAT) and two salivary proteomic biomarkers (IL-8 and thioredoxin) has been found to have good sensitivity and specificity for oral cancer detection [112].

Optical nanobiosensors are based on the application of optical concepts to convert biochemical interactions into appropriate output signals. The fiber-optic equipment enables non-invasive study of intracellular components such as cytochrome C, which is essential for cellular energy production and a well-known protein involved in programmed cell death [110, 115].

Nanotechnologies already allow for intracellular imaging by attaching quantum dots or synthetic chromophores to specific molecules, such as proteins, or by introducing naturally occurring fluorescent proteins that, when combined with optical techniques like confocal microscopy and correlation imaging, allow for direct investigation of intracellular biochemical processes [116]. Despite the fact that microarray/biochip approaches based on the detection of specific biomolecular interactions have become an essential tool for molecular diagnostics, there are certain drawbacks. Some of the drawbacks of biochip technology are being addressed via nanotechnology [117, 118]. Nanophosphor scintillators can be utilized in digital dental imaging procedures to obtain high-quality images while lowering the radiation exposure [119].

4.4.3 Local anaesthesia

Injections of local anaesthetics are frequently used in dental treatment. However, the conventional methods have several drawbacks such as a prolonged period of action, different degrees of efficacy, and patient discomfort [120]. Some popular alternatives which include transcutaneous electronic nerve stimulation, topical, transmucosal or intraosseous techniques, cell demodulated electronic targeted anaesthesia, have all shown to have little clinical efficacy.

One advancement in nanodentistry outlined that dental nanorobots are able to travel between the teeth and the surrounding tissues. Their movement and function can be controlled by a pre-programmed computer. They can be used to perform the intended purpose and then be deactivated when no longer needed. Therefore, the ongoing nanodentistry research is focusing on colloidal suspensions and dental nanorobots. This yielded a promising procedure for making oral anaesthesia in dental practice today, which involved administration of colloidal suspension by the dental professionals, whereby colloidal nanoparticles are loaded with millions of active analgesic dental nanorobots in the patient's gingiva. The ambulating nanorobots reach the dentin by migrating into the gingival sulcus and passing painlessly through the lamina propria or the 1 to 3-micron thick layer of loose tissue at the cementodentinal junction after contacting the surface of the crown or mucosa [99]. When the nanorobots hit the dentin, they traverse 1 to 4 micron-sized dentinal tubules apertures and advance toward the pulp, steered by a mix of chemical gradients, temperature differentials, and positional guidance, all under the direction of the dentist's onboard nanocomputer. Many pathways are available for selection, near the cemento-enamel junction (CEJ), midway between the junction and the pulp, and close to the pulp [99]. Once installed in the pulp and under control of nerve impulse traffic, the dentist can instruct the analgesic dental nanorobots to turn off all sensitivity in any particular tooth that requires treatment. The targeted tooth becomes numb as soon as the hand-held controller display is activated. Following the completion of the dental operations, the dentist instructs the nanorobots to restore full sensation, re-establish nerve traffic control, and egress, followed by aspiration [120]. Nanorobotic analgesics provide greater patient comfort and anxiety reduction as no needles are involved, greater analgesic selectivity and controllability, fast and completely reversible switchable action, and avoidance of most side effects and complications [121].

4.4.4 Endodontics

Since the introduction of applications of nanoparticles in endodontics over the last decade, the overall understanding of endodontics discipline has expanded substantially, as new findings are constantly adding to what we already know. The complex anatomy of the root canal makes cleaning and shaping to eliminate bacteria in the canal extremely difficult, and some of pathogens have developed resistance to the irrigants and medicaments used [122]. Hence, the hallmarks of successful endodontic treatment are effective disinfection and sealing of the root canal system. Due to the favourable features of nanoparticles which include their nanoscale dimensions, nanoparticles can improve the treatment of endodontic infections by increasing antibacterial activity, enhancement of reactivity, and their ability be functionalized with other reactive compounds [123]. For instance, antimicrobial nanoparticle-based irrigants and intracanal medicaments can improve the elimination of endodontic infections. Similarly, photoactive nanoparticles could be employed as a potent adjunct to root canal disinfection [123]. Nanoparticle applications in regenerative endodontics also show promise, such as facilitating the release of bioactive molecules and improving the biophysical properties of scaffolds. Nanotechnology allows for the construction of experimental scaffolds made of natural nanofibers (Collagen-I) to aid in endodontic regeneration, along with microporous and nanofibrous membranes [124, 125]. Incorporation of bio-ceramic nanoparticles such as bioglass, glass ceramics, and zirconiain endodontic sealers, represents another application of nanotechnology in endodontics [126]. A team of investigators reported experimental pulp regeneration material. They developed nanostructured poly-L-lysine polymers dendrigraft with acid poly-glutamic coatings that encouraged pulp fibroblast proliferation, making them highly promising for endodontic applications [127].

A recently published review article showed that nanoparticles may be produced for a variety of applications in endodontics, including photodynamic therapy, disinfection tactics, regenerative procedures, and obturation materials [123]. However, according to the review, nanoparticles should not be used in a one-size-fits-all manner, because different materials, formulas, and combinations will produce diverse qualities, both positive and detrimental. Thus, nanoparticle applications in endodontics offer a lot of promise, but there is still more work to be done before basic research is translated into clinical investigations [123].

4.4.5 Drug delivery

The positive role of nanomedicine on drug delivery is undeniable. Many medicines such as conventional oral formulations that cannot be effectively used due to poor bioavailability are now finding new therapeutic prospects, thanks to nanotechnology. Reformulation of those medicines with a reduced particle size may improve oral bioavailability among other benefits. Nanoformulations preserve agents that are prone to degradation or denaturation in extreme pH environments, as well as extending the length of a drug's exposure by enhancing the formulation's retention through bioadhesion. Theoretically, all nanoformulations would improve the drug's absorption, stability, and therapeutic concentration within the target tissue, and allow for their repeatable and long-term release at the desired site of action. Therefore, the ability to control and target drug release, improve pharmacokinetics, increase drug bioavailability and selectivity, and thus treatment effectiveness are the most significant benefits of using drug carriers. Integration of nanotechnology into the process of drug development enables scientists to design and develop nanoscale pharmaceuticals which satisfy these requirements that are critically needed for safe and effective drug delivery [103]. Controlled drug release and improved local drug delivery can be achieved by encapsulating therapeutics compounds in nanocapsules or loading them onto nanospheres. These nanostructures undergo slow enzymatic breakdown over a period of time, and the drug is released as the membrane disintegrates. The initial burst release typically happens early after the administration process, and then the drug or other active substance progressively released over time [75].

The success of delivery systems is heavily reliant on the design of delivery carriers. The therapeutic molecules can be mixed within the 3D structure or on the surface of the scaffolds when 3D scaffolds are used as carriers for the molecules to be delivered. As a result, therapeutic molecules are released into the surrounding tissues, where they control cellular activities. They can also be employed as tissue engineering matrices for stem cell differentiation and culture. The inclusion of signaling molecules should be carefully studied in these circumstances based on scaffold nanochemistry, such as molecular bonding between scaffold networks and drug [104]. Nanoparticles, on the other hand, are the most common delivery carriers since they are frequently meant to infiltrate the cell membrane and, in some cases, the nucleus.

Periodontal disease is a major problem in dentistry that requires a special therapeutic molecule delivery technique. Periodontal disease is caused by a bacterial infection that can potentially destroy the host periodontal tissues, resulting in tooth loosening and loss [128]. The disease is generally treated using two basic approaches: bacterial eradication to prevent disease development and regenerative therapy, which may allow for the regeneration of damaged tissue [13, 129]. Bone morphogenetic protein has been widely used in this regard since it is involved in the osteogenic differentiation of stem cells and could regenerate periodontal ligament, cementum, and alveolar bone when administered via nanomaterials [130, 131]. Drug delivery to the periodontal pockets can be accomplished using triclosan-impregnated nanoparticles [16, 37]. Inflammation was reduced after triclosan particles were applied to the targeted location [20].

Calcium carbonate nanoparticles can have good oral surface retention due to their colloidal particle size and capability for calcium ion delivery. They serve as a vehicle for the slow and steady release of high amounts of calcium ions into the oral fluids. Calcium carbonate nanoparticles have the ability to raise the pH of the surrounding fluid. When added into an experimental tooth dentifrice, calcium carbonate nanoparticles were successful in remineralizing incipient enamel defects [44].

4.4.6 Dental restorative

Nanodental approaches for major tooth repair may have to go through multiple stages of engineering, beginning with genetic engineering, tissue engineering and regeneration, and then progression to the *in vitro* growth and implantation of the entire new teeth. Eventually, the nanorobotic fabrication and installation of a biologically autologous whole-replacement tooth that encompasses both cellular and mineral components—that is, complete dentition replacement therapy—should be possible within the time and financial constraints of a normal medical visit, thanks to the use of a low-cost desktop manufacturing facility that would fabricate the new tooth due to nanodentistry innovations. Many nanodentistry innovations continue to emerge for dental restoration.

In a recent development, Sanaei-rad *et al* (2021), fabricated poly(L-lactide) (PLLA) nanofibers via electrospinning method and coated them with mineral trioxide aggregate (MTA), and then investigated attachment, proliferation and odontogenic differentiation of dental pulp stem cells on the developed nanofibrous scaffolds [50]. The electrospun nanofibers were extensively evaluated. The results of this study showed that the fabricated MTA-coated PLLA nanofibers showed nano-fibrous morphology with an average diameter of 756 ± 157 nm. It also exhibited interconnected pores, interesting mechanical strength, and biocompatible properties. The fabricated MTA-coated PLLA nanofibers were also found to support the attachment and proliferation of dental pulp stem cells. In addition, there was observed biomineralization effect that was attributed to the developed nanofibers. It was concluded that these electrospun nanofibers play a vital role in dental therapeutics by enhancing the odontogenic differentiation of dental pulp stem cells, and thus demonstrated the desired characteristics of a pulp capping material [50].

In another development, dental resin nanocomposites were developed for dental restorative application [93]. The role of the presence of functional groups in the chemical structure of the nanoclay organic modifier on the mechanical and/or the physicochemical properties of the dental resin nanocomposites was investigated. Result of the study revealed the synthesized nanocomposites with intercalated structures and mostly with high distribution of filler into the polymer matrix. The presence of nanoclays resulted in limitation of polymerization shrinkage to about 50%. On the other hand, setting contraction was retained at low levels due to the absence of reactive groups in the organ modified montmorillonite nanofiller structure. At high filler loadings, an improvement in flexural modulus was observed, primarily due to the use of clay nanoparticles decorated with methacrylated groups, as well as a decrease in flexural strength. The nanocomposites with organomodified montmorillonite containing double bonds had the best overall performance. The authors concluded that their study is significant because it provides new knowledge regarding chemical interactions between nanofillers and the organic matrix, which can be used to improve tooth restorative materials [93].

The integration of nanotechnology with fluoralumino-silicate technology was found to drastically enhance the wear resistance, aesthetics, and polish ability of glass ionomer materials. Core build-ups, class I, III, and V restorations, sandwich method, and primary tooth restorative treatment are all clinical indications for this material [112, 120].

4.4.7 Tissue engineering

Tissue engineering is the science of designing and manufacturing new tissues to replace missing portions that were lost as a result of diseases such as cancer and trauma. Nanodentistry is crucial in natural tooth repair since it allows for genetic and tissue engineering, and regeneration, and also the creation and installation of an entire new tooth *in vitro* [132]. Signals for morphogenesis, extracellular matrix scaffolding, and stem cells for reacting to morphogens are the three fundamental ingredients for tissue engineering. Cell therapy and gene therapy have been developed in preclinical studies for a variety of tissues and organs, including the bone, heart, liver, and kidney, as a means of delivering growth factors, morphogens, or cytokines with stem/progenitor cells in a scaffold to the sites of tissue injury to accelerate and/or induce natural biological regeneration [99, 133]. New developments in stem cell biology and tissue engineering are paving the way for creation of cutting-edge dental treatments for both tooth repair and replacement. Periodontitis, dental caries, and pulpitis raise health-care expenses and, as a result, reduce economic productivity. They eventually lead to early tooth loss, lowering one's quality of life [99]. Advances in vital pulp therapy using pulp stem/progenitor cells may provide the impetus for the regeneration of the dentin–pulp complex without removing the entire pulp.

A team of researchers reported the development of nanofiber scaffolds fabricated using multiple biomaterials, namely, poly-ε-caprolactone (PCL), starch, and collagen by electrospinning technique for bone tissue engineering [134]. The developed nanofiber scaffolds were found to be monodispersed nanosized diameter (60 nm), have negative zeta potential (−13.5 mV), and uniform network shaped morphology. They also demonstrated attractive physicochemical and interconnected properties that are desirable in bone tissues engineering. The authors believed that the developed nanofibers could match the speed of tissue regeneration by incorporating some materials that can decrease the nanofibers' degradation rate, hence making them relevant as membrane biomaterials, like guided bone regeneration membrane [134]. In a similar but expanded study, nanofibers loaded with nano-hydroxyapatite (nHA) and zeolite were synthesized using a biodegradable and biocompatible blend of polymers, namely, poly(lactic acid) (PLA) and PCL [49]. The nanofibers were aimed for potential application in dental tissue engineering by stimulating and causing the proliferation of human dental pulp-derived stem cells. The nHA and zeolite-loaded nanofibers were prepared via electrospinning method and were subsequently characterized using field emission-scanning electron microscopy (FE-SEM) (for morphological features), x-ray diffraction (for crystals properties), FTIR spectrometry (to provide information about functional groups for the nanofiber structures), and water contact angle measurement techniques. The researchers also investigated the viability and adhesion of human dental pulp-derived stem cells on the loaded electrospun nanofibers using MTT assay and FE-SEM, respectively. The results revealed that the viability of human dental pulp-derived stem cells on the nanofiberscaffolds was significantly enhanced following 1, 7 and 14 days of culturing, and the nanofibers also improve the cell adhesion rate when compared with free nHA-Zeolite scaffolds. The authors concluded that nHA and zeolite-loaded nanofibers turn out to be a suitable scaffold for bone and tooth tissue engineering applications [49].

4.5 Conclusion

Through nanotechnology innovations, researchers were able to improve the physical, chemical, and mechanical properties of existing materials. The area of dentistry has been influenced by nanotechnology, which led to the creation of intelligent nanomaterials that provide better prophylaxis, diagnostic, and treatment regimens, and also regenerate tissue for the improvement and maintenance of oral health. Nanomedicines have demonstrated promising results in a variety of current and potential future applications in different areas of dentistry. They demonstrated antimicrobial effect helping in the preventive and treatment stages. The nanomaterials such as nanoamorphous calcium phosphate, nanosized calcium phosphate, nanocrystals of carbonate hydroxyapatite, and nanoparticulate bioactive glass have remineralizing potential that is desirable in bone restoration and tissue engineering. With continuous better understanding and the introduction of new nanomaterials, it is possible that the science of dental materials will also significantly change in the future. In the coming decade, it is hoped that promising nanomedicines will provide an avenue for achieving complete oral health. However, further research is needed to determine whether there are any potential dangers posed by dental nanomaterials to human health and the environment as a result of their rapid proliferation.

References

[1] Aminu N, Bello I, Umar N M, Tanko N, Aminu A and Audu M M 2020 The influence of nanoparticulate drug delivery systems in drug therapy *J. Drug Deliv. Sci. Technol.* **60** 101961

[2] Florence A T 2009 Pharmaceutical aspects of nanotechnology *Modern Pharmaceutics* ed A T Florence and J Siepmann 5th edn (New York: Informa Healthcare USA, Inc.) pp 453–92

[3] Zafar M S, Khurshid Z, Najeeb S, Zohaib S and Rehman I U 2017 Therapeutic applications of nanotechnology in dentistry *Nanostructures for Oral Medicine* (Amsterdam: Elsevier) pp 833–62

[4] Williams R O III and Vaughn J M 2007 Nanoparticle engineering *Encyclopedia of Pharmaceutical Technology* ed J Swarbrick 3rd edn (New York: Informa Healthcare USA, Inc.) pp 2384–98

[5] Perrie Y 2013 Pharmaceutical nanotechnology and nanomedicines *Aulton's Pharmaceutics the Design and Manufacture of Medicines* ed M E Aulton and K M G Taylor 4th edn (London: Elsevier) pp 777–96

[6] AlKahtani R N 2018 The implications and applications of nanotechnology in dentistry: a review *Saudi Dent. J.* **30** 107–16

[7] 2021 United States Budget https://nano.gov/2021budgetsupplement)

[8] Narang R S and Narang J K 2015 Nanomedicines for dental applications-scope and future perspective *Int. J. Pharm. Investig.* **5** 121–3

[9] Abiodun-Solanke I, Ajayi D and Arigbede A 2014 Nanotechnology and its application in dentistry *Ann. Med. Health Sci. Res.* **4** 171–7

[10] Hua S, de Matos M B C, Metselaar J M and Storm G 2018 Current trends and challenges in the clinical translation of nanoparticulate nanomedicines: pathways for translational development and commercialization *Front. Pharmacol.* **9** 1–14

[11] Kankala R K 2018 Nanomedicine: the next-generation therapeutics *J. Nanomed.* **2** 1–5

[12] Kirtane A R, Wong H L, Guru B R, Lis L G, Georg G I and Gurvich V J *et al* 2015 Reformulating tylocrebrine in epidermal growth factor receptor targeted polymeric nanoparticles improves its therapeutic index *Mol. Pharm.* **12** 2912–23

[13] Aminu N and Toh S M 2017 Applicability of nanoparticles-hydrogel composite in treating periodontal diseases and beyond. *Asian J. Pharm. Clin. Res.* **10** 65–70

[14] Khurshid Z, Zafar M, Qasim S, Shahab S, Naseem M and AbuReqaiba A 2015 Advances in nanotechnology for restorative dentistry *Materials* **8** 717–31

[15] Adeola H A, Sabiu S, Adekiya T A, Aruleba R T, Aruwa C E and Oyinloye B E 2020 Prospects of nanodentistry for the diagnosis and treatment of maxillofacial pathologies and cancers *Heliyon* **6** e04890

[16] Aminu N, Chan S Y and Toh S M 2018 Formulation design and optimization of triclosan loaded nanoparticles for enhanced drug delivery across gingival sulcus by Resolution IV modeling of Design-Expert® *J. Biomed. Clin. Sci.* **3** 83–90

[17] Aminu N, Chan S Y, Khan N H and Toh S M 2018 Concurrent determination of triclosan and flurbiprofen by high-performance liquid chromatography in simulated saliva and its application in dental nanogel formulation *Acta Chromatogr.* **30** 219–24

[18] Pramod K, Aminu N and Ali J 2014 Targeted drug delivery systems for the treatment of periodontal infections *Biotechnology Volume 8: Novel Drug Delivery* ed B Singh, O P Katare and J N Govil (Houston, TX: Studium Press LLC) pp 97–128

[19] Foong L K, Foroughi M M, Mirhosseini A F, Safaei M, Jahani S and Mostafavi M *et al* 2020 Applications of nano-materials in diverse dentistry regimes *RSC Adv.* **10** 15430–60

[20] Aminu N, Yam M-F, Chan S-Y, Bello I, Umar N M and Nuhu T *et al* 2020 The evaluation of healing effect of triclosan and flurbiprofen-loaded nanogels in experimental periodontitis in rats by morphometric analysis *Saudi Dent. J.* in press

[21] Amini H, Shenkin J D and Chi D L 2019 Dental public health issues in pediatric dentistry *Pediatric Dentistry* (Amsterdam: Elsevier) pp 159–68

[22] Gokulgandhi M R, Patel A, Cholkar K, Barot M and Mitra A K 2014 Nanoscale drug delivery systems *Advanced Drug Delivery* ed A K Mitra, C H Lee and K Cheng (Hoboken, NJ: Wiley) pp 140–55

[23] Pelaz B, Alexiou C, Alvarez-Puebla R A, Alves F, Andrews A M and Ashraf S *et al* 2017 Diverse applications of nanomedicine *ACS Nano.* **11** 2313–81
[24] Elkassas D and Arafa A 2017 The innovative applications of therapeutic nanostructures in dentistry *Nanomed. Nanotechnol. Biol. Med.* **13** 1543–62
[25] Hubbs A F, Sargent L M, Porter D W, Sager T M, Chen B T and Frazer D G *et al* 2013 Nanotechnology: toxicologic pathology *Toxicol. Pathol.* **41** 395–409
[26] Ledet G and Mandal T K 2012 Nanomedicine: emerging therapeutics for the 21st century https://uspharmacist.com/article/nanomedicine-emerging-therapeutics-for-the-21st-century)
[27] Caruthers S D, Wickline S A and Lanza G M 2007 Nanotechnological applications in medicine *Curr. Opin. Biotechnol.* **18** 26–30
[28] Odiba A, Ottah V, Ottah C, Anunobi O, Ukegbu C and Edeke A *et al* 2017 Therapeutic nanomedicine surmounts the limitations of pharmacotherapy *Open Med.* **12** 271–87
[29] Chandra Hembram K, Prabha S, Chandra R, Ahmed B and Nimesh S 2016 Advances in preparation and characterization of chitosan nanoparticles for therapeutics. *Artif. Cells Nanomed. Biotechnol.* **44** 305–14
[30] Raura N, Garg A, Arora A and Roma M 2020 Nanoparticle technology and its implications in endodontics: a review *Biomater. Res.* **24** 21
[31] Najeeb S, Khurshid Z, Agwan A, Zafar M, Alrahabi M and Bin Qasim S *et al* 2016 Dental applications of nanodiamonds *Sci. Adv. Mater.* **8** 2064–70
[32] Fernandez C C, Sokolonski A R, Fonseca M S, Stanisic D, Araújo D B and Azevedo V *et al* 2021 Applications of silver nanoparticles in dentistry: advances and technological innovation *Int. J. Mol. Sci.* **22** 1–21
[33] Adam R Z and Khan S B 2021 Antimicrobial efficacy of silver nanoparticles against *Candida albicans*: a systematic review protocol *PLoS One* **16** 1–13
[34] Fattah Z, Jowkar Z and Rezaeian S 2021 Microshear bond strength of nanoparticle-incorporated conventional and resin-modified glass ionomer to caries-affected dentin. *Int. J. Dent.* **2021** 1–8
[35] Bouropoulos N and Moradian–Oldak J 2003 Analysis of hydroxyapatite surface coverage by amelogenin nanospheres following the langmuir model for protein adsorption *Calcif. Tissue Int.* **72** 599–603
[36] Vasiliu S, Racovita S, Gugoasa I A, Lungan M-A, Popa M and Desbrieres J 2021 The benefits of smart nanoparticles in dental applications *Int. J. Mol. Sci.* **22** 1–23
[37] Aminu N, Baboota S, Pramod K, Singh M, Dang S and Ansari S H *et al* 2013 Development and evaluation of triclosan loaded poly-ε-caprolactone nanoparticulate system for the treatment of periodontal infections *J. Nanoparticle Res.* **15** 2075
[38] Aminu N, Chan S Y and Toh S M 2018 Development and validation of a stability-indicating HPLC-UV method for the simultaneous determination of flurbiprofen and triclosan in dental nanogel formulations *J. Phys. Sci.* **29** 1–7
[39] Toledano M, Osorio R, Osorio E, Medina-Castillo A L, Toledano-Osorio M and Aguilera F S 2017 Ions-modified nanoparticles affect functional remineralization and energy dissipation through the resin-dentin interface *J. Mech. Behav. Biomed. Mater.* **68** 62–79
[40] Moradpoor H, Safaei M, Mozaffari H R, Sharifi R, Imani M M and Golshah A *et al* 2021 An overview of recent progress in dental applications of zinc oxide nanoparticles *RSC Adv.* **11** 21189–206
[41] Akram Z, Aati S, Ngo H and Fawzy A 2021 pH-dependent delivery of chlorhexidine from PGA grafted mesoporous silica nanoparticles at resin-dentin interface *J. Nanobiotechnol.* **19** 1–16

[42] Enan E T, Ashour A A, Basha S, Felemban N H and Gad El-Rab S M F 2021 Antimicrobial activity of biosynthesized silver nanoparticles, amoxicillin, and glass-ionomer cement against *Streptococcus mutans* and *Staphylococcus aureus Nanotechnology* **32** 215101

[43] Kim T-H, Eltohamy M, Kim M, Perez R A, Kim J-H and Yun Y-R *et al* 2014 Therapeutic foam scaffolds incorporating biopolymer-shelled mesoporous nanospheres with growth factors *Acta Biomater.* **10** 2612–21

[44] Nakashima S, Yoshie M, Sano H and Bahar A 2009 Effect of a test dentifrice containing nano-sized calcium carbonate on remineralization of enamel lesions *in vitro J. Oral Sci.* **51** 69–77

[45] Priyadarsini S, Mukherjee S and Mishra M 2018 Nanoparticles used in dentistry: a review *J. Oral Biol. Craniofac. Res.* **8** 58–67

[46] Aminu N, Chan S Y, Yam M F and Toh S M 2019 A dual-action chitosan-based nanogel system of triclosan and flurbiprofen for localised treatment of periodontitis *Int. J. Pharm.* **570** 118659

[47] Bottino M C, Yassen G H, Platt J A, Labban N, Windsor L J and Spolnik K J *et al* 2015 A novel three-dimensional scaffold for regenerative endodontics: materials and biological characterizations *J. Tissue Eng. Regen. Med.* **9** E116–23

[48] Zafar M, Najeeb S, Khurshid Z, Vazirzadeh M, Zohaib S and Najeeb B *et al* 2016 Potential of electrospun nanofibers for biomedical and dental applications *Materials* **9** 1–21

[49] Mohandesnezhad S, Pilehvar-Soltanahmadi Y, Alizadeh E, Goodarzi A, Davaran S and Khatamian M *et al* 2020 *In vitro* evaluation of Zeolite-nHA blended PCL/PLA nanofibers for dental tissue engineering *Mater. Chem. Phys.* **252** 123152

[50] Sanaei-rad P, Jamshidi D, Adel M and Seyedjafari E 2021 Electrospun poly(L-lactide) nanofibers coated with mineraltrioxide aggregate enhance odontogenic differentiation ofdental pulp stem cells *Polym. Adv. Technol.* **32** 402–10

[51] Ainola M, Tomaszewski W, Ostrowska B, Wesolowska E, Wagner H D and Swieszkowski W *et al* 2016 A bioactive hybrid three-dimensional tissue-engineering construct for cartilage repair *J. Biomater. Appl.* **30** 873–85

[52] Tang J, Han Y, Zhang F, Ge Z, Liu X and Lu Q 2015 Buccal mucosa repair with electrospun silk fibroin matrix in a rat model *Int. J. Artif. Organs* **38** 105–12

[53] Joshi D, Garg T, Goyal A and Rath G 2015 Development and characterization of novel medicated nanofibers against periodontitis *Curr. Drug Deliv.* **12** 564–77

[54] Ranjbar-Mohammadi M, Zamani M, Prabhakaran M P, Bahrami S H and Ramakrishna S 2016 Electrospinning of PLGA/gum tragacanth nanofibers containing tetracycline hydrochloride for periodontal regeneration *Mater. Sci. Eng.* C **58** 521–31

[55] Qasim S B, Najeeb S, Delaine-Smith R M, Rawlinson A and Ur Rehman I 2017 Potential of electrospun chitosan fibers as a surface layer in functionally graded GTR membrane for periodontal regeneration *Dent. Mater.* **33** 71–83

[56] Samprasit W, Kaomongkolgit R, Sukma M, Rojanarata T, Ngawhirunpat T and Opanasopit P 2015 Mucoadhesive electrospun chitosan-based nanofibre mats for dental caries prevention *Carbohydr. Polym.* **117** 933–40

[57] Aadil K R, Nathani A, Sharma C S, Lenka N and Gupta P 2019 Investigation of poly (vinyl) alcohol-gellan gum based nanofiber as scaffolds for tissue engineering applications *J. Drug Deliv. Sci. Technol.* **54** 101276

[58] Zinger A, Adir O, Alper M, Simon A, Poley M and Tzror C *et al* 2018 Proteolytic nanoparticles replace a surgical blade by controllably remodeling the oral connective tissue *ACS Nano.* **12** 1482–90

[59] Ghazal M, Albashaireh Z S and Kern M 2008 Wear resistance of nanofilled composite resin and feldspathic ceramic artificial teeth *J. Prosthet. Dent.* **100** 441–8

[60] Nagar N and Vaz A C 2013 Comparison of shear bond strengths of conventional orthodontic composite and nano-ceramic restorative composite: an *in vitro* study *Indian J. Dent. Res.* **24** 713–8

[61] Yang Y, Xu Z, Guo Y, Zhang H, Qiu Y and Li J *et al* 2021 Novel core–shell CHX/ACP nanoparticles effectively improve the mechanical, antibacterial and remineralized properties of the dental resin composite *Dent. Mater.* **37** 636–47

[62] Mutar M A and Mahdi M S 2019 Synthesis and characterization of novel nanocomposites with nanofillers particles and their applications as dental materials *Period. Eng. Nat. Sci.* **7** 1512–38

[63] Zaki D Y, Safwat E M, Nagi S M, Salem H N, Hamdy T M and Moharam L M *et al* 2021 A novel dental re-mineralizing blend of hydroxyethyl-cellulose and cellulose nanofibers oral film loaded with nepheline apatite glass: preparation, characterization and *in vitro* evaluation of re-mineralizing effect *Carbohydr. Polym. Technol. Appl.* **2** 100035

[64] Yushau U S, Almofeez L and Bozkurt A 2020 Novel polymer nanocomposites comprising triazole functional silica for dental application *Silicon* **12** 109–16

[65] Yushau U S, Almofeez L and Bozkurt A 2019 Aminotriazole functional silica incorporated BisGMA/TEGDMA resins as dental nanocomposites *Polym. Polym. Compos.* **27** 488–95

[66] Jaymand M, Lotfi M, Barar J, Eskandani M and Maleki H 2018 Novel dental nanocomposites: fabrication and investigation of their physicochemical, mechanical and biological properties *Bull. Mater. Sci.* **41** 84

[67] Elshereksi N, Muchtar A and Azhari C 2021 Effects of nanobarium titanate on physical and mechanical properties of poly(methyl methacrylate) denture base nanocomposites *Polym. Polym. Compos.* **29** 484–96

[68] Kanaparthy R and Kanaparthy A 2011 The changing face of dentistry: nanotechnology *Int. J. Nanomed.* **6** 2799–804

[69] Kasaj A, Willershausen B, Reichert C, Röhrig B, Smeets R and Schmidt M 2008 Ability of nanocrystalline hydroxyapatite paste to promote human periodontal ligament cell proliferation *J. Oral Sci.* **50** 279–85

[70] Gupta K C, Haider A, Choi Y and Kang I 2014 Nanofibrous scaffolds in biomedical applications *Biomater. Res.* **18** 5

[71] Wehlage D, Blattner H, Mamun A, Kutzli I, Diestelhorst E and Rattenholl A *et al* 2020 Cell growth on electrospun nanofiber mats from polyacrylonitrile (PAN) blends *AIMS Bioeng.* **7** 43–54

[72] Kamsani N H, Haris M S, Pandey M, Taher M and Rullah K 2021 Biomedical application of responsive 'smart' electrospun nanofibers in drug delivery system: A minireview *Arab. J. Chem.* **14** 103199

[73] Liang D, Hsiao B S and Chu B 2007 Functional electrospun nanofibrous scaffolds for biomedical applications *Adv. Drug. Deliv. Rev.* **59** 1392–412

[74] Perán M, García M A, López-Ruiz E, Bustamante M, Jiménez G and Madeddu R *et al* 2012 Functionalized nanostructures with application in regenerative medicine *Int. J. Mol. Sci.* **13** 3847–886

[75] Baghaban-Eslaminejad M, Oryan A, Kamali A and Moshiri A 2017 The role of nanomedicine, nanotechnology, and nanostructures on oral bone healing, modeling, and remodeling *Nanostructures for Oral Medicine* (Amsterdam: Elsevier) pp 777–832

[76] He X, Cheng L, Zhang X, Xiao Q, Zhang W and Lu C 2015 Tissue engineering scaffolds electrospun from cotton cellulose *Carbohydr. Polym.* **115** 485–93

[77] Ardeshirzadeh B, Anaraki N A, Irani M, Rad L R and Shamshiri S 2015 Controlled release of doxorubicin from electrospun PEO/chitosan/graphene oxide nanocomposite nanofibrous scaffolds *Mater. Sci. Eng.* C **48** 384–90

[78] Qasim S B, Zafar M S, Najeeb S, Khurshid Z, Shah A H and Husain S *et al* 2018 Electrospinning of chitosan-based solutions for tissue engineering and regenerative medicine *Int. J. Mol. Sci.* **19** 1–26

[79] Cavalcanti B N, Zeitlin B D and Nör J E 2013 A hydrogel scaffold that maintains viability and supports differentiation of dental pulp stem cells *Dent. Mater.* **29** 97–102

[80] Park C H, Rios H F, Jin Q, Bland M E, Flanagan C L and Hollister S J *et al* 2010 Biomimetic hybrid scaffolds for engineering human tooth-ligament interfaces *Biomaterials* **31** 5945–52

[81] Zafar M S, Alnazzawi A A, Alrahabi M, Fareed M A, Najeeb S and Khurshid Z 2019 Nanotechnology and nanomaterials in dentistry *Advanced Dental Biomaterials* (Amsterdam: Elsevier) pp 477–505

[82] Saunders S A 2009 Current practicality of nanotechnology in dentistry. Part 1: Focus on nanocomposite restoratives and biomimetics *Clin. Cosmet. Investig. Dent.* **1** 47–61

[83] Kukulka E C, Sato T P, Archangelo K C, Santos J D, De Souza J R and Borges A L S 2021 Development of electrospun-based polyetherimide fibers and diameter analysis for potential use in dental materials *Brazil. Dent. Sci.* **24** 1–5

[84] Feng L, Li S, Li H, Zhai J, Song Y and Jiang L *et al* 2002 Super-hydrophobic surface of aligned polyacrylonitrile nanofibers *Angew Chem. Int. Ed.* **41** 1221–3

[85] Bhattarai R, Bachu R, Boddu S and Bhaduri S 2018 Biomedical applications of electrospun nanofibers: drug and nanoparticle delivery *Pharmaceutics.* **11** 5

[86] Nugroho D A, Nuryono W, Asmara W, Rahmasari A and Dyota E 2021 Tensile strength and microscopic adhesion observation of dental composite restorative material manufactured from Sisal nanofiber as filler *Adv. Heal. Sci. Res.* **33** 268–72

[87] Antonova O Y, Kochetkova O Y and Shlyapnikov Y M 2021 ECM-mimetic nylon nanofiber scaffolds for neurite growth guidance *Nanomaterials* **11** 516

[88] Chaughule R, Raorane D, Pednekar S and Dashaputra R 2018 Nanocomposites and their use in dentistry *Dental Applications of Nanotechnology* (Cham: Springer) pp 59–79

[89] Sen M 2020 Nanocomposite materials *Nanotechnology and the Environment* (London: IntechOpen) pp 1–12

[90] Nikolaidis A K, Koulaouzidou E A, Gogos C and Achilias D S 2021 Synthesis of novel dental nanocomposite resins by incorporating polymerizable, quaternary ammonium silane-modified silica nanoparticles *Polymers* **13** 1682

[91] Alamgir M, Mallick A and Nayak G C 2021 Mechanical and thermal behavior of pHEMA and pHEMA nanocomposites targeting for dental materials *Appl. Nanosci.* **11** 1257–65

[92] Wang W, Sun X, Huang L, Gao Y, Ban J and Shen L *et al* 2014 Structure–property relationships in hybrid dental nanocomposite resins containing monofunctional and multifunctional polyhedral oligomeric silsesquioxanes *Int. J. Nanomed.* **9** 841–52

[93] Nikolaidis A K, Koulaouzidou E A, Gogos C and Achilias D S 2019 Synthesis and characterization of dental nanocomposite resins filled with different clay nanoparticles *Polymers* **11** 730

[94] Gawdzinska K, Paszkiewicz S, Piesowicz E, Bryll K, Irska I and Lapis A *et al* 2019 Preparation and characterization of hybrid nanocomposites for dental applications *Appl. Sci.* **9** 1381

[95] Calisir M 2019 Nanotechnology in dentistry: past, present, and future *Nanomaterials for Regenerative Medicine Stem Cell Biology and Regenerative Medicine* (Cham: Springer) pp 197–216
[96] Wilson D J 2018 https://news-medical.net/health/What-is-Nanodentistry.aspx
[97] Pokrowiecki R, Pałka K and Mielczarek A 2018 Nanomaterials in dentistry: a cornerstone or a black box? *Nanomedicine* **13** 639–67
[98] Dogra S, Gupta A, Goyal V, Chib A S and Kataria V 2020 Recent trends, therapeutic applications, and future trends of nanomaterials in dentistry *Nanomaterials in Diagnostic Tools and Devices* (Amsterdam: Elsevier) pp 257–92
[99] Moezizadeh M 2013 Future of dentistry, nanodentistry, ozone therapy and tissue engineering *J. Dev. Biol. Tissue Eng.* **5** 1–6
[100] Freitas R A 2005 Nanotechnology, nanomedicine and nanosurgery *Int. J. Surg.* **3** 243–6
[101] Sousa M G C, Maximiano M R, Costa R A, Rezende T M B and Franco O L 2020 Nanofibers as drug-delivery systems for infection control in dentistry *Expert Opin. Drug. Deliv.* **17** 919–30
[102] Barot T, Rawtani D and Kulkarni P 2021 Nanotechnology-based materials as emerging trends for dental applications *Rev. Adv. Mater. Sci.* **60** 173–89
[103] Makvandi P, Josic U, Delfi M, Pinelli F, Jahed V and Kaya E *et al* 2021 Drug delivery (nano)platforms for oral and dental applications: tissue regeneration, infection control, and cancer management *Adv. Sci.* **8** 1–28
[104] Abou Neel E A, Bozec L, Perez R A, Kim H-W and Knowles J C 2015 Nanotechnology in dentistry: prevention, diagnosis, and therapy *Int. J. Nanomed.* **10** 6371–94
[105] Neves P B A, das, Agnelli J A M, Kurachi C and Souza CWO de 2014 Addition of silver nanoparticles to composite resin: effect on physical and bactericidal properties *in vitro Braz. Dent. J.* **25** 141–5
[106] Li F, Weir M D, Fouad A F and Xu H H K 2014 Effect of salivary pellicle on antibacterial activity of novel antibacterial dental adhesives using a dental plaque microcosm biofilm model *Dent. Mater.* **30** 182–91
[107] Allaker R P and Ian Douglas C 2015 Non-conventional therapeutics for oral infections *Virulence.* **6** 196–207
[108] Besinis A, De Peralta T and Handy R D 2014 Inhibition of biofilm formation and antibacterial properties of a silver nano-coating on human dentine *Nanotoxicology* **8** 745–54
[109] Mielczarek A, Gedrange T and Michalik J 2015 An *in vitro* evaluation of the effect of fluoride products on white spot lesion remineralization *Am. J. Dent.* **28** 51–6
[110] Tschoppe P, Zandim D L, Martus P and Kielbassa A M 2011 Enamel and dentine remineralization by nano-hydroxyapatite toothpastes *J. Dent.* **39** 430–7
[111] Hernández-Sierra J F, Ruiz F, Cruz Pena D C, Martínez-Gutiérrez F, Martínez A E and de Jesús Pozos Guillén A *et al* 2008 The antimicrobial sensitivity of *Streptococcus mutans* to nanoparticles of silver, zinc oxide, and gold. *Nanomed. Nanotechnol. Biol. Med.* **4** 237–40
[112] Kasimoglu Y, Tabakcilar D, Guclu Z A, Yamamoto-Nemoto S, Tuna E. B. and Ozen B *et al* 2020 Nanomaterials and nanorobotics in dentistry: a review *J. Dent. Indones.* **27** 77–84
[113] Parvanian S, Mostafavi S M and Aghashiri M 2017 Multifunctional nanoparticle developments in cancer diagnosis and treatment *Sens. Bio.-Sensing Res.* **13** 81–7
[114] Aeran H, Kumar V, Uniyal S and Tanwer P 2015 Nanodentistry: is just a fiction or future *J. Oral. Biol. Craniofac. Res.* **5** 207–11
[115] Punyadeera C and Slowey P D 2019 Saliva as an emerging biofluid for clinical diagnosis and applications of MEMS/NEMS in salivary diagnostics *Nanobiomaterials in Clinical Dentistry.* (Amsterdam: Elsevier) pp 543–65

[116] Li K C P, Pandit S D, Guccione S and Bednarski M D 2004 Molecular imaging applications in nanomedicine *Biomed. Microdevices* **6** 113–6

[117] Jain K 2003 Nanodiagnostics: application of nanotechnology in molecular diagnostics *Expert Rev. Mol. Diagn.* **3** 153–61

[118] Jain K K 2005 Nanotechnology in clinical laboratory diagnostics *Clin. Chim. Acta* **358** 37–54

[119] Mupparapu M 2006 New nanophosphor scintillators for solid-state digital dental imagers *Dentomaxillofac. Radiol.* **35** 475–6

[120] Chandki R, Kala M, Kiran Kumar N, Brigit B, Banthia P and Banthia R 2012 'Nanodentistry': exploring the beauty of miniature *J. Clin. Exp. Dent.* **4** 119–24

[121] Schleyer T L 2000 Nanodentistry *J. Am. Dent. Assoc.* **131** 1567–8

[122] Chandak P G, Ghanshyamdasj M, Chandak C, Relan K N, Chandak M and Rathi C *et al* 2021 Nanoparticles in endodontics—a review *J. Evol. Med. Dent. Sci.* **10** 976–82

[123] Wong J, Zou T, Lee A H C and Zhang C 2021 The potential translational applications of nanoparticles in endodontics *Int. J. Nanomed.* **16** 2087–106

[124] Yamauchi N, Yamauchi S, Nagaoka H, Duggan D, Zhong S and Lee S M *et al* 2011 Tissue engineering strategies for immature teeth with apical periodontitis *J. Endod.* **37** 390–7

[125] Albuquerque M T P, Valera M C, Nakashima M, Nör J E and Bottino M C 2014 Tissue-engineering-based strategies for regenerative endodontics *J. Dent. Res.* **93** 1222–31

[126] Utneja S, Nawal R R, Talwar S and Verma M 2015 Current perspectives of bio-ceramic technology in endodontics: calcium enriched mixture cement—review of its composition, properties and applications *Restor. Dent. Endod.* **40** 1

[127] Fioretti F, Mendoza-Palomares C, Avoaka-Boni M C, Ramaroson J, Bahi S and Richert L *et al* 2011 Nano-odontology: nanostructured assemblies for endodontic regeneration *J. Biomed. Nanotechnol.* **7** 471–5

[128] Aminu N, Chan S Y and Toh S M 2017 Roles of nanotechnological approaches in periodontal disease therapy *J. Appl. Pharm. Sci.* **7** 234–42

[129] Pragati S, Ashok S and Kuldeep S 2009 Recent advances in periodontal drug delivery systems *Int. J. Drug Deliv.* **1** 1–14

[130] Ripamonti U, Crooks J, Petit J-C and Rueger D C 2001 Periodontal tissue regeneration by combined applications of recombinant human osteogenic protein-1 and bone morphogenetic protein-2. A pilot study in Chacma baboons (Papio ursinus) *Eur. J. Oral Sci.* **109** 241–8

[131] Wikesjo U M E, Sorensen R G, Kinoshita A, Jian Li X and Wozney J M 2004 Periodontal repair in dogs: effect of recombinant human bone morphogenetic protein-12 (rhBMP-12) on regeneration of alveolar bone and periodontal attachment. A pilot study *J. Clin. Periodontol.* **31** 662–70

[132] Ozak S T and Ozkan P 2013 Nanotechnology and dentistry *Eur. J. Dent.* **7** 145–51

[133] Lacerda-Pinheiro S, Marchadier A, Donãs P, Septier D, Benhamou L and Kellermann O *et al* 2008 An *in vivo* model for short-term evaluation of the implantation effects of biomolecules or stem cells in the dental pulp *Open Dent. J.* **2** 67–72

[134] Ghavimi M A, Negahdari R, Shahabadi A B, Sharifi S, Kazeminejad E and Shahi S *et al* 2020 Preparation and study of starch/collagen/polycaprolactone nanofiber scaffolds for bone tissue engineering using electrospinning technique *Eur. Chem. Commun.* **2** 122–7

Chapter 5

Nanotechnology applications in managing periodontal diseases

Rana Anju and Kaur Arundeep

Science was never static and will never be. It is a continuously evolving event and so are the various techniques for treating inflammatory and destructive diseases. Periodontium is a highly specialised structure comprising hard and soft tissues. Any deviation in this organised structure may result in various periodontal diseases including periodontitis and peri-implantitis resulting in loss of periodontium and eventually tooth loss and implant failure. To overcome this, an alternative counterpart for the biomechanical properties of the periodontium in terms of physicochemical properties and bio-mimetic features is required. Even after several decades of research, the complete regeneration of periodontium seems a challenge to the periodontist. But now there can be a ray of scope through the vision of nanotechnology which uses various nanomaterials and nanofibers for regeneration. Nanotechnology seems to be the future of dentistry and in this chapter, we are going to discuss the recent nanomaterials and nanoscaffolds that can deliver drugs, signaling molecules, growth factors, and cells for tissue engineering in managing and treating periodontal and implant diseases.

NOTE: Parts of this chapter have been adapted from [5], with permission from MDPI.

5.1 Introduction

The periodontium is defined as the supporting structures that comprise two soft fibrous tissues i.e. gingiva and periodontal ligament and two hard tissues which include cementum and alveolar bone. In November 2017, the World Workshop of Periodontology was held at Chicago, USA and a new periodontal disease and peri-implant diseases classification was given [1]. For the first time in history, gingival health, periodontal health and peri-implant health was included in the classification.

doi:10.1088/978-0-7503-3671-0ch5

Periodontal disease is a microbial dysbiosis which results in disturbances within the host immune balance. The National Oral Health Survey conducted by Indian Govt. and Indian Dental Council in collaboration with the World Health Organization reported that almost 85% of children and 95%–100% adult population suffer from periodontal disease once in their lifetime. The prevalence in different age groups up to 12, 15, 35–44 and 65–74 years was reported as 57%, 67.7%, 89.6% and 79.9%, respectively [2]. This disease prevalence increases with the age, leading to silent periodontal bone and tissue destruction and eventually loss of teeth in the oral cavity.

This prevalence of periodontal disease has lead to the search for more effective treatment options. The main goal of treatment focuses on ideally reconstructing the original healthy periodontium including new cementum, new bone and periodontal ligament. Various techniques have been employed in the field of periodontics for regeneration such as flap surgery, bone grafting and guided tissue regeneration but ideal regeneration still remains a challenge.

5.2 What are nanosubstances or nanostructures?

Substances less than 100 nm in at least one dimension are known as nanosubstances or nanostructures [3]. Various fibres, materials, grains, films, powder, scaffold, gels or membranes less than 100 nm come under this category.

$1 \text{ nm} = 10^{-6} \text{ mm}$

Nanotechnology is the control of matter at a scale of approximately 1 to 100 nanometres (nm), where due to nanosize, novel properties and function of the material are seen (figure 5.1) [4].

Advantages of nano-structures include:

1. High colloidal stability;

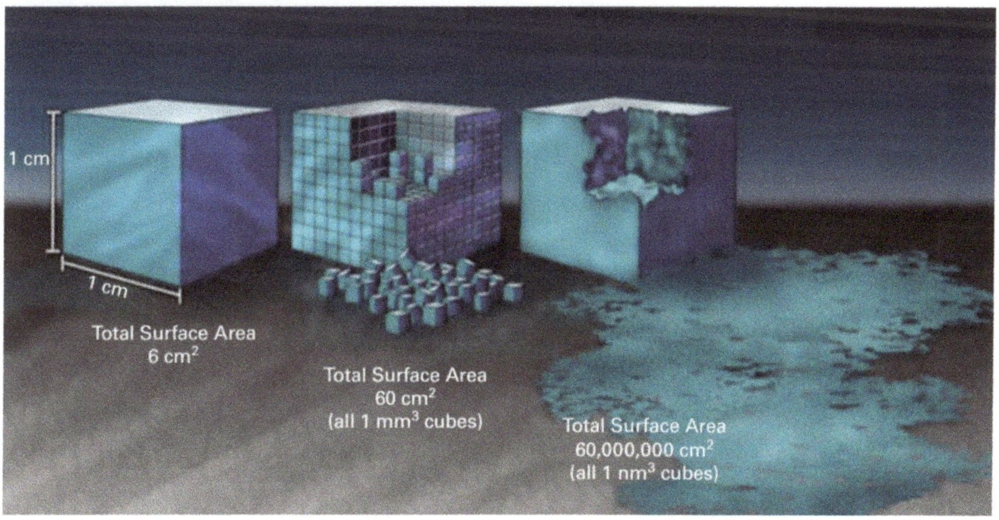

Figure 5.1. Graphic representation of increased surface area provided by nanotexturing. (Adapted from [4] CC BY 3.0.)

2. High dispersibility;
3. Easy to manipulate into nano-phases/nano-domains.

Various types of nano-materials (figure 5.2):
1. Organic/natural nano-materials;
2. Synthetic nano-materials.

5.3 Application of nanotechnology in periodontal disease management

1. Local drug delivery;
2. Host modulation and regeneration;
3. Managing implant conditions and disease.

5.4 Application in local drug delivery

Drugs play an important role in inhibiting growth of bacteria and inflammatory response after a good phase I therapy. There are two ways of administration of drug, i.e., locally administered and systemic administration. Since systemic administration of drug has several disadvantages like drug resistance, dysbacteriosis, systemic side effects and lack of site specificity, local drug delivery (LDD) is the mostly preferred route in periodontal and peri-implant cases. LDD provides more precise targeting therapy. These systems can release and transport the therapeutic agents and bioactive substances to the required sites at required rates.

Bio-adhesive carriers like chips and gels were used for the drug delivery system (DDS) in the past. They used to adhere onto the tooth surface for sustained release of drug. Antibiotics or antibacterial agents like chlorhexidine were loaded in the chips for local biofilm inhibition. Today, newer advances, micro/nano-scaled

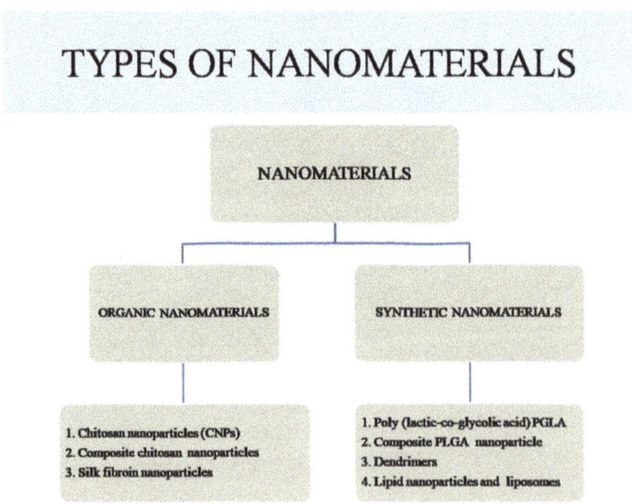

Figure 5.2. Schematic representation of various types of nanomaterials.

carriers, are reported which are more comfortable and multifunctional. These carriers have been designed and modified for the loading of anti-bacterial agents, anti-inflammatory drugs, and biomolecules like proteins and growth factors (figure 5.3).

Various carriers of DDS [5] (figure 5.4) include:

1. Microparticles/nanoparticles;
    a. Microsphere/nanosphere
    b. Nanofibres;
    c. Nanocapsules.
2. Hydrogel;
3. Dendrimer.

Studies by different researchers in the field of nanotechnology have shown advancement in local drug delivery carriers for periodontitis and peri-implantitis cases. Seneviratne *et al* [6], Chen *et al* [7], Moura *et al* [8], Phaechamud *et al* [9], Marziyeh *et al* [10], Kimura *et al* [11], and Yu *et al* [12] are a few of the researchers who documented their work on the significance of nanoparticles in periodontal disease.

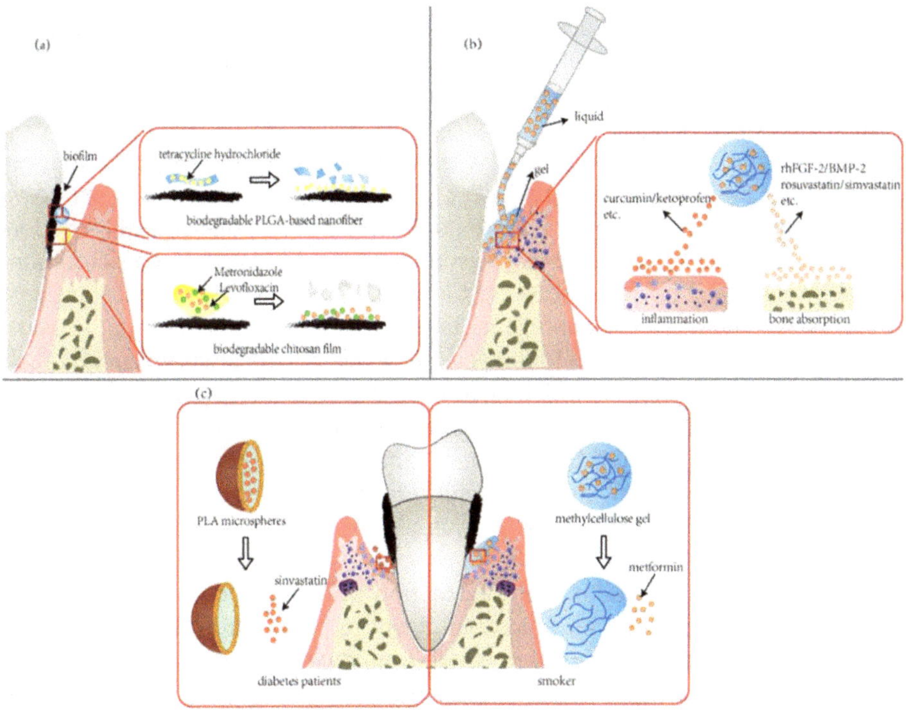

Figure 5.3. DDS used in periodontitis. (a) Antimicrobial agents (like metronidazole and levofloxacin) can be loaded in drug delivery devices, such as fibers and films. (b) Drugs for immunomodulation and alveolar bone regeneration, i.e., anti-inflammation agents, bone growth factors and osteogenesis drugs used in injectable drug delivery systems. (c) DDS used in periodontitis with risk factors (such as smoking and diabetes). (Adapted from [5], copyright (2020), with permission from Molecular Diversity Preservation International (CC BY 4.0).)

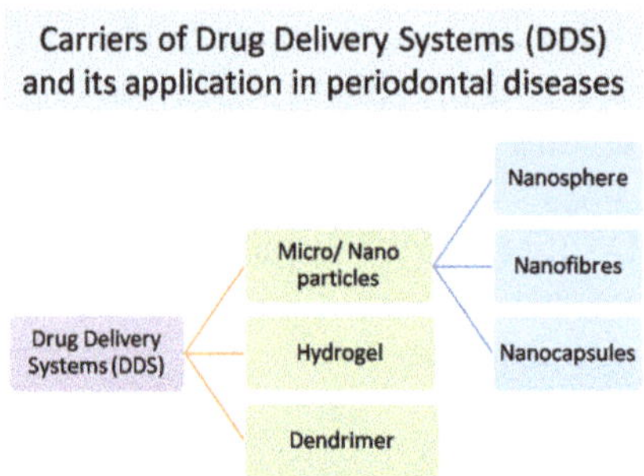

Figure 5.4. Schematic representation of various carriers of DDS available.

Seneviratne *et al* in 2014, synthesized a Nano-CHX which was a mesoporous silica nanoparticle-encapsulated pure chlorhexidine agent. This Nano-CHX demonstrated a potent antibacterial effect on planktonic bacteria and mono-species biofilms at concentrations of 50–200 μg mL^{-1} against *Streptococcus mutans, Aggregatibacter actinomycetemcomitans, Fusobacterium nucleatum*, and *Enterococccus faecalis* [6]. Moreover, Nano-CHX effectively suppressed *S. mutans, F. nucleatum, A. actinomycetemcomitans* and *Porphyromonas gingivalis* up to 72 h. Thus, it was acting as a promising anti-biofilm agent in periodontal diseases. Similarly Chen *et al* formulated triclosan-loaded tooth-binding micelles and proved their capability of reducing the viability of preformed biofilm by 4-log CFU/HA disc compared to untreated control biofilm [7].

Moura *et al* in their clinical study, assessed the release of locally delivered doxycycline by poly (l-lactide-co-glycolide) (PLGA) microspheres in the periodontal pocket of patients with chronic periodontitis, treated by non-surgical periodontal therapy [8]. A significant decrease in gingival crevicular fluid (GCF) concentration of drug (19.69 ± 4.70 μg ml^{-1}) was observed after the 20th day, thus implicating potent DOX delivery system in a successfully sustained release in the periodontal pockets.

Phaechamud T *et al* formulated doxycycline hyclate loaded with bleached shellac *in situ* forming microparticles which effectively inhibited *Porphyromonas gingivalis, Staphylococcus aureus* and *Streptococcus mutans* and showed a sustainable drug release for up to 47 days [9]. Marziyeh *et al* fabricated another drug delivery device by incorporation of tetracycline hydrochloride into core–shell nanofibers. In this drug delivery device, drug release was sustained for 75 days with only 19% of burst release within the first two hours [10].

Kimura *et al* reported ofloxacin loaded poly (methacrylic acid) and hydroxypropyl- cellulose strips as an antimicrobial agent in periodontal therapy [11].

Another study done by Yu *et al* in which a tinidazole loaded *in situ* gel forming system using a diblock copolymer, monomethoxy poly(ethylene glycol)–block-poly (D,L-lactide) (mPEG–PDLLA) was designed. This *in situ* gel forming system showed sustained tinidazole release over 192 h with low burst effect (around 7% in the first 8 h) in their *in vitro* and *in vivo* study [12]. Other products commercially available include perioChip insert (chlorhexidine gluconate, crosslinked hydrolyzed gelatin) and the injectable formulations attridox (doxycycline, poly-DL-lactide), arestin

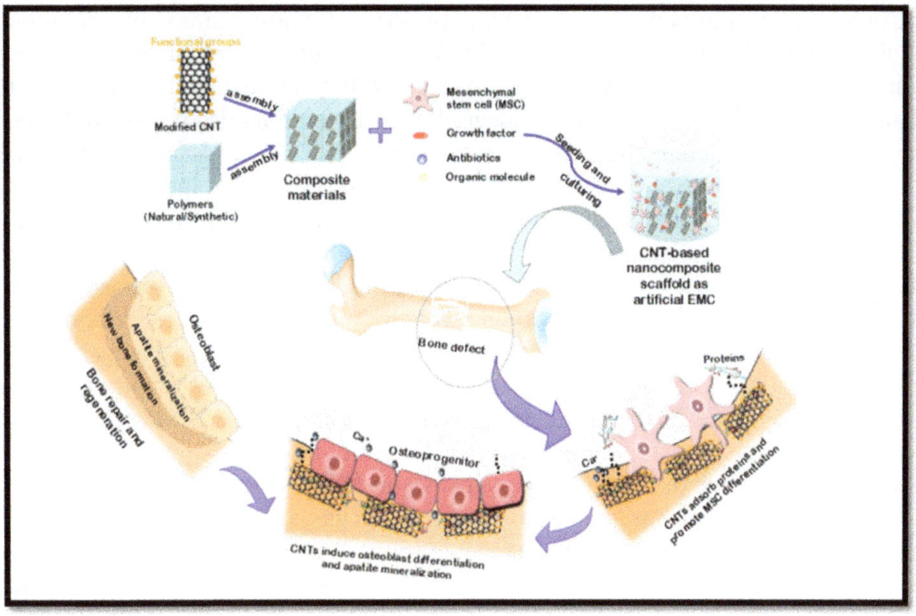

Figure 5.5. Role of carbon nanotubes (CNTs) as scaffold composition in bone engineering and regeneration. (Adapted from [26], copyright (2019), with permission from Molecular Diversity Preservation International. (CC BY 4.0).)

(minocycline, polyglycolide-co-DL-lactide), and elyzole (metronidazole benzoate in glyceryl mono-oleate/sesame oil) [5].

One of the DDS carriers includes hydrogel. Since hydrogels easily respond to certain stimuli, Liang *et al* documented that these polymeric structures could also be used to control antimicrobial agents' release behavior leading to desired therapeutic outcomes [5].

Different mechanisms during the disease initiation which are based on redox potential, pH, or enzymatic changes, could interfere with the drug mechanism. To overcome this action de Avila E *et al* in 2020 documented that the chemical and structural LbL (layer by layer) modifications allowed slow release of local antimicrobial agents over time (figure 5.5) [13]. For example, to achieve slower release kinetics, pH-responsive polymeric capsules were assembled combining poly(2-(diisopropylamino)ethyl methacrylate) (PDPA) and poly(methacrylic acid) (PMA) via LbL assembly and cross-linked by click chemistry. Effectively, PDPA undergoes a charge-shifting transition from hydrophobic to hydrophilic when in an acidic environment. With this strategy, the capsules swell, allowing degradation of the unstable disulfide bond moieties, and release the drug into the required site at the desired rate.

5.5 Application in host modulation and regeneration

Curcumin, non-steroid anti-inflammatory drugs, statins and different growth factors are already used for host modulation and alveolar bone regeneration in the form of films and injectable gels.

Curcumin has antioxidant, antiseptic, antimicrobial, anti-inflammatory, immunostimulant, and antimutagenic properties. Different formulations of curcumin delivered by nanotextured drug delivery systems are reported for the treatment of periodontitis. Chauhan *et al* in 2018, prepared a biodegradable curcumin loaded crosslinked gelatin film which had a sustained delivery up to seven days [14]. Also, Nasra *et al* developed *in situ* gel formulations of curcumin using 30% of pluronic F127, and 1% of carbopol P934. This formulation was later also verified in clinical research [15].

Ketoprofen, a non-steroid anti-inflammatory drug (NSAIDS) is another anti-inflammatory agent used for host modulation. It inhibits the cyclooxygenase enzyme, which is indeed related to prostaglandins secretion and increases host inflammatory response. Scrivastava *et al* in 2016 created an *in situ* injectable nano-emulgel with the sustained release of ketoprofen into the periodontal pocket [16]. Sundararaj *et al* formulated a multiple DDS, which can load four types of drugs. The drugs used were metronidazole to eliminate infection, ketoprofen for inhibiting inflammation, doxycycline to prevent tissue destruction, and simvastatin for bone regeneration [17].

Some therapeutic drugs such as osteogenesis drugs and bone factors have also been clinically proven to help in bone regeneration. A randomized controlled clinical trial conducted in 2015 by Santana *et al* concluded that the intrabony periodontal defects when treated with a biological hydrogel of recombinant human fibroblast growth factor type 2 (rhFGF-2) in combination with hyaluronic acid (HA) significantly improved the clinical parameters of periodontal wound healing [18]. These growth factors were also used with bone grafts to enhance their regenerative potential. Oortgiesen *et al* formulated an injectable macroporous calcium phosphate graft (CaP) in combination with bone morphogenetic protein-2 (BMP-2) or fibroblast growth factor-2 (FGF-2) [19]. Oortgiesen *et al* in another formulation combined enamel matrix derivative with an injectable calcium phosphate graft for regeneration of bone in previously diseased periodontium [20].

Another category of drugs known as Statins such as simvastatin and rosuvastatin stimulated *in vivo* bone formation with increase in the new bone volume. Pradeep *et al* formulated 1.2% rosuvastatin gel incorporated into a methylcellulose vehicle for the treatment of intrabody defect sites. It showed a greater reduction in probing depth along with increased gain in clinical attachment level [21]. Also, the efficacy of 1.2% simvastatin gel as a local drug delivery on Interleukin-6 (IL-6) and Interleukin-8 (IL-8) levels of GCF was evaluated and proved to be an effective agent for the management of chronic periodontitis cases [22].

Smokers have a higher risk of periodontal disease. Therefore, Rao *et al* in 2013, developed an indigenously prepared, controlled-release gel which was biodegradable and loaded with metformin [23]. It was used for the treatment of vertical defects in smokers with generalized chronic periodontitis with better clinical therapeutic effect. Also, simvastatin use in smokers showed significant improvement in the bone fill.

Diabetes, a systemic disease is also closely related to periodontitis. A double-masked clinical trial was designed by Pradeep *et al* in 2012, to investigate the effectiveness of local drug delivery systems, like 1.2% simvastatin *in situ* gel (methylcellulose) and 1% alendronate gel (carbopol 934P and triethanolamine) in diabetic patients [24]. Both drug delivery systems, showed a significant increase in

PD reduction, CAL gain, and improved bone fill. Also, Li *et al* in 2013, prepared 25-hydroxyvitamin D3-loaded polylactic acid (PLA) microspheres and established a diabetic periodontitis model. They used the microspheres to treat rats, reporting that alveolar bone loss was inhibited along with osteoid formation in murine models [25], thus implicating the use of nano-delivered drugs in providing an effective approach for the treatment of diabetic periodontitis.

CNTs have also been used for bone tissue engineering. The structure of CNTs comprises only carbon atoms exclusively bonded to each other through sp2 bonds. A CNT is one graphene sheet rolled up into a hollow cylindrical nanostructure. They are commonly divided into two subtypes: single-walled carbon nanotubes (SWCNTs) formed from a single-layer of tubular graphene and multi-walled carbon nanotubes (MWCNTs) composed of multiple concentric tubular graphene layers. SWCNTs usually present in a closely arranged bundle of hexagons, approximately 1 nm in diameter and up to a millimeter or more in length. MWCNTs have a higher degree of structure similar to hollow graphite fiber but with larger diameter than SWCNTs (2–100 nm).

For bone tissue regeneration and engineering, CNTs in composites can be simply considered as biomimetic nanocomposites of collagen fibers on a cell hierarchy level. They have a multiple effect which includes stimulating cell adhesion via interaction with cell-binding proteins, regulates cell morphology and accelerates stem cell differentiation. Also, they have been shown to promote differentiation of osteoblasts and apatite mineralization leading to new bone regeneration (figure 5.6) [26]. Li *et al* demonstrated that MWCNTs can adsorb and concentrate more proteins (i.e., rhBMP-2), induce the expression of alkaline phosphatase (ALP) and the cbfa1 and COLIA1 genes, and also promote osteogenic differentiation of the human adipose-derived mesenchymal stem cells (MSCs) *in vitro*.

5.6 Application in managing implant conditions and disease

Nano-technology is the control of matter at a scale of approximately 1–100 nanometres (nm), where novel properties and function occur because of size [4]. Thus, for the surface of an implant truly to possess nanotechnology, aspects of shape and structure must be formed at the nano-scale which specifically enhance the properties of the implant. The impact of nano-texturising on the implant surface can be illustrated by imagining the macro, micro, and nano-surface area of a 1 cm × 1 cm × 1 cm cubic implant. When the implant cubic dimensions are 1 cm × 1 cm × 1 cm, then the macro surface area is 6 cm^2. If the same implant is divided into 1 mm cubes, the total surface area increases to 60 cm^2 and on further decreasing the dimensions to 1 nm cubes, the surface area increases to 60 000 000 cm^2 (figure 5.1) [4]. Therefore, increasing the surface area of the implant surface by using nanotubes leads to increased surface area for osseointegration between the nanotextured implant surface and the bone progenitor cells.

Nano-scaled DDS are also combined with dental implants for anti-bacterial modification of the implant surface without affecting their basic properties. Already available products like arestin, perioChip, and atridox used for periodontitis have also being effective in peri-implantitis. Recently, dual function DDS with

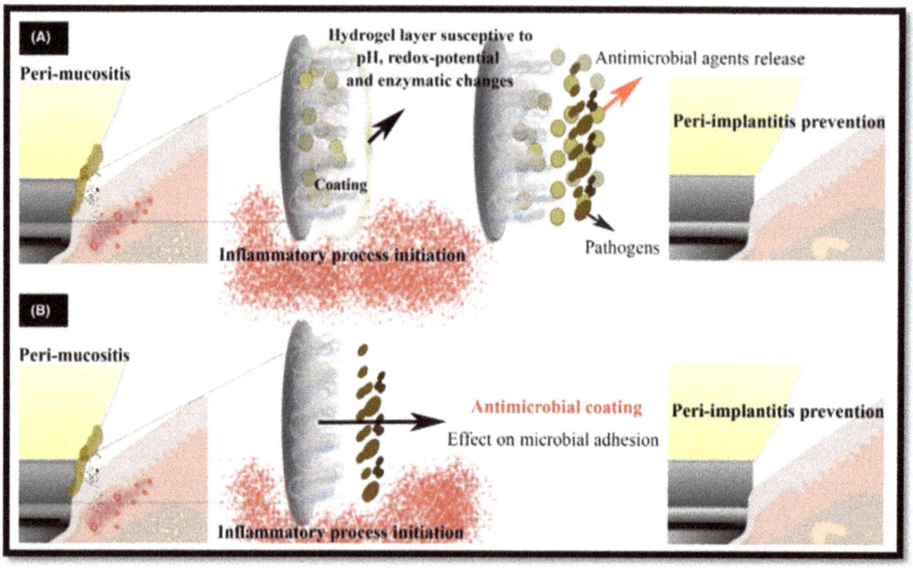

Figure 5.6. Proposed antimicrobial mechanism for the prevention of peri-implantitis. Biomaterial coating onto Ti substrate causing slow release of antimicrobial substances after the initiation of an inflammatory process. (Adapted from [13], copyright (2019) The Authors. *Journal of Periodontal Research* published by John Wiley & Sons Ltd.)

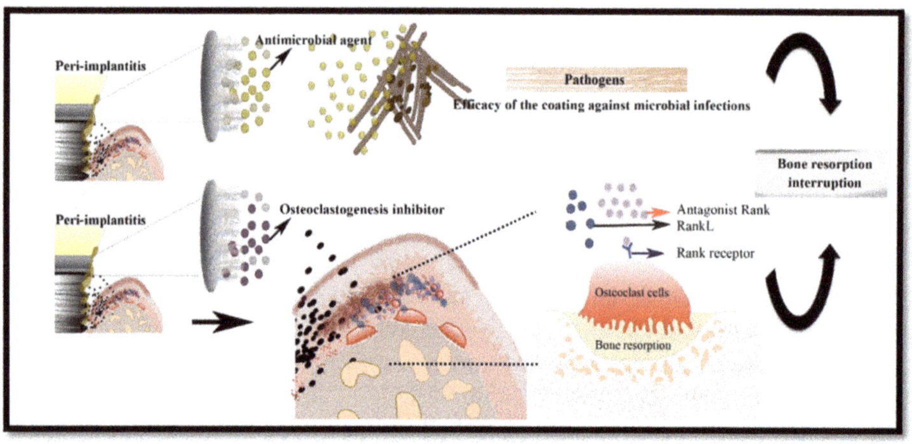

Figure 5.7. Proposed antimicrobial mechanism for the treatment of peri-implantitis. Biomaterial coating onto Ti substrate releases a sustained high dose of antimicrobial agent into the affected inflamed sites. (Adapted from [13], copyright (2019) The Authors. *Journal of Periodontal Research* published by John Wiley & Sons Ltd.)

antibacterial along with bone regeneration properties have been reported by Diniz *et al.* They formulated silver lactate loaded hydrogel having a dual effect, i.e., inducing osteogenic differentiation of gingival mesenchymal stem cells and inhibiting *A. actinomycetemcomitans* for up to two weeks [27].

Anti-bacterial materials, antibiotics, and other bioactive agents loaded DDS were coated on an implant surface by topographical modifications and biodegrading polymer incorporation. Fisher *et al* in 2016 developed *in situ* implants to deliver doxycycline and simvastatin [28]. Anti-bacterial peptide, chlorhexidine, antibiotics like doxycycline and tetracycline were incorporated into dental implants by covalent conjugation. Nanoparticles mostly used in coating dental implants are zinc oxide and silver nanoparticles. Electrodeposition technology is used to incorporate these metal nanoparticles into titanium and abutments. Nanomaterials having bioactive and biodegradable properties such as gelatin, chitosan, and hydroxylapatite, help in bone regeneration. Besinis *et al* in 2017 documented a dual-layered nanocoated implant with silver, titanium dioxide, and hydroxyapatite [29]. Also, Bottino *et al* in the same year, fabricated tetracycline incorporated gelatin nanofiber as a dental implant modifier which induced osteo-precursor cells growth and anti-bacterial property [30]. Yang *et al* in 2018, reported a PLGA(Ag-Fe_3O_4)-coated dental implant which proved the safety and efficacy of paramagnetic dental implants. The magnetic field induced the osteogenic potential of precursor osteoblastic cells and inhibited bacterial adherence [31].

Further clinical research on the treatment modalities for peri-implantitis using nanoparticles biomolecules/drugs incorporated into an LbL system is still ongoing. Although efficacy of biomedical application of LbL self-assembly has been derived from *in vitro* studies, the ability of the LbL system to incorporate drugs and chemical properties of polyelectrodes for coating dental implants requires validation (figure 5.7) [13].

References

[1] Papapanou P N, Sanz M, Buduneli N, Dietrich T, Feres M and Fine D H *et al* 2018 Periodontitis: consensus report of Workgroup 2 of the 2017 World Workshop on the Classification of Periodontal and Peri-implant Diseases and Conditions *J. Periodontol.* **89** S173–S82 Suppl 1

[2] Shah N, Pandey R M, Duggal R, Mathur V P and Rajan K 2007 Oral health in India: a report of the multi centric study, Directorate General of Health Services, Ministry of Health and Family Welfare, Government of India and World Health Organisation Collaborative Program

[3] Manjunath R G and Rana A 2015 Nanotechnology in periodontal management *J. Adv. Oral Res.* **6** 1–8

[4] Su E P, Justin D F, Pratt C R, Sarin V K, Nguyen V S, Oh S and Jin S 2018 Effects of titanium nanotubes on the osseointegration, cell differentiation, mineralisation and anti-bacterial properties of orthopaedic implant surfaces *Bone Joint J.* **100-B** 9–16 (1 Suppl. A)

[5] Liang J, Peng X, Zhou X, Zou J and Cheng L 2020 Emerging applications of drug delivery systems in oral infectious diseases prevention and treatment *Molecules* **25** 516

[6] Seneviratne C J, Leung K C, Wong C H, Lee S F, Li X and Leung P C *et al* 2014 Nanoparticle-encapsulated chlorhexidine against oral bacterial biofilms *PLoS One* **9** e103234

[7] Chen F, Rice K C, Liu X M, Reinhardt R A, Bayles K W and Wang D 2010 Triclosan-loaded tooth-binding micelles for prevention and treatment of dental biofilm *Pharm. Res.* **27** 2356–64

[8] Moura L A, Ribeiro F V, Aiello T B, Duek E A, Sallum E A and Nociti Junior F H *et al* 2015 Characterization of the release profile of doxycycline by PLGA microspheres adjunct to non-surgical periodontal therapy *J. Biomater. Sci. Polym. Ed.* **26** 573–84

[9] Phaechamud T, Chanyaboonsub N and Setthajindalert O 2016 Doxycycline hyclate-loaded bleached shellac *in situ* forming microparticle for intraperiodontal pocket local delivery *Eur. J. Pharm. Sci.* **93** 360–70

[10] Marziyeh R M, Zamani M, Prabhakaran M P, Bahrami S H and Ramakrishna S 2016 Electrospinning of PLGA/gum tragacanth nanofibers containing tetracycline hydrochloride for periodontal regeneration *Mater. Sci. Eng. C: Mater. Biol. Appl.* **58** 521–31

[11] Kimura S, Toda H, Shimabukuro Y, Kitamura M, Fujimoto N and Miki Y *et al* 2010 Topical chemotherapy in human periodontitis using a new controlled-release insert containing ofloxacin. I. Microbiological observation *J. Periodontal. Res.* **26** 33–41

[12] Tian Y, Shen Y and Jv M 2016 Synthesis, characterization and evaluation of tinidazole-loaded mPEG–PDLLA (10/90) *in situ* gel forming system for periodontitis treatment *Drug Deliv.* **23** 2726–35

[13] de Avila E, van Oirschot B and van den Beucken J 2020 Biomaterial-based possibilities for managing peri-implantitis *J. Periodont. Res.* **55** 165–73

[14] Chauhan S, Bansal M, Khan G, Yadav S K, Singh A K and Prakash P *et al* 2018 Development, optimization and evaluation of curcumin loaded biodegradable crosslinked gelatin film for the effective treatment of periodontitis *Drug Dev. Ind. Pharm.* **44** 1212–21

[15] Nasra M M, Khiri H M, Hazzah H A and Abdallah O Y 2017 Formulation, in-vitro characterization and clinical evaluation of curcumin *in situ* gel for treatment of periodontitis *Drug Deliv.* **24** 133–42

[16] Srivastava M, Kohli K and Ali M 2016 Formulation development of novel *in situ* nanoemulgel (NEG) of ketoprofen for the treatment of periodontitis *Drug Deliv.* **23** 154–66

[17] Sundararaj S C, Thomas M V, Peyyala R, Dziubla T D and Puleo D A 2013 Design of a multiple drug delivery system directed at periodontitis *Biomaterials* **34** 8835–42

[18] De Santana R B and de Santana C M 2015 Human intrabony defect regeneration with rhFGF-2 and hyaluronic acid—a randomized controlled clinical trial *J. Clin. Periodontol.* **42** 658–65

[19] Oortgiesen D A, Walboomers X F, Bronckers A L, Meijer G J and Jansen J A 2014 Periodontal regeneration using an injectable bone cement combined with BMP-2 or FGF-2 *J. Tissue Eng. Regen. Med.* **8** 202–9

[20] Oortgiesen D A, Meijer G J, Bronckers A L, Walboomers X F and Jansen J A 2013 Regeneration of the periodontium using enamel matrix derivative in combination with an injectable bone cement *Clin. Oral Investig.* **17** 411–21

[21] Pradeep A R, Karvekar S, Nagpal K, Patnaik K, Guruprasad C N and Kumaraswamy K M 2015 Efficacy of locally delivered 1.2% rosuvastatin gel to non-surgical treatment of patients with chronic periodontitis: a randomized, placebo-controlled clinical trial *J. Periodontol.* **86** 738–45

[22] Gunjiganur Vemanaradhya G, Emani S, Mehta D S and Bhandari S 2017 Effect of 1.2% of simvastatin gel as a local drug delivery system on Gingival Crevicular Fluid interleukin-6 & interleukin-8 levels in non surgical treatment of chronic periodontitis patients *Arch. Oral Biol.* **82** 55–61

[23] Rao N S, Pradeep A R, Bajaj P, Kumari M and Naik S B 2013 Simvastatin local drug delivery in smokers with chronic periodontitis: a randomized controlled clinical trial *Aust. Dent. J.* **58** 156–62

[24] Pradeep A R, Anuj S, Rao N S, Pavan B, Naik S B and Minal K 2012 Local drug delivery of alendronate gel for the treatment of patients with chronic periodontitis with diabetes mellitus: a double-masked controlled clinical trial *J. Periodontol.* **83** 1322–8

[25] Li H, Wang Q, Xiao Y, Bao C and Li W 2013 25-Hydroxyvitamin D 3-loaded PLA microspheres: *in vitro* characterization and application in diabetic periodontitis models *Aaps Pharmscitech* **14** 880–9

[26] Pei B, Wang W, Dunne N and Li X 2019 Applications of carbon nanotubes in bone tissue regeneration and engineering: superiority, concerns, current advancements, and prospects *Nanomaterials (Basel)* **9** 1501

[27] Diniz I M, Chen C, Ansari S, Zadeh H H, Moshaverinia M and Chee D *et al* 2016 Gingival Mesenchymal Stem Cell (GMSC) delivery system based on RGD-coupled alginate hydrogel with antimicrobial properties: a novel treatment modality for peri-implantitis *J. Prosthodont.* **25** 105–15

[28] Fisher P D, Clemens J, Zach Hilt J and Puleo D A 2016 Multifunctional poly(beta-amino ester) hydrogel microparticles in periodontal *in situ* forming drug delivery systems *Biomed. Mater.* **11** 025002

[29] Besinis A, Hadi S D, Le H R, Tredwin C and Handy R D 2017 Antibacterial activity and biofilm inhibition by surface modified titanium alloy medical implants following application of silver, titanium dioxide and hydroxyapatite nanocoatings *Nanotoxicology* **11** 327–38

[30] Bottino M C, Munchow E A, Albuquerque M T P, Kamocki K, Shahi R and Gregory R L *et al* 2017 Tetracycline-incorporated polymer nanofibers as a potential dental implant surface modifier *J. Biomed. Mater Res. B Appl. Biomater.* **105** 2085–92

[31] Yang Y, Ren S, Zhang X, Yu Y, Liu C and Yang J *et al* 2018 Safety and efficacy of PLGA (Ag-Fe_3O_4)-coated dental implants in inhibiting bacteria adherence and osteogenic inducement under a magnetic field *Int. J. Nanomed.* **13** 3751–62

IOP Publishing

Nanotechnology for Dentistry Applications

Ashutosh Kumar Shukla

Chapter 6

Drug nanocarriers: alternative therapy to fight oral biofilms

Douglas Roberto Monteiro, Anne Caroline Morais Caldeirão, Beatriz Díaz Fabregat, Laís Salomão Arias, Heitor Ceolin Araujo, Wilmer Ramírez Carmona and Juliano Pelim Pessan

Biofilms are highly organized microbial communities, and their formation on different surfaces of the oral cavity is considered one of the main virulence factors of oral microorganisms. The formation of biofilms is also associated with the emergence of several oral diseases, such as candidiasis, periodontal disease, dental caries and others. In order to overcome the issue of biofilms' resistance to conventional drugs used in dentistry, the employment of nanocarriers has emerged as an interesting alternative strategy. Nanocarriers have potential to transport drugs to their targets, reducing their concentration of use and their side effects on the tissues of the human body. Thus, this chapter addresses the main nanocarriers used in several areas of dentistry, including the different types of nanoparticles and compounds for their functionalization, the main drugs conjugated to the core–shell compound, as well as their effects on oral biofilms, toxic effects on eukaryotic cells and other applications of dental interest.

6.1 Introduction

In the oral cavity can be found more than 700 microbial species, which are organized in extremely complex and organized communities, namely biofilms [1, 2]. These can adhere to any oral surface, as mucosal tissues, gingiva or dental structures, in addition to the surface of restorative materials and prosthetic structures. They are covered by an extracellular matrix of polysaccharides that provides nutrition, support and defence for the microbial community. Different oral diseases appear as a result of the interactions between the species that make up the oral microbiome, a process that is conditioned by the susceptibility of the host and its immune response [1, 2].

doi:10.1088/978-0-7503-3671-0ch6

The dysbiosis of the oral microbiome can result in the development of oral diseases such as dental caries, periodontitis, peri-implantitis, candidiasis and others, which often implies the need to use drug therapies to treat these diseases. Over the years, biofilms have been changing their defence mechanisms, becoming more resistant to drugs applied to fight oral diseases. Among the resistance mechanisms of biofilms, the tolerance to oxidative stress, the role of efflux pumps, and the protection provided by the extracellular matrix are of special relevance [3]. In addition, the low bioavailability of some drugs in the oral environment, the limited ability to penetrate into the deeper layers of biofilms and their cytotoxicity, have stimulated the development of alternative therapies for the control of oral biofilms [4].

Nanotechnology is shown as one of the alternative therapies for the use of drug delivery in nanostructures, called nanocarriers, which are known to facilitate the penetration of drugs into biofilms [5, 6]. Carrier nanoparticles can be coated with several compounds to prevent their agglomeration and help stabilizing the nanocomposite. Currently, nanocarriers are being used to control/combat biofilms in several areas of dentistry, including stomatology, periodontology, endodontics, orthodontics and cariology. Moreover, these nanostructures have been explored in bone regeneration, in restorative dentistry (dentin adhesives), in anaesthesiology and in anti-tumor therapies.

In light of the above, this chapter aims to present a review on the use of drug nanocarriers in the control of oral biofilms or in other applications of dental interest.

6.2 Nanocarriers in stomatology to fight oral candidiasis

Oral fungal infections are predominantly caused by *Candida* species, which colonize the oral cavity as commensal and opportunistic pathogens [7]. Oral candidiasis is common in extremes of age and in patients with low immunity [7–9]. A study conducted in northern Sweden showed a prevalence of candidiasis in infants' first year of life between 11 and 15% [10]. Besides, data showed clinical and microbiological signs of oral candidiasis in about 30% of patients with advanced cancer [9]. Notably, the prevalence of candidiasis in elderly users of removable dental prosthesis is higher compared to non-users, while non-*albicans Candida* species is increasingly isolated from patients affected by this disease [8].

In the last few years, multiple studies have shown a direct correlation between the increasing incidence of oral and systemic fungal infections and the rise in fungal strains resistant to conventional antimicrobials [11, 12]. In this sense, the design of controlled drug delivery systems using nanocarriers has emerged as an alternative in order to optimize the effect of antimicrobials, while reducing the clinical concentration of drug used and slowing the emergence of mutations and problems of microbial resistance [13, 14]. Moreover, the use of nanocarriers can provide a number of other advantages, such as improved bioavailability, biocompatibility, reduction of drugs' side effects and reduction of toxicity [13–16].

Certain nanocarriers have shown prominence in the development of these nanosystems, such as iron oxide nanoparticles (IONPs). Used in the last decades in hyperthermia and image resonance therapies, IONPs have also proved to be interesting for use in controlled drug delivery systems due to their easy multi-functionalization, biocompatibility, size and magnetic properties, which allow one to explore the kinetic control of drug delivery [13]. In addition, lipid nanocarriers have been extensively explored due to their ability to encapsulate hydrophobic drugs, increasing their diffusion through biological barriers [15–17], while natural and synthetic polymeric nanocarriers have been investigated in different *in vitro* and *in vivo* studies due to their increased biocompatibility [14, 15, 18]. Other types of carriers such as micelles and liposomes can also be employed [14, 15].

The preparation of antifungal nanosystems usually involves the encapsulation, emulsion or direct binding of the drug to the nanocarrier [13]. However, this design may have disadvantages for some combinations of drugs and nanoparticles. As a solution, functionalizing agents can be used as a coating (shell) to bind the drug to the nanocarrier (core), reducing problems of particle agglomeration, biocompatibility and nanosystem stability [13]. Chitosan (CS) is a natural polymer with antimicrobial, hemostatic, mucoadhesive and biocompatibility properties, being the most utilized compound in the functionalization of IONPs (figure 6.1(a)) in studies focused on antifungal nanotherapies [19–21]. Literature shows that other agents can be used for the same purpose, such as other types of polymers, organic surfactants, inorganic compounds and bioactive molecules [13–15].

Due to their extensive clinical use, polyenes and azoles are the most investigated compounds in the manufacture of antifungal nanotherapies. In this sense, the functionalization of IONPs with CS (IONPs-CS) has been especially interesting, since it allows the binding of different drugs such as nystatin, chlorhexidine (CHX), cetylpyridinium chloride (CPC), miconazole (MCZ) and fluconazole (FLZ) [19–23]. Figure 6.1(b) illustrates the binding of MCZ to IONPs-CS, forming

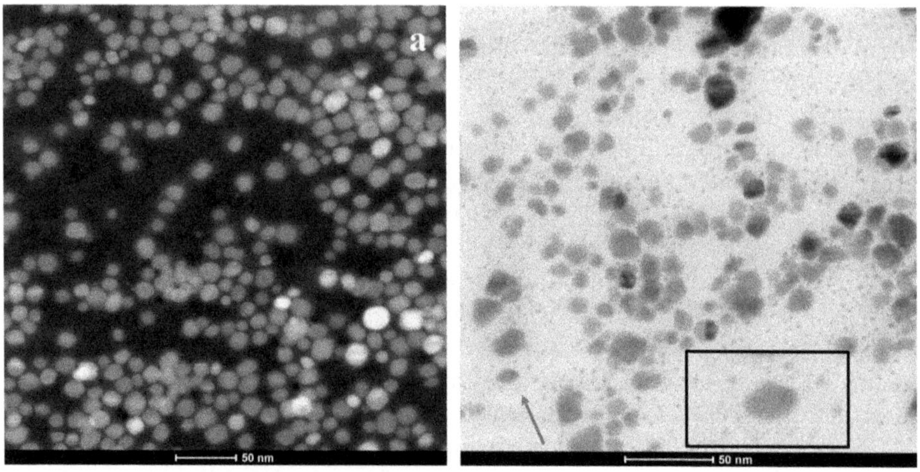

Figure 6.1. Transmission electron microscopy images for chitosan-coated iron oxide nanoparticles (a) and for the nanocarrier of miconazole (b). The area marked in black in figure 'b' represents an increase of the region indicated with the arrow, showing the miconazole particles bound to chitosan-coated iron oxide nanoparticles.

the IONPs-CS-MCZ nanocarrier. On the other hand, drugs such as amphotericin B, which are hydrophobic in nature and difficult to solubilize, are commonly encapsulated in lipid nanocarriers, and sometimes combined with other materials, which can bring even more advantages in their optimization [17]. An example is the association with sodium alginate, which is a pH-dependent polymer capable of forming hydrogel particles that can absorb large amounts of water [17, 18]. Such absorption leads to swelling and, consequently, to the appearance of pores in the particle that allow the drug to escape [17, 18]. In addition, the literature reports antifungal potential in the association of nanocarriers with alternative drugs such as miltefosine [18], curcumin [24, 25] and osthol, a compound derived from the *Cnidium monnieri* (L.) Cusson plant, the latter being widely used in Eastern medicine to treat vaginal fungal infections [26].

Understanding the phenomena of drug and nanocarrier interaction is an important step towards the improvement of a nanosystem geared for clinical use. In IONPs-CS associations, studies suggest that CS adsorbs on the nanocarrier surface through glycosidic interactions and interacts with the negatively charged groups of the magnetic nanoparticles [20–22]. In turn, the chemical interaction of CS with different drugs varies according to the nature of each compound, interfering with its antimicrobial capacity. The hypotheses suggest that MCZ and nystatin bind to CS through electrostatic interaction of the polymer's amine groups with the negative charge of the drugs [20, 23]. Further, hydrogen bonds are also responsible for the conjugation of NH_3^+ groups from CS with negative groups from FLZ and CPC [19, 22]. Finally, it is likely that covalent bonds link CHX to IONPs functionalized with CS [21].

The methodology of synthesis and assembly of a nanocarrier directly affects its biological effect. As an example, the association of nystatin with IONPs-CS can maintain the antifungal action of the compound [23], while the same antifungal drug associated with a silicate nanocarrier may lose its antimicrobial effectiveness [27]. These results reinforce that the types of chemical interactions in the compound can affect its molecular integrity and bioavailability. On the other hand, the real benefits of nanocarriers must be considered in view of their clinical use. At the same concentration ($\leqslant 3.9\ \mu g\ ml^{-1}$), despite CHX carried by IONPs-CS does not have its antifungal effect potentiated against *Candida albicans* and *Candida glabrata* [21, 28], its cytotoxicity is significantly reduced compared to drug alone (without nanocarrier) [28]. The inverse occurs when IONPs-CS is carrying CPC. In this case, the presence of the nanocarrier potentiates the effect of CPC on *Candida* biofilms without altering its cytotoxicity on murine fibroblasts [19].

The effect of a nanocarrier can also vary depending on the fungal species, type of strain, biofilms' development stage and growing conditions of the investigated biofilm. The association of IONPs-CS with MCZ is synergistic and quite effective against *C. albicans* biofilms, but has reduced effects against *C. glabrata* due to its specific resistance mechanism based on efflux pumps and also to the different mechanisms of action that the nanocarrier presents against each cell type [20]. Faced with even more complex models of pathogenic biofilms developed *in vitro*, the IONPs-CS-MCZ nanocarrier maintains its antifungal action while affecting biofilms

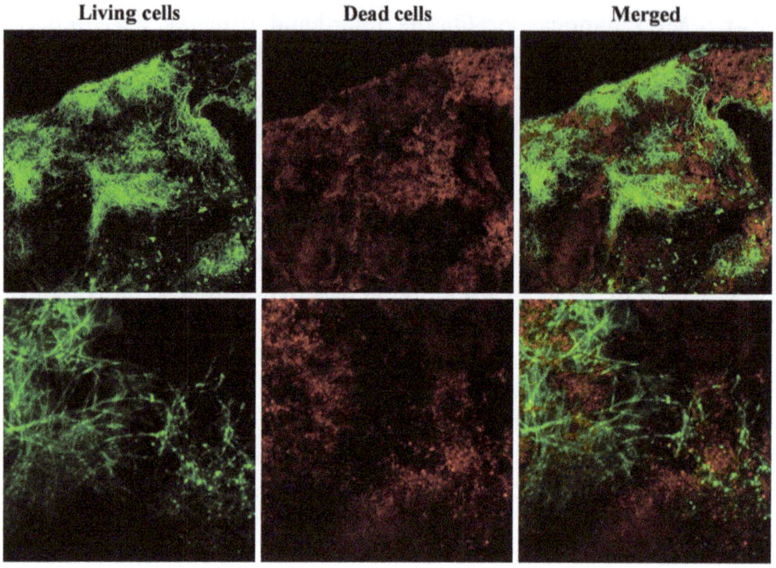

Figure 6.2. Confocal laser scanning microscopy images of 96-h microcosm biofilms developed on glass coverslips and treated for 24 h with chitosan-coated iron oxide nanoparticles carrying miconazole at 156 μg mL^{-1}. Green and red fluorescence indicate living and dead cells, respectively. Magnification: 20x.

differently, which occurs due to the different conditions of development and cellular interactions within the biofilm [29]. Regarding the effect on salivary microcosm biofilms, IONPs-CS-MCZ carrying MCZ at 156 μg mL^{-1} is significantly more effective than MCZ alone (156 μg mL^{-1}) in reducing mutans streptococci, *Lactobacillus* spp., *C. albicans* and *C. glabrata*. Interestingly, this nanocarrier shows a more prominent antibiofilm effect on bacterial and yeast cells than on *C. albicans* hyphae, as demonstrated by confocal laser scanning microscopy (figure 6.2).

Nanocarriers of FLZ based on IONPs-CS show synergistic antifungal effect against planktonic cells of *C. albicans* and *C. glabrata*, despite not enhancing the effects of FLZ alone when tested against biofilms of these species [22]. Scanning electron microscopy (SEM) images show that treatment with both FLZ alone and IONPs-CS-FLZ generate *Candida* biofilms with less robust architecture compared to untreated biofilms (figure 6.3). IONPs also act efficiently as a carrier of FLZ, since fluorine (characteristic of FLZ) and iron (characteristic of IONPs) atoms have been shown to be distributed throughout the biofilm sample (figure 6.4). In turn, in rats with invasive candidiasis, the polymeric nanocarrier Eudragit S100 is capable of potentiating the antifungal effect of osthol [26]. However, it is important to consider that, in addition to synergism in antifungal efficacy, advantages of a different nature can be obtained when using different nanocarriers, such as increased bioavailability, biocompatibility and reduced toxicity and biodegradability of the drug.

Synergism occurs when there are different mechanisms of action involved between the components that form the nanocarrier. As an example, MCZ acts by producing an accumulation of reactive oxygen species (ROS) in the fungal cytoplasm, leading to cell death [20, 29]. It also acts on the actin of the fungal

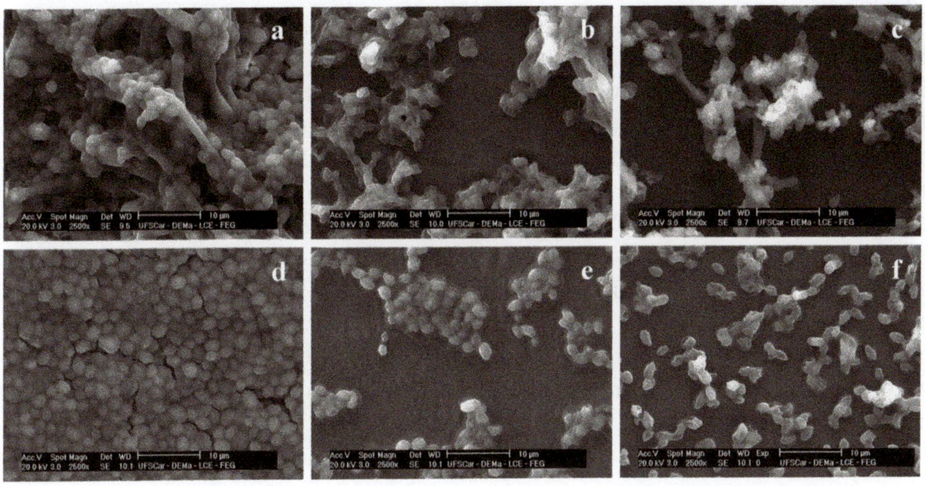

Figure 6.3. Scanning electron microscopy images of mono-species biofilms of *C. albicans* (a–c) and *C. glabrata* (d–f) grown for 48 h within 96-well plates and treated during 24 h with 1250 μg mL^{-1} fluconazole (b, e) and chitosan-coated iron oxide nanoparticles carrying fluconazole at 1250 μg mL^{-1} (c, f). Images 'a' and 'd' represent untreated biofilms. Magnification: 2500x. Bars: 10 μm.

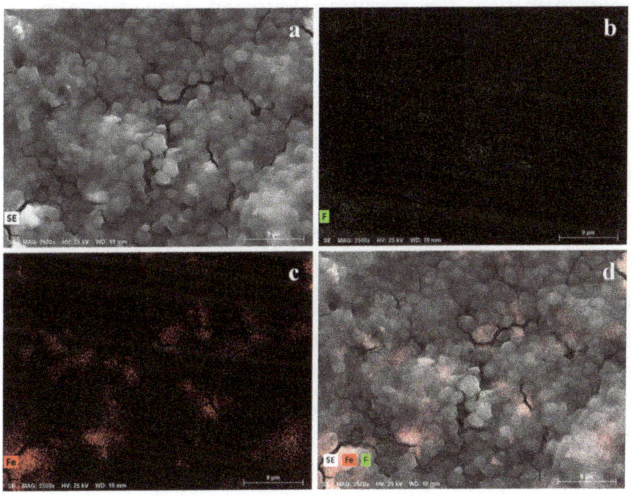

Figure 6.4. Energy dispersive spectroscopy mapping of *C. glabrata* mono-species biofilm formed for 48 h within 96-well plates and treated during 24 h with chitosan-coated iron oxide nanoparticles carrying fluconazole at 500 μg ml^{-1} (a). Images 'b' and 'c' display fluorine and iron atoms distributed throughout the sample, respectively. Image 'd' depicts overlap of the images 'a', 'b' and 'c'. Magnification: 2500x. Bars: 9 μm.

cytoskeleton, creating microtubules in the fungal mitochondria [20, 29]. In turn, CS acts by antagonizing the SAGA complex, which coordinates Ada2 and transport genes (CDR1 and MDR1), causing changes in the integrity of the cell surface [29]. Thus, when combined within a nanocarrier, such compounds have the ability to potentiate their antifungal effects[20, 29]. The same trend occurs for the IONPs-CS-FLZ nanocarrier.

FLZ inhibits the enzyme C14-α-lanosterol demethylase, causing depletion of ergosterol in the fungal cell membrane and accumulation of toxic C14-α-methylesterols. In turn, IONPs-CS can generate microbial toxicity via depolarization of the membrane, release of metal ions that affect cell homeostasis, ROS production and inhibition of mRNA transcription and protein synthesis via DNA binding [22]. Therefore, the combined action of these compounds may explain the lower values of minimum inhibitory concentration found for IONPs-CS-FLZ against *C. albicans* and *C. glabrata* cells compared to FLZ alone [22]. Finally, the resulting charge of a nanosystem is capable of interfering with the compound's mechanism of action and antifungal capacity. An anionic curcumin nanocarrier was shown to be ineffective against *C. albicans* compared to a cationic nanocarrier, probably because the negative charge of the formulation promotes repulsion with the negative charge of the biofilm matrix and the cell wall of the fungal cell [24].

6.3 Nanocarriers in periodontology

Periodontal disease begins with the invasion of the gingival sulcus by pathogenic microorganisms (e.g. Gram-negative anaerobic bacteria) and the subsequent multiplication and coaggregation of new species, leading to the development of complex polymicrobial communities. Cooperation between species makes microbial communities more resistant to traditional drugs used in periodontal therapies to control biofilm. Therefore, one of the alternatives to overcome the limitations of conventional drugs is nanotechnology, which uses nanocarriers as vehicles in the release of drugs to cause changes in the composition of multispecies biofilms.

IONPs have been studied for that purpose. These nanoparticles coated with CS bind by electrostatic interactions to MCZ, generating a nanocarrier with dimensions around 50 nm. When the IONPs-CS-MCZ nanocarrier was evaluated in a polymicrobial biofilm model of gingivitis, the most important change in the biofilm composition was the lowest percentage of *Fusobacterium* spp. in relation to the results observed for MCZ in its free form [29]. At the same time, the nanocarrier reduced the number of viable cells, the total biomass and the metabolic activity of the biofilm without a significant difference in relation to the free MCZ. These changes in composition affect the structure of the extracellular matrix, exposing the biofilm to a greater susceptibility to the action of drugs.

For the drug to act against pathogenic microorganisms in the oral cavity, it needs to remain available long enough before it is absorbed. Drugs used in periodontal therapy, such as metronidazole, have difficulty crossing the membranes of bacterial cells due to resistance mechanisms such as efflux pumps. This scenario leads to the need for the use of combined therapies, given that drugs transported in nanocarriers to bacterial cells result in greater therapeutic efficacy compared with their free form. In this sense, a nanocarrier of metronidazole with size ranging from 2 to 4 nm, formed by carbon quantum dots derived from chlorophyll, showed an antimicrobial effect on *Porphyromonas gingivalis* 72% more effective than that found for the free drug. Furthermore, it had no cytotoxic effect on the epithelial cell line H413 (derived from a human oral squamous cell carcinoma), even at concentrations of

1000 mg ml^{-1} after 4–24 h of incubation [4]. Another example is that presented by a nanosystem that used polylactide-glycolic acid co-polymer (PLGA) as nanocarrier combined with CS in the nanosphere form, encapsulating metronidazole and N-phenacylthiazolium bromide, with a final size of 499 nm [30]. In this *in vivo* study, the progression of periodontal bone loss, inflammatory cell infiltration, and collagen deposition were assessed. Both nanocarriers had effects in improving the evaluated parameters, and the N-phenacylthiazolium bromide nanocarrier modulated inflammation and promoted long-term repair, overcoming the effects promoted by the metronidazole nanocarrier.

The effect of modulating the inflammatory response is another aspect to be considered in periodontal therapy. Turmeric, whose active component is curcumin, has been explored for its anti-inflammatory properties, although with low oral bioavailability. A nanogel called pluronic F127 loaded with 2% curcumin has been clinically evaluated in patients with periodontal pockets in order to improve the effect of scaling and root planing (SRP) therapy [31]. On the other hand, SRP has limitations in relation to the difficult access to certain anatomical areas, being combined with other drug therapies. The clinical parameters evaluated (plaque index, gingival index, and clinical attachment level) at 21 and 45 days after receiving the intervention showed no differences between the group treated with the curcumin nanocarrier and the group treated with CHX gel 1%, both therapies adjuvant to SRP therapy. This clearly shows that the curcumin nanocarrier may be a promising treatment for periodontal pockets, due to its similar effects to a conventional drug (CHX), without the inconveniences related to unpleasant taste, tooth staining and dysbiotic effects.

During the inflammatory process from the immune system in response to pathogenic microorganisms, large amounts of oxygen free radicals are produced. As the antioxidant cellular response is insufficient due to the overproduction of these radicals, this condition contributes to the destruction of periodontal tissues. In order to solve this clinical problem, a micellar nanocarrier of coenzyme Q_{10} as an antioxidant agent has contributed clinically to reduce the damage to periodontal tissues [32]. This nanocarrier can reach up to 154 nm and improves both periodontal parameters and antioxidant activity compared to SRP alone. It is used in a gel formulation that is applied in syringe with a traumatic needle inserted in the deepest point of the periodontal pockets after SRP treatment. The application is performed each alternate day for a week. In the sixth week, the nanosystem is capable of decreasing the gingival index (57%), the plaque index (61%), probing pocket depth (41%), bleeding on probing (70%), malondialdehyde (42%), in addition to increasing clinical attachment level (42%) and total antioxidant capacity (52%). Thus, nanotechnology moves towards the development of periodontal therapies, providing greater clinical efficacy to therapeutic procedures and enhancing the action of drugs.

6.4 Nanocarriers in bone regeneration

Several nanocarriers are applied in bone regeneration in order to stimulate osteo/odontogenesis, including ions, genes, biopolymer fiber systems, collagen matrices and double layer membranes. Mesoporous bioglass nanoparticle is one of the most used nanocarriers in the delivery of ions, genes, growth factors and drugs in bone regeneration, since it plays an important role in mineralization due to its ability to release compounds, in addition to adjustable pore size, surface functionalization capacity and biocompatibility [33].

When structurally coupled to a mesoporous bioglass nanoparticle of 60 nm (pore size of only 6 nm), the Sr ion (in addition to Ca and Si) has the ability to penetrate mesenchymal cells by macropinocytosis and increase the intracellular levels of Sr, Ca and Si, with subsequent stimulation of bone formation. This nanosystem externally charged with phenamil favours the activation of small molecules of bone morphogenetic protein 2 signaling, resulting in effects on the expression of osteo/odontogenic genes, alkaline phosphatase activity and cell mineralization. Bone regeneration is favored due to the synergism between the effect of the Sr ion and the phenamil drug *in vivo* under conditions of mal-calcification [34].

This nanocarrier is also used for the delivery of genetic material, such as silencing systems (e.g, siRNA) that inhibit osteoclastic functions, favouring the increase in bone mass. Specifically, siRNA acts by inhibiting the expression of the nuclear factor kappa B (RANK) activating receptor to suppress osteoclastogenesis, with limitations in life span due to its rapid degradation and little penetration through membranes. However, when coupled with the 90 nm mesoporous bioglass nanoparticle, siRNA presents sustained release for up to 4 days, with 70% intracellular uptake and negative regulation of osteoclastogenesis. On the other hand, doses higher than 320 μg mL^{-1} may have cytotoxic effects [33].

Another example related to nanocarrier engineering is the polyethylene oxide/polycaprolactone biopolymer fiber system, which is internally constituted by electrospinning of fibroblast growth factor (FGF)-2 and mesoporous bioglass nanoparticle of 90 nm loaded with FGF-18. A fiber loaded with FGF-2 aims to release this factor quickly to stimulate mitosis, angiogenesis and cell proliferation. In turn, the FGF-18 contained in the mesoporous bioglass nanoparticle is gradually released, stimulating osteogenesis from the positive regulation of bone morphogenetic protein 2. The nanosystem as a whole favours prolonged cell proliferation, alkaline phosphatase induction and cell mineralization [35]. Mesoporous bioglass nanoparticle (pore size of only 10.3 nm) loaded with FGF-18 is also used in collagen matrix, to which it binds by fibrillar interactions. The nanocarrier has shown improvements in osteogenic differentiation of mesenchymal cells, possibly due to the role of FGF-18 in bone regeneration [36].

Nanotechnology is also applied to guide bone regeneration in the synthesis of membranes. These are used to prevent the invasion of cells from epithelial and connective tissues, in addition to maintaining a stable space for the internal growth of bone tissue. Thus, scientific research is developed to improve the regenerative potential of these membranes. One way to stimulate the bioactivity of membranes has been to include nanocarriers in their structure, forming a double layer membrane. In this sense, a structurally porous layer favours cell adhesion and infiltration, being composed by conjugated electrospun of PLGA/gelatin nanofibers incorporating mesoporous silica nanoparticles loaded with dexamethasone (layer size of approximately 0.080 mm). The second layer is dense and acts as a barrier composed of PLGA/nanofibers loaded with the antibiotic doxycycline hyclate (layer size of approximately 0.305 mm) [37]. The results of this double-layer membrane show good biological compatibility, sustained release of drugs and osteoinductive capacity derived from increased activity of alkaline phosphatase, calcium deposition and increased expression of osteocalcin [37].

6.5 Nanocarriers in endodontics

The disinfection process of root canals is one of the biggest challenges during endodontic treatments. The anatomical complexity of the root canal system (presence of isthmus, accessory channels and dentinal tubules) hinders the eradication of pathogenic microorganisms and biofilms housed therein during mechanical and chemical preparation. Nanotechnology has provided new alternatives to make endodontic treatment more effective in the disinfection process, either through its antibacterial activity or through the drug delivery system [38].

6.5.1 Irrigants

The application of nanoparticles in irrigation systems is an alternative explored to improve the disinfection of root canal and dentinal tubules. Hydrogen peroxide (H_2O_2), a compound commonly used as a disinfectant, has been carried by IONPs. This association used in the *in vitro* disinfection of dentinal tubules for 5 min exhibits high antimicrobial activity on *Enterococcus faecalis* (one of the most resistant microorganisms in endodontic treatments), exceeding 5 to 30 times the effect obtained with conventional irrigants such as sodium hypochlorite (NaOCl) and CHX. These favorable results are attributed to the activation of IONPs and the beginning of a catalytic process, leading to faster release of free radicals from H_2O_2, and causing the death of microorganisms. In addition, the IONPs' nanoscale size allows access to more restricted areas, such as regions of dentinal tubules with a diameter ranging from 2.4 to 2.9 μm [38].

The removal of smear layer after mechanical instrumentation is another important step for the successful treatment of root canals [39]. Silver nanoparticles (AgNPs) have been incorporated into irrigating materials in order to improve the results obtained with the use of conventional irrigation. These nanoparticles display broad spectrum antibacterial activity and their application has been associated with treatments for resistant microorganisms. Due to their greater propensity to oxidation, surface coatings are applied for stabilization of AgNPs. In this sense, AgPNs coated with mesoporous silica (SiO_2—a biocompatible material and able to perform the controlled release of the core particles) have a spherical shape and mean diameter of 119 ± 29 nm, and have been incorporated into two irrigation systems for smear layer removal. Their application was evaluated by conjugating NaOCl and carrying sodium phytate (SP) or ethylene glycol-bis (β- aminoethyl ether) N, N, N′, N′—tetraacetic acid (EGTA), obtaining a final product of NaOCl and AgNPs + SiO_2 + Tris + EGTA or SP. These compounds demonstrate potential in preventing biofilm growth for at least seven days, while AgNPs alone lose their antimicrobial effect almost immediately after use. Moreover, the use of AgNPs + SiO_2 becomes more relevant due to the cytotoxicity to human gingival fibroblasts being reduced compared with the solutions that are conventionally used (NaOCl 3% and EDTA 17%) [40].

The smear layer may limit the penetration of disinfectants, interfering in the treatment of root canals. For this reason, mesoporous silica nanoparticles (MSNs) were fluorescently marked (RITC-NPs) and evaluated for their infiltration profile in the dentinal tubules of extracted permanent teeth containing smear layer. RITC-NPs were ellipsoid and spherical in shape, with two different diameters. RITC-NPs with an average size of 384.0 ± 86.0 nm are accumulated on the root canal, while particles with about 100 nm infiltrate through the dentinal tubules. These characteristics reinforce the potential of nanoparticles as carriers of antimicrobials [39].

6.5.2 Intracanal medication

Debridement of root canals is not effective only with instrumentation, requiring the use of intracanal medications. Bacteria, mainly *E. faecalis*, can lodge in the dentinal tubules up to 1500 μm from the root canal and protect themselves in biofilms, resisting to nutritional deprivation. Conventional intracanal medications have limitations such as difficulty in penetrating biofilms and dentinal tubules [41]. Thus, CS nanoparticles [42], PLGA nanoparticles [41], AgNPs [43] and mesoporous calcium silicate nanoparticles [44] have carried these medications as ways to overcome the limitations observed in current treatments. Carrier nanoparticles have greater surface area and charge density than microparticles, which favours direct interaction with the opposite charge of microorganisms. This enhances the antimicrobial effect and promotes drug delivery within the dentinal tubules [43].

Calcium hydroxide (CH) is a material commonly used in the treatment of apical lesions and in the disinfection of root canals, and its action depends on the release of calcium ions and the alkalinity of the environment. Nanocarriers of CH based on gelatin-coated CS nanoparticles are being developed. CS nanoparticles are biocompatible, have antimicrobial effect and promote the growth of osteoblasts and wound healing. In turn, gelatin has excellent biocompatibility and adhesiveness, increasing the possibility of chemical bonds. This nanocarrier is capable of promoting controlled and sustained release of CH ions, allowing future tests in endodontic treatments [42]. CH has also been combined with AgNPs in an attempt to eliminate *E. faecalis*, considering that CH's antimicrobial activity is associated with the vehicle used. In this case, the AgNPs-CH nanocarrier effectively eliminates *E. faecalis* after 1 day of exposure, while CH alone is unable to eliminate this microorganism after 1 week. This result indicates that the AgNPs-CH nanocarrier can be used as an intracanal medication in the short term [43].

CS-coated PLGA nanoparticles (339.6 nm ± 9.8 nm) are also used in endodontics to carry ciprofloxacin. These nanoparticles have a high capacity to permeate cell membranes, which also contributes to the compound's antimicrobial effect. This nanocarrier has shown controlled release of ciprofloxacin (78.3% within 72 h) and greater antimicrobial and antibiofilm effects on *E. faecalis* compared to the drug alone [41].

CHX is another compound used as an irrigant or intracanal medication during endodontic treatments. Although it has a good antimicrobial effect, CHX is cytotoxic at high concentrations, which contraindicates its direct contact with tissues around the root apex, in addition to being unable to neutralize some bacterial inflammatory metabolic products. In turn, calcium and silicate ions are effective in promoting osteogenesis, but have low antimicrobial activity. Thus, mesoporous calcium silicate nanoparticles (MCSNs) have been loaded with CHX (M-CHX) in an attempt to improve the limitations found in each material. SEM images show the formation of a spherical nanocarrier with a diameter of 78.6 ± 11.1 nm. M-CHX has an alkaline pH, favorable for inhibiting the growth of bacteria. This nanocarrier acts by releasing CHX and favouring the antimicrobial effect against *E. faecalis*, besides acting on the sustained release of Ca^{2+}/SiO_3^{2-} ions, promoting bone regeneration [44].

6.5.3 Sealing materials

The final stage of root canal treatment comprises filling the canal using materials that will seal the space previously occupied by the dental pulp. Gutta-percha (GP) is the standard material used in this step and has advantages such as biocompatibility, low cost and ease of removal. However, GP can show flaws in the sealing of root canals, creating empty spaces that allow the growth of microorganisms and the formation of biofilms [45].

In order to avoid the possibilities of reinfection of the root canal, a nanodiamond-based compound (ND) has been conjugated to amoxicillin and incorporated into GP (NDGP). ND are biocompatible carbon nanoparticles (4–6 nm in diameter) that enable the conjugation of several compounds to their surface, favouring the drug delivery system, in addition to presenting antimicrobial activity. The NDGP compound shows greater hardness and tensile strength compared to unmodified GP, characterizing ease of handling during the clinical process. Furthermore, NDGP significantly reduces *Staphylococcus aureus* growth compared to unmodified GP. The mechanism of antimicrobial action of NDGP is associated with the release of amoxicillin or contact-mediated inhibition after bacterial deposition on its surface [45].

6.6 Nanocarriers in orthodontics

In orthodontics, the use of nanocarriers has emerged as an alternative for controlling biofilm-associated oral pathologies [46, 47]. Nanocarriers have been incorporated into orthodontic materials for sustained drug release, as a way of controlling the dependence on patient collaboration in the control of oral hygiene in conventional treatments [46].

Among the problems observed during orthodontic treatments, enamel white spot lesions (i.e., non-cavitated caries lesions) are commonly reported. These lesions originate from biofilm formation in retentive niches created by orthodontic appliances or from excessive acid conditioning for the adherence of orthodontic brackets. In this context, the application of bioactive glass nanoparticles (BAG) shows good results as a fluorine carrier (FBAG) and is incorporated into orthodontic adhesives. FBAG has shown greater antibacterial and anti-demineralization activities than those observed for the pure Transbond Low Flow (LV) orthodontic adhesive, 24 and 48 h after its application. The ions released from the FBAG compound lead to changes in osmotic pressure and increased intracellular pH, causing microbial cell death. As for cytotoxicity, FBAG has no difference in cell viability for human gingival fibroblasts compared to LV, which is already used clinically. In addition, FBAG has physical and chemical stability, without interfering with the fixation of the orthodontic appliance to the dental enamel [46].

Another association between orthodontics and nanotechnology is the incorporation of nanocarriers to chemically activated polymethylmethacrylate (PMMA) used in removable orthodontic appliances. This material has porosities that facilitate the emergence of oral bacterial and fungal infections associated with biofilm formation [47]. In order to avoid this clinical problem, the incorporation of MSNs into PMMA has been evaluated due to their favourable characteristics, such as biocompatibility, high stability, good durability, low cost and the possibility of acting as a carrier, besides the slow release of ions [48]. MSNs (85 nm) carrying amphotericin B were incorporated into PMMA in order to inhibit the adhesion of *Streptococcus oralis* and *Candida albicans*. The observed mechanism is an anti-adherent effect related to hydrophilic surface energy and long-term release of silicon ions. Furthermore, the amphotericin B release occurs quickly in the first seven days (approximately 30%) and decelerates over time (10%–15% every seven days), maintaining an effective dosage in inhibiting microbial growth. After the incorporation of the nanocarrier, the mechanical properties (modulus of elasticity and hardness) of the PMMA are increased, being another favourable aspect of the material [47]. MSNs (20 nm) carrying silver sulfadiazine also reduce microbial adhesion (*S. oralis* and *C. albicans*) for up to 14 days when incorporated into PMMA, in addition to increasing the mechanical properties (flexural strength and hardness) of the acrylic resin. In this case, silver sulfadiazine acts by denaturing the microbial genetic material, inhibiting cell replication or enzymes necessary for cell wall formation [48]. Interestingly, the incorporation of the nanocarriers described above into PMMA does not show significant cytotoxicity when evaluated on oral keratinocytes, showing 70% cell viability [47, 48].

6.7 Nanocarriers in dentin adhesives

The failure of adhesive restorations and/or the reduction of their longevity are related to enzymatic degradation and hydrolysis at the dentin-restoration interface [49, 50]. Thus, nanotechnology has been shown to be a promising alternative, considering that carrier nanoparticles may preserve the structural integrity and, consequently, increase the longevity of aesthetic restorations [51, 52].

Polymeric nanoparticles (PolymP-***n***Active) composed of 2-hydroxyethyl methacrylate, ethylene glycol dimethacrylate and methacrylic acid have been loaded with zinc and applied to demineralized dentin, before or after application of the Single Bond (SB) adhesive (3M ESPE) [51]. The nanocarrier generated shows 350 nm in diameter, and the polymeric nanoparticles do not agglomerate and are able to penetrate the dentinal tubules and retain calcium. The zinc-carrying nanoparticles may bind to the demineralized dentin and remain in the hybrid layer, even after composite resin infiltration [51]. Furthermore, the nanocarrier inhibits dentin collagen degradation mediated by matrix metalloproteinases and remineralizes hard tissues [53], since this nanocarrier may supply calcium ions for mineral deposition at the resin–dentin interface [51]. Regarding mechanical properties, the bond strength at the dentin-restoration interface is unaffected when the carrier nanoparticles are applied before or after using the adhesive compared to the adhesive without prior application of the nanocarrier [51].

Copper-loaded mesoporous bioactive glass nanoparticles have been incorporated into experimental adhesives composed of bisphenol A diglycidyl ether dimethacrylate, 2-hydroxylethyl methacrylate, ethyl N, N-dimethyl-4-aminobenzoate and camphorquinone. The use of these copper-carrying nanoparticles (size ranging from 50–85 nm) favours the release of ions (copper, calcium and silicon) that promote the remineralization of the adhesive interface and inactivation of metalloproteinases. In addition to these properties, the experimental adhesive system containing the nanocarrier promotes the mineralization of human dental pulp stem cells, is cytocompatible for that same cell line and has good bond strength compared to the experimental control adhesives (without nanocarrier). Given these properties, the experimental adhesive system containing mesoporous bioactive glass nanoparticles has potential to increase the longevity of the resin–dentin interface [52].

6.8 Nanocarriers in cariology

Dental caries is the most prevalent oral disease [54, 55], affecting 60 to 90% of the population, regardless of age [56–58]. Its development is associated with the interaction among host (dental surfaces and saliva), diet (sugars and fermentable carbohydrates) and microorganisms. Among the microorganisms with cariogenic potential, the bacterium *Streptococcus mutans* stands out, which is acidogenic and aciduric, and synthesizes exopolysaccharides in dental biofilms [59]. Considering its important role in dental caries etiology, *S. mutans* has been the target of several studies with nanocarriers in cariology.

In this context, the chloraluminiumphthalocyanine photosensitizer has been encapsulated in CS mucoadhesive nanoparticles, generating a 350 nm nanocarrier. This nanocarrier combined with a diode laser (660 nm, 100 J cm^{-2}) is able to significantly reduce the number of viable *S. mutans* biofilm cells formed on bovine dentine blocks compared to the negative control group (phosphate buffered saline). It is believed that these favourable results are related to the size of CS at the nanoscale, which increases its antimicrobial potential and improves the electrostatic interaction of the photosensitizer with bacterial cells and with the extracellular polymeric matrix of biofilms [60].

Silver (Ag)-coated MSNs have been used as carriers of CHX (average size of the nanocarrier of 100 nm). This nanocarrier (Ag-MSNs-CHX) at 12.5 μg mL^{-1} is able to inhibit the growth of *S. mutans* in the planktonic state, while its minimal bactericidal effect occurs at 25 μg mL^{-1}. For biofilms of the same species, a minimum inhibitory concentration of the nanocarrier of 50 μg mL^{-1} has been found after 24 h of exposure. Furthermore, the Ag-MSNs-CHX nanocarrier has shown greater antibiofilm effects on *S. mutans* than Ag-MSNs or CHX alone, when evaluated by MTS assay and by images of confocal microscopy and SEM. The aforementioned nanocarrier also acts responsively to pH and antioxidant activity, since in acidic pH (5.5) and in the presence of glutathione (GSH; non-enzymatic antioxidant) it exhibits greater release of CHX and silver ions in comparison to the release of these compounds at pH 7.4 and in the absence of GSH. This behaviour may contribute to the degradation of the extracellular matrix of *S. mutans* biofilm. Interestingly, this nanocarrier of CHX is less cytotoxic at 50 $\mu g\ mL^{-1}$ for immortalized oral epithelial cells compared to pure CHX, and the compound Ag-MSNs shows no abnormal effects on blood biochemical tests and on mouse organs when administered orally at 50 $mg\ kg^{-1}$ [61].

CHX has also been carried by CS-coated IONPs, forming a nanocarrier only 33.6 nm in diameter. For *S. mutans* planktonic cells, the IONPs-CS-CHX nanocarrier does not differ from free CHX, both showing the same minimum inhibitory concentration value (0.78 $\mu g\ mL^{-1}$). On the other hand, the nanocarrier containing CHX at 39 $\mu g\ mL^{-1}$ shows reducing effects on the quantification of cultivable cells, total biomass and metabolic activity of *S. mutans* biofilms similar to those found for free CHX at 78 $\mu g\ mL^{-1}$. This result reinforces the possibility of using a lower concentration of CHX in nanocarriers to reduce side effects associated with this drug [21].

6.9 Nanocarriers in anaesthesiology

Anaesthesiology is an area of dentistry in which different polymeric and lipid nanocarriers have been tested [62–64]. Among the polymeric nanocarriers are the poly (ethyleneglycol)-poly (ol-caprolactone) nanocapsules (PEG-PCL) and alginate/CS nanospheres (AG/CS) with a diameter of 100–300 nm, both carrying the anaesthetic articainehydrochloride (ATC). The connection between nanoparticles and polymers is believed to occur as result of electrostatic interactions after the disappearance of ATC endothermic peaks. ATC associated with PEG-PCL nanocapsules has less free anesthetic available compared to ATC without nanocarrier. In addition, the ATC release values from both AG/CS nanospheres and PEG-PCL nanocapsules are higher than those observed for ATC alone, indicating that both formulations are able to alter the permeation profile of the anaesthetic. The capsulation of ATC to PEG-PCL nanocapsules also reduces toxicity to mouse BALB/c-3T3 cells compared to free anaesthetic [62].

Nanocarriers such as nanostructured lipids (NLC) based on cetylpalmitateand-capric/caprylictriglycerides, and solid lipid nanoparticles (SLN) have also been explored [63, 64]. The nanostructures sizes range from 153 to 389 nm for NLC and 120 nm for SLN. These nanocarriers have been used to transport the topical anaesthetics lidocaine (LID) and prilocaine (PRI), and the electrostatic interactions between the amine groups and the free groups available in the NLC may be responsible for the chemical binding mechanism. Structural lipids act as a physical barrier, protecting LID and PRI against hydrolysis in aqueous environments such as the oral mucosa. This scenario allows the release of anaesthetics with a sustained profile, improving their analgesic effects and skin penetration [63, 64]. NLC containing 5% LID or PRI displays satisfactory sustained release profiles of 59% and 66% up to 20 h, respectively [63]. SLN and NLC loaded with LID and/or PRI show greater and longer effectiveness than free anaesthetics (without nanocarrier). Finally, the cytotoxicity evaluated on mouse embryonic fibroblasts (BALB/c-3T3 cells) is lower after encapsulation and protection of LID and/or PRI by NLC compared to free anaesthetics [64].

6.10 Nanocarriers in chemotherapy

The use of nanocarriers has also been explored as a way of treating neoplastic diseases related to dentistry. Several *in vitro* studies in this specific area have used different illnesses, including *Hela* cancer cell, head and neck cancer cells and oral squamous cells [65–67]. Three groups of nanocarriers have been used in the *in vitro* treatment of oral neoplasms, highlighting (i) fluorinated graphene (FG) coated by CS, (ii) poly (3,4-dihydroxy-L-phenylalanine) (PDOPA) with a layer of PEG and assembled with calcium phosphate (CaP), as well as (iii) polyethyleneimine (PEI)-cholesterol (PC) coated with berberine (BER). The sizes of the nanocomposites vary according to the type of nanocarrier and drug carried, with an average of approximately 60 nm for FG carrying doxorubicin (DOX) and camptothecin (CPT), 154.3 nm for PDOPA carrying DOX and up to 193 nm for PC nanoparticle carrying miR-122 (Mr). Covalent bonds [65] and electrostatic interactions [67] have been described as mechanisms of connection between nanocarrier and drug.

Regarding biological effects, the coating of FG with CS improves the photo thermal effects of the compound and its loading capacity for mixed drugs (DOX and CPT). On the other hand, CaP favours the conservation of the DOX-CaP-nanoparticles structure, inhibiting and enhancing DOX release in the extracellular space and in the endosomal pH, respectively. In addition, Mr-BER-PC assists the endosomal leakage of Mr and improves its delivery compared to Mr alone [65–67]. Functionalized FG also exhibits greater cytotoxicity on *Hela* cells than CPT or DOX alone. DOX-CaP-nanoparticles outperform the therapeutic efficacy of free DOX, while mr-BER-PC shows significant effects on invasion and migration cell compared to treatments with isolated drugs [65–67]. For normal human cells, FG-CS has little toxicity, DOX-CaP-nanoparticles have relatively low toxicity (mean inhibitory concentration of 6.59 μg mL^{-1} compared to free DOX of 1.22 μg ml^{-1}), while mr-BER-PC is not cytotoxic [65–67].

6.11 Conclusions

The use of nanocarriers is a promising alternative strategy, capable of increasing the efficacy of antimicrobial agents, facilitating their penetration into biofilms and tissue barriers, and promoting an increase in its stability, bioavailability, delivery control and reduction of toxicity. Nanocarriers have also been shown to enhance the effects of conventional therapies in other fields, including bone regeneration, dentin adhesives, anaesthesiology and chemotherapy. Thus, when well designed, nano-carriers can bring solutions to several clinical dental challenges. However, most of studies on nanocarriers in dentistry are still in basic research, so that intensive investigation is needed before the pharmaceutical industry can incorporate them into commercial formulations.

References

[1] Foster J S and Kolenbrander P E 2004 Development of a multispecies oral bacterial community in a saliva-conditioned flow cell *Appl. Environ. Microbiol.* **70** 4340–8

[2] Verma D, Garg P K and Dubey A K 2018 Insights into the human oral microbiome *Arch. Microbiol.* **200** 525–40

[3] Van Acker H, Van Dijck P and Coenye T 2014 Molecular mechanisms of antimicrobial tolerance and resistance in bacterial and fungal biofilms *Trends Microbiol.* **22** 326–33

[4] Ardekani S M, Dehghani A, Ye P, Nguyen K A and Gomes V G 2019 Conjugated carbon quantum dots: potent nano-antibiotic for intracellular pathogens *J. Colloid Interface Sci.* **552** 378–87

[5] Liu Y, Shi L, Su L, van der Mei H C, Jutte P C, Ren Y and Busscher H J 2019 Nanotechnology-based antimicrobials and delivery systems for biofilm-infection control *Chem. Soc. Rev.* **48** 428–46

[6] Gupta A, Mumtaz S, Li C H, Hussain I and Rotello V M 2019 Combatting antibiotic-resistant bacteria using nanomaterials *Chem. Soc. Rev.* **48** 415–27

[7] Patil S, Rao R S, Majumdar B and Anil S 2015 Clinical appearance of oral *Candida* infection and therapeutic strategies *Front. Microbiol.* **6** 1391

[8] Pires de Carvalho Bianchi C M, Bianchi H A, Tadano T, Paula C R, Hoffmann-Santos H D, Leite D P Jr and Hahn R C 2006 Factors related to oral candidiasis in elderly users and non-users of removable dental prostheses *Rev. Inst. Med. Trop. Sao Paulo.* **58** 17

[9] Davies A N, Brailsford S R and Beighton D 2006 Oral candidosis in patients with advanced cancer *Oral Oncol.* **42** 698–702

[10] Stecksen-Blicks C, Granstrom E, Silfverdal S A and West C E 2015 Prevalence of oral *Candida* in the first year of life *Mycoses* **58** 550–56

[11] Lima S L, Colombo A L and de Almeida Junior J N 2019 Fungal cell wall: emerging antifungals and drug resistance *Front. Microbiol.* **10** 2573

[12] Song J, Zhou J, Zhang L and Li R 2020 Mitochondria-mediated azole drug resistance and fungal pathogenicity: opportunities for therapeutic development *Microorganisms* **8** 1574

[13] Arias L S, Pessan J P, Vieira A P M, Lima T M T, Delbem A C B and Monteiro D R 2018 Iron oxide nanoparticles for biomedical applications: a perspective on synthesis, drugs, antimicrobial activity, and toxicity *Antibiotics (Basel)* **7** 46

[14] Waghule T, Sankar S, Rapalli V K, Gorantla S, Dubey S K, Chellappan D K, Dua K and Singhvi G 2020 Emerging role of nanocarriers based topical delivery of anti-fungal agents in combating growing fungal infections *Dermatol. Ther.* **33** e13905

[15] Garg A, Sharma G S, Goyal A K, Ghosh G, Si S C and Rath G 2020 Recent advances in topical carriers of anti-fungal agents *Heliyon* **6** e04663

[16] Verma S and Utreja P 2019 Vesicular nanocarrier based treatment of skin fungal infections: potential and emerging trends in nanoscale pharmacotherapy *Asian J. Pharm. Sci.* **14** 117–29

[17] Senna J P, Barradas T N, Cardoso S, Castiglione T C, Serpe M J, Silva K and Mansur C R E 2018 Dual alginate-lipid nanocarriers as oral delivery systems for amphotericin B *Colloids Surf.* B **166** 187–94

[18] de Bastiani F, Spadari c c, de Matos J K R, Salata G C, Lopes L B and Ishida K 2019 Nanocarriers provide sustained antifungal activity for amphotericin B and miltefosine in the topical treatment of murine vaginal candidiasis *Front. Microbiol.* **10** 2976

[19] Araujo H C, Arias L S, Caldeirao A C M, Assumpção L C F, Morceli M G, de Souza Neto F N, de Camargo E R, Oliveira S H P, Pessan J P and Monteiro D R 2020 Novel colloidal nanocarrier of cetylpyridinium chloride: antifungal activities on *Candida* species and cytotoxic potential on murine fibroblasts *J. Fungi (Basel)* **6** 218.

[20] AriasL S, Pessan J P, de Souza Neto F N, Lima B H R, de Camargo E R, Ramage G, Delbem A C B and Monteiro D R 2020 Novel nanocarrier of miconazole based on chitosan-coated iron oxide nanoparticles as a nanotherapy to fight *Candida* biofilms *Colloids Surf.* B **192** 111080

[21] VieiraA P M, Arias L S, de Souza Neto F N, Kubo A M, Lima B H R, de Camargo E R, Pessan J P, Delbem A C B and Monteiro D R 2019 Antibiofilm effect of chlorhexidine-carrier nanosystem based on iron oxide magnetic nanoparticles and chitosan *Colloids Surf.* B **174** 224–31

[22] de Lima T M, Arias L S, Afanaci L F, Ferraresse R F, Neto F N S, de Lima B H, Straioto F G, de Camargo E R, Pessan J P and Monteiro D R 2020 Assembly and antifungal effect of a new fluconazole-carrier nanosystem *Future Microbiol.* **15** 273–85

[23] Hussein-Al-Ali S H, El Zowalaty M E, Kura A U, Geilich B, Fakurazi S, Webster T J and Hussein M Z 2014 Antimicrobial and controlled release studies of a novel nystatin conjugated iron oxide nanocomposite *BioMed Res. Int.* **2014** 651831

[24] Sakima V T, Barbugli P A, Cerri P S, Chorilli M, Carmello J C, Pavarina A C and Mima E G O 2018 Antimicrobial photodynamic therapy mediated by curcumin-loaded polymeric nanoparticles in a murine model of oral candidiasis *Molecules* **23** 2075

[25] Trigo Gutierrez J K, Zanatta G C, Ortega A L M, Balastegui M I C, Sanita P V, Pavarina A C, Barbugli P A and Mima E G O 2017 Encapsulation of curcumin in polymeric nanoparticles for antimicrobial photodynamic therapy *PLoS One* **12** e0187418

[26] Li L P *et al* 2018 Antifungal activity of osthol *in vitro* and enhancement in vivo through Eudragit S100 nanocarriers *Virulence* **9** 555–62

[27] Pupe C G, Villardi M, Rodrigues C R, Rocha H V, Maia L C, de Sousa V P and Cabral L M 2011 Preparation and evaluation of antimicrobial activity of nanosystems for the control of oral pathogens *Streptococcus mutans* and *Candida albicans Int. J. Nanomed.* **6** 2581–90

[28] Araujo H C, da Silva A C G, Paiao L I, Magario M K W, Frasnelli S C T, Oliveira S H P, Pessan J P and Monteiro D R 2020 Antimicrobial, antibiofilm and cytotoxic effects of a colloidal nanocarrier composed by chitosan-coated iron oxide nanoparticles loaded with chlorhexidine *J. Dent.* **101** 103453

[29] Arias L S, Brown J L, Butcher M C, Delaney C, Monteiro D R and Ramage G 2020 A nanocarrier system that potentiates the effect of miconazole within different interkingdom biofilms *J. Oral Microbiol.* **12** 1771071

[30] Lin J H, Feng F, Yu M C, Wang C H and Chang P C 2017 Modulation of periodontitis progression using pH-responsive nanosphere encapsulating metronidazole or N-phenacylthialzolium bromide *J. Periodontal Res.* **53** 2–8

[31] Guru S R, Reddy K A, Rao R J, Padmanabhan S, Guru R and Srinivasa T S 2020 Comparative evaluation of 2% turmeric extract with nanocarrier and 1% chlorhexidine gel as an adjunct to scaling and root planing in patients with chronic periodontitis: a pilot randomized controlled clinical trial *J. Indian Soc. Periodontol.* **24** 244–52

[32] Shaheen M A, Elmeadawy S H, Bazeed F B, Anees M and Saleh N M 2020 Innovative coenzyme Q10-loaded nanoformulation as an adjunct approach for the management of moderate periodontitis: preparation, evaluation, and clinical study *Drug Deliv. Transl. Res.* **10** 548–64

[33] Kim T H, Singh R K, Kang M S, Kim J H and Kim H W 2015 Inhibition of osteoclastogenesis through siRNA delivery with tunable mesoporous bioactive nanocarriers *Acta Biomater.* **29** 352–64

[34] Lee J H, Mandakhbayar N, El-Fiqi A and Kim H W 2017 Intracellular co-delivery of Sr ion and phenamil drug through mesoporous bioglass nanocarriers synergizes BMP signaling and tissue mineralization *Acta Biomater.* **15** 93–108

[35] Kang M S, Kim J H, Singh R K, Jang J H and Kim H W 2015 Therapeutic-designed electrospun bone scaffolds: mesoporous bioactive nanocarriers in hollow fiber composites to sequentially deliver dual growth factors *Acta Biomater.* **16** 103–16

[36] Mahapatra C, Singh R K, Kim J J, Patel K D, Perez R A, Jang J H and Kim H W 2016 Osteopromoting reservoir of stem cells: bioactive mesoporous nanocarrier/collagen gel through slow-releasing FGF18 and the activated BMP signaling *ACS Appl. Mater. Interfaces* **8** 27573–84

[37] Lian M, Sun B, Qiao Z, Zhao K, Zhou X, Zhang Q, Zou D, He C and Zhang X 2019 Bi-layered electrospun nanofibrous membrane with osteogenic and antibacterial properties for guided bone regeneration *Colloids Surf.* B **176** 219–29

[38] Bukhari S, Kim D, Liu Y, Karabucak B and Koo H 2018 Novel endodontic disinfection approach using catalytic nanoparticles *J. Endod.* **44** 806–12

[39] Li X, Li X, Wang S, Leung K C F, Zhang C and Jin L 2018 Infiltration and profiles of mesoporous silica nanoparticles in dentinal tubules *ACS Biomater. Sci. Eng.* **4** 1428–36

[40] Ertem E, Gutt B, Zuber F, Allegri S, Ouya B L, Mefti S, Formentin K, Stellacci F and Ren Q 2017 Core–shell silver nanoparticles in endodontic disinfection solutions enable long-term antimicrobial effect on oral biofilms *ACS Appl. Mater. Interfaces* **9** 34762–72

[41] Arafa M G, Mousa H A and Afifi N N 2020 Preparation of PLGA-chitosan based nanocarriers for enhancing antibacterial effect of ciprofloxacin in root canal infection *Drug Deliv.* **27** 26–39

[42] Farhadian N, Godiny M, Moradi S, Azandaryani A H and Shahlaei M 2018 Chitosan/gelatin as a new nano-carrier system for calcium hydroxide delivery in endodontic applications: Development, characterization and process optimization *Mater. Sci. Eng. C: Mater. Biol. Appl.* **92** 540–46

[43] Javidi M, Afkhami F, Zarei M, Ghazvini K and Rajabi O 2014 Efficacy of a combined nanoparticulate/calcium hydroxide root canal medication on elimination of *Enterococcus faecalis Aust. Endod. J.* **40** 61–5

[44] Fan W, Li Y, Sun Q, Ma T and Fan B 2016 Calcium-silicate mesoporous nanoparticles loaded with chlorhexidine for both anti-*Enterococcus faecalis* and mineralization properties *J. Nanobiotechnol.* **14** 72

[45] Lee D K *et al* 2015 Nanodiamond-guttapercha composite biomaterials for root canal therapy *ACS Nano.* **9** 11490–501

[46] Nam H J, Kim Y M, Kwon Y H, Yoo K H, Yoon S Y, Kim I R, Park B S, Son W S, Lee S M and Kim Y I 2019 Fluorinated bioactive glass nanoparticles: enamel demineralization prevention and antibacterial effect of orthodontic bonding resin *Materials (Basel)* **12** 1813

[47] Lee J H, El-Fiqi A, Jo J K, Kim D A, Jun S K, Kim H W and Lee H H 2016 Development of long-term antimicrobial poly (methyl methacrylate) by incorporating mesoporous silica nanocarriers *Dent. Mater.* **32** 1564–74

[48] Jo J K, Fiqi A E, Lee J H, Kim D A, Kim H W and Lee H H 2017 Rechargeable microbial anti-adhesive polymethylmethacrylate incorporating silver sulfadiazine-loaded mesoporous silica nanocarriers *Dent. Mater.* **33** e361–72

[49] Helling A L, Tsekoura E K, Biggs M, Bayon Y, Pandit A and Zeugolis D I 2017 *In vitro* enzymatic degradation of tissue grafs and collagen biomaterials by matrix metalloproteinases: improving the collagenase assay *ACS Biomater. Sci. Eng.* **3** 1922–32

[50] Aguda A H, Panwar P, Du X, Nguyen N T, Brayer G D and Brömme D 2014 Structural basis of collagen fiber degradation by cathepsin K *Proc. Natl Acad. Sci. USA* **111** 17474–9

[51] Osorio R, Osorio E, Medina-Castillo A L and Toledano M 2014 Polymer nanocarriers for dentin adhesion *J. Dent. Res.* **93** 1258–63

[52] Jun S, Yang S A, Kim Y, El-Figi A and Mandkhbayer N 2018 Functional nano-adhesive releasing therapeutic ions for MMP-deactivation and remineralization *Sci. Rep.* **8** 5663

[53] Hoppe A, Güldal N S and Boccaccini A R 2011 A review of the biological response to ionic dissolution products from bioactive glasses and glass-ceramics *Biomaterials* **32** 2757–74

[54] Mathur V P and Dhillon J K 2018 Dental caries: a disease which needs attention *Indian J. Pediatr.* **85** 202–6

[55] Tafere Y, Chanie S, Dessie T and Gedamu H 2018 Assessment of prevalence of dental caries and the associated factors among patients attending dental clinic in Debre Tabor general hospital: a hospital-based cross-sectional study *BMC Oral Health* **18** 119

[56] Petersen P E and Ogawa H 2016 Prevention of dental caries through the use of fluoride—the WHO approach *Community Dent. Health* **33** 66–8

[57] Salunke S, Shah V, Ostbye T, Gandhi A, Phalgune D, Ogundare M O and Sable V 2019 Prevalence of dental caries, oral health awareness and treatment—seeking behavior of elderly population in rural Maharashtra *Indian J. Dent. Res.* **30** 332–6

[58] Carvalho J C and Schiffner U 2019 Dental caries in european adults and senior citizens 1996–2016: ORCA Saturday Afternoon Symp. in Greifswald, Germany—Part II *Caries Res.* **53** 242–52

[59] Meng Y, Wu T, Billings R, Kopycka-Kedzierawski D T and Xiao J 2019 Human genes influence the interaction between *Streptococcus mutans* and host caries susceptibility: a genome-wide association study in children with primary dentition *Int. J. Oral Sci.* **11** 19

[60] Cavalcante L L R, Tedesco A C, Takahashi L A U, Curylofo-Zotti F A, Souza-Gabriel A E and Corona S A M 2020 Conjugate of chitosan nanoparticles with chloroaluminiumphthalocyanine: Synthesis, characterization and photoinactivation of *Streptococcus mutans* biofilm *Photodiagnosis Photodyn. Ther.* **30** 101709

[61] Lu M M *et al* 2018 Redox/pH dual-controlled release of chlorhexidine and silver ions from biodegradable mesoporous silica nanoparticles against oral biofilms *Int. J. Nanomed.* **13** 7697–709

[62] de Melo N F S, Campos E V R, Gonçalves C M, de Paula E, Pasquoto T, de Lima R, Rosa A H and Fraceto L F 2014 Development of hydrophilic nanocarriers for the charged form of the local anestheticarticaine *Colloids Surf.* B **121** 66–73

[63] Ribeiro L N, Franz-Montan M, Breitkreitz M C, Alcântara A C, Castro S R, Guilherme V A, Barbosa R M and de Paula E 2016 Nanostructured lipid carriers as robust systems for topical lidocaine-prilocaine release in dentistry *Eur. J. Pharm. Sci.* **93** 192–202

[64] You P, Yuan R and Chen C 2017 Design and evaluation of lidocaine- and prilocaine-coloadednanoparticulate drug delivery systems for topical anesthetic analgesic therapy: a comparison between solid lipid nanoparticles and nanostructured lipid carriers *Drug Des. Devel. Ther.* **11** 2743–52

[65] Gong P, Zhao Q, Dai D, Zhang S, Tian Z, Sun L, Ren J and Liu Z 2017 Functionalized ultrasmall fluorinated graphene with high NIR absorbance for controlled delivery of mixed anticancer drugs *Chemistry* **23** 17531–41
[66] Nam H Y, Min K H, Kim D E, Choi J R, Lee H J and Lee S C 2017 Mussel-inspired poly (L-DOPA)-templated mineralization for calcium phosphate-assembled intracellular nano-carriers *Colloids Surf.* B **157** 215–22
[67] Li L *et al* 2018 Synergistic anticancer effects of nanocarrier loaded with berberine and *miR-122 Biosci. Rep.* **38** BSR20180311

IOP Publishing

Nanotechnology for Dentistry Applications

Ashutosh Kumar Shukla

Chapter 7

Nanotechnology in orthodontics: recent trends and future

Ra'ed Al-Dboush and Tarek El-Bialy

Nanotechnology has helped in improvement of many dental materials like dental fillings and implants and facilitating many dental procedures including orthodontic treatment. Friction between the brackets and arch wires is an inevitable physical phenomenon that could burn out anchorage and impede orthodontic tooth movement. However, coating of orthodontic brackets and wires with nanoparticles could reduce the friction and ease the orthodontic tooth movement. Development of white spot lesions on teeth after orthodontic treatment increases the risk of dental caries. These side effects of orthodontic treatment could be reduced through the antimicrobial effects of nanoparticles-coated brackets or nanoparticles-containing orthodontic cement and the enamel remineralizing effects of nanoparticles-filled pastes. Future orthodontic applications of nanotechnology include acceleration of orthodontic treatment through nanorobotics, reducing post-treatment relapse in cleft lip palatal expansion using an injectable estrogen-nano-diamond hydrogel, reducing orthodontic induced root resorption using nano-low intensity pulsed ultrasound (LIPUS) devices, and using smart nano-mechanical sensors to provide instant feedback about the applied orthodontic forces and moments.

7.1 Introduction

The topic of nanotechnology was introduced by a meeting talk entitled 'there's plenty of room at the bottom' that was given by American physicist Richard Feynman in 1959 [1, 2]. However, the first definition of the word 'nanotechnology' was given by Norio Taniguchi at the Tokyo Science University in 1974 [1]. Taniguchi defined it as follows 'Nano-technology mainly consists of the processing of separation, consolidation, and deformation of materials by one atom or one molecule' [3]. The prefix nano had originated from the Greek word nanos, which means dwarf [4]. Nanotechnology involves the fabrication and operating of materials and devices in the range of 0 to 100 nanometers [5]. This range is similar to the size scale of

doi:10.1088/978-0-7503-3671-0ch7

intracellular structures and molecules [5]. There are two approaches of nanomaterials construction: top-down approach which means breaking bulk material into smaller pieces using mechanical, chemical or other forms of energy; and bottom-up approach which means constructing the material from molecular or atomic particles through chemical reactions. Nanotechnology is advancing well in many sciences including medicine, pharmacy, and dentistry [6]. Nano-dentistry could be defined as the science and technology of maintaining near-perfect oral health through the use of nanomaterials including tissue engineering and nanorobotics [7]. Current applications of nano-dentistry include the use of nano adhesives (nanocomposite and nano light-curing glass-ionomer restorative) [8], improving the handling and physical properties of endodontic sealers [8], and enhancing the osteointegration of dental implants through surface modifications [9]. Interest in applying nanotechnology in the orthodontic field is increasing every day. In this chapter we will discuss the current and the future applications of nano-orthodontics.

7.2 Recent trends

7.2.1 Nanotechnology to reduce frictional forces during orthodontic treatment

Friction is considered a clinical challenge to the treating orthodontist, particularly with orthodontic sliding mechanics, as high levels of friction could burn out anchorage and impede orthodontic tooth movement [10]. It is defined as a force which is tangential to the shared boundary of two materials in contact that resists the motion or tendency to motion of one relative to the other [11]. Several studies have been done to analyze the effects of material, size and shape of wires and brackets as well as type of ligation on friction in different conditions and temperatures [12]. The impact of material of wires and brackets compromises two aspects, the alloy and surface characteristics [12]. Nanoparticles of metal dichalcogenide with fullerene-like structure were discovered in the early 90s [13, 14]. These nanoparticles work as excellent dry solid lubricants that have the capability to minimize friction between two materials sliding against each other without the need for liquid media [15, 16]. It was found that orthodontic stainless wires coated with self-lubricating coating containing inorganic fullerene-like tungsten disulfide (IF-WS_2) showed a 54% reduction of frictional forces compared to the non-coated wire [14, 17]. The coating process is done by inserting the wires into the electroless solution of the nanoparticles for about 30 min [18]. Different types of nanoparticles are used to reduce frictional forces, which include IF-WS_2, copper with IF-WS_2, carbon nitride, and molybdenum disulfide. [14, 17–19]

7.2.2 Nanotechnology to control oral flora during orthodontic treatment

White spot lesions are a common iatrogenic effect during orthodontic treatment [20]. Accumulation of a complex biofilm organization called dental plaque is considered one of the reasons that cause white spot lesions, dental caries, and periodontal disease [20, 21]. One of the risk factors for plaque accumulation is orthodontic treatment using fixed appliances [22]. The presence of fixed appliances could alter the microbiological composition of dental plaque with increased quantity and quality of oral microbiota [23]. Previous studies showed that quantity of colonies of acidogenic *Streptococci mutans* and *Lactobacilli* in saliva and dental biofilm were significantly elevated during active orthodontic treatment [24–26]. Nanoparticles are characterized by a greater surface-to-volume ratio compared to non-nanoscale particles, thus they interact more closely with microbial membranes and provide considerably larger surface area for antimicrobial activity [27, 28]. Specific nanoparticles have biocidal or anti-adhesive effects [26]. Hence, these nanoparticles have been incorporated into orthodontic adhesives to control the oral biofilm and to minimize the development of white spot lesions around the orthodontic fixed applainces [26]. Metal nanoparticles in the size range of 1–10 nm have particularly shown the greatest biocidal activity against bacteria [28]. These nanoparticles include titanium dioxide (TiO_2), silver (Ag), Gold (Au), and Silica (SiO_2) [26]. Cao *et al* used brackets coated with nitrogen-doped titanium dioxide (TiO_2) [29]. The nitrogen doping and modification enable titanium dioxide to exhibit catalytic activity within the visible-light region [29]. The activation leads to the formation of hydroxyl free radicals ($^{\bullet}OH$), superoxide ions (O_2), peroxyl radicals (HO_2) and hydrogen peroxide (H_2O_2) [29]. These chemicals, through a series of oxidation reactions, react with biological molecules such as lipids, proteins, enzymes and nucleic acids, damage biological cell structures, but also exert antimicrobial activity [29]. Cao *et al* found that bracket coated with the TiO_2 thin film showed high antimicrobial activity against *S. mutans, L. acidophilus, A. viscous*, and *C. albicans* [29]. They found that TiO_2-coated brackets showed strong anti-adhesive properties against *S. mutans* [29]. The findings of the Cao *et al* study demonstrated that TiO_2-coated brackets may be effective for the prevention of white spot lesions and gingivitis during orthodontic treatment [29]. Ahn *et al* found that the addition of silver nanoparticles to an experimental composite adhesive containing silica nanofillers resulted in less bacterial adhesion compared to conventional adhesives [30]. Thus, these adhesives can help prevent enamel demineralization around orthodontic brackets without compromising physical properties [30]. Poosti *et al* found that adding TiO_2 nanoparticles to light cure orthodontic composite paste (Transbond XT) enhances its antibacterial effects without compromising the shear bonding strength [31]. Elsaka *et al* found that the addition of 3% (w/w) TiO_2 nanoparticles to the conventional glass-ionomer restorative resulted in improved mechanical and antibacterial properties [32]. Lin *et al* assessed the fluoride release properties and the effect on bond strength of two experimental adhesive cements that contained nano-fluorapatite (nano-FA) or nano-fluorohydroxyapatite (nano-FHA) that were incorporated into a resin-modified glass ionomer cement (Fuji Ortho LC) [33]. The results showed that cements containing nano-FA or nano-FHA resulted in improved fluoride release properties, however, significant reduction in shear bond strength was reported [33]. The optimal concentration of added nano-FA and nano-FHA for maximum fluoride release was 25 wt.%, which nearly tripled fluoride release after 70 days compared with the control group. However, shear bond strength was reduced [33]. Enan *et al* found that modification of the banding glass ionomer cement with 15% nano-hydroxyapatite (nano-HA) resulted in significant reduction of microleakage around orthodontic bands when compared to those cemented with nano-HA-modified GIC [34]. A recent study by Cheng *et al* found that calcium phosphate nanoparticles have the ability of minimizing demineralization and increasing remineralization by releasing calcium and phosphate ions [35]. Cheng *et al* concluded in their study that the incorporation of nanoparticles into dental composites and adhesives could have double benefits: antibacterial capability and remineralization of tooth lesions [35].

7.3 Future orthodontic applications of nanotechnology

7.3.1 Acceleration of orthodontic treatment through nanorobotics

Several studies were done to build biosensors and nano-kinetics devices that are needed to control nanorobotics operation and locomotion [36]. Orthodontic nano-robots could direct manipulation of the periodontal tissues, including gingivae, periodontal ligament, cementum and alveolar bone, which could allow rapid and painless correction of crowded teeth, rotations and vertical repositioning within minutes to hours instead of the regular lengthy treatment [7].

7.3.2 Reducing post-treatment relapse in cleft lip palatal expansion

Cleft lip and/or palate is the most common congenital anomaly involving the craniofacial region with a prevalence of 1 in 700 births[37]. Maxillary expansion to correct transverse deficiency in these patients contributes in improvement of mastication, speech, nasal breathing, facilitation of permanent teeth eruption, and enhanced esthetics leading to greater self-esteem [38, 39]. However, expanded arches tend to relapse in the long term, especially in the canine width [40]. In their comparative study, Nicholson *et al* found that only 20% of cleft lip and palate patients retained the maxillary expansion in the molar region. Hence, relapse of maxillary expansion can be considered a clinical challenge during treatment of cleft and lip patients as it necessitates additional expansion and surgical procedures which could lead to increased complications and morbidities [39]. Using an animal model, Hong *et al* developed an innovative method to combat this clinical challenge by utilizing estrogen, 17β-estradiol (E2), which has proven bone-building properties [39]. They synthesized an injectable system with photo–crosslink able hydrogel (G) and nanodiamond (ND) technology to facilitate the targeted and sustained delivery of E2 to promote bone formation [39]. Their results showed that E2/ND/G system reduced post expansion relapse by threefold compared with the control group through enhancements in sutural remodeling [39]. Also, E2/ND/G injectable system resulted in a twofold increase of bone volume and a higher osteoblast number by threefold when compared to a control group [39]. They concluded that the E2/ND/G platform is an applicable E2 delivery platform that maximized the beneficial effects of E2 through its extended release with superior efficacy and safety at the local level so it could be a promising adjunctive therapy in the craniofacial care of cleft lip and palate patients [39].

7.3.3 Reducing orthodontic induced root resorption using nano-low intensity pulsed ultrasound (LIPUS) devices

One of the iatrogenic side effects of orthodontic treatment is orthodontically induced inflammatory root resorption (OIIRR). Recent studies found that daily application of low intensity pulsed ultrasound (LIPUS) for 20 min has the ability to reduce OIIRR during orthodontic treatment [41, 42]. However, there are some challenges in using

LIPUS intraorally which include that commercially available LIPUS transducers have large size and difficult adjustment [6]. In addition, the outcome of LIPUS use depends largely on patient compliance. Hence, the need for noncompliant LIPUS devices was the main interest for researchers recently. There are two positive applications that nanotechnology could impose on the use of noncompliant LIPUS devices. The first one is the construction of a nano-scale operation circuit and the second one is the fabrication of nano-scale batteries to run these circuits [6]. By these improvements, the intraoral LIPUS device would be independent of power supply or patient compliance [6]. Using nano LIPUS devices, LIPUS can be targeted to specific tooth or teeth as would be easily mounted on a bracket or even a plastic removable crown [27]. In addition, a nano-energy sensor could be used to ensure that the nano LIPUS devices are working properly by measuring the LIPUS power in the targeted area [27].

7.3.4 Using smart nano-mechanical sensors to provide instant feedback about the applied orthodontic forces and moments

In order to move the teeth efficiently without traumatic adverse effects, it is necessary to assess the orthodontic force that is applied on the teeth. Nanosensors to detect three-dimensional force-moment systems can be manufactured and built into the base of orthodontic brackets that aim to provide real-time feedback about the applied orthodontic forces [27]. This real-time feedback allows the orthodontist to adjust the applied force to be within a biological range to efficiently move the teeth with minimal side effects [27]. Lapaki *et al* constructed a 'smart' bracket with an integrated microelectronic chip equipped with multiple piezoresistive stress sensors for quantitative determination of three-dimensional force-moment systems applied to the bracket [43, 44]. Development of a nano system chip that can be incorporated into small low-profile contemporary bracket systems will allow clinical testing of the utilization of this technology [6].

7.4 Conclusions

Nanotechnology has positive impact on many aspects of dental care including orthodontic treatment. There is a need for more research projects in this field. The use of these nanotechnology materials would increase the cost of orthodontic treatment. Hence, affordability of these materials by the patients is an important aspect that should be taken into consideration while constructing them to facilitate their implantation in daily orthodontic practice.

References

[1] Kovvuru S, Mahita V, Manjunatha B S and Buma S 2012 Nanotechnology: the emerging science in dentistry *J Orofacial Res.* **2** 33–6

[2] Feynman R P 1960 There's plenty of room at the bottom *Calif. Inst. Technol. J. Eng. Sci.* **4** 23–6

[3] Taniguchi N 1974 On the basic concept of 'nano-technology' *Proc. Intl. Conf. Prod. Eng. Tokyo, Part II* (Japan Society of Precision Engineering)

[4] Shashirekha G, Jena A and Mohapatra S 2017 Nanotechnology in dentistry: clinical applications, benefits, and hazards *Compend. Contin. Educ. Dent.* **38** e1–4

[5] Zarbin M A, Montemagno C, Leary J F and Ritch R 2010 Nanotechnology in ophthalmology *Can. J. Ophthalmol.* **45** 457–76

[6] El-Bialy T 2019 Nanotechnology in orthodontics—2: facts and possible future applications *Nanobiomaterials in Clinical Dentistry* ed K Subramani and W Ahmed 2nd edn (Amsterdam: Elsevier) pp 299–308

[7] Freitas R A 2000 Nanodentistry *J. Am. Dent. Assoc.* **131** 1559–65

[8] Govind S, Jena A and Mohapatra S 2017 Nanotechnology in dentistry: clinical applications, benefits, and hazards *Compend. Contin. Educ. Dent.* **38** e1–4

[9] Lavenus S, Louarn G and Layrolle P 2010 Nanotechnology and dental implants *Int. J. Biomater.* **2010** 915327

[10] Rossouw P E 2003 Friction: an overview *Semin. Orthod.* **9** 218–22

[11] Drescher D, Bourauel C and Schumacher H A 1989 Frictional forces between bracket and arch wire *Am. J. Orthod. Dentofacial. Orthop.* **96** 397–404

[12] Pizzoni L, Ravnholt G and Melsen B 1998 Frictional forces related to self-ligating brackets *Eur. J. Orthod.* **20** 283–91

[13] Tenne R, Margulis L, Genut M and Hodes G 1992 Polyhedral and cylindrical structures of tungsten disulphide *Nature* **360** 444–6

[14] Katz A, Redlich M, Rapoport L, Wagner H D and Tenne R 2006 Self-lubricating coatings containing fullerene-like WS_2 nanoparticles for orthodontic wires and other possible medical applications *Tribol. Lett.* **21** 135–9

[15] Rapoport L, Fleischer N and Tenne R 2005 Applications of WS_2 (MoS_2) inorganic nanotubes and fullerene-like nanoparticles for solid lubrication and for structural nanocomposites *J. Mater. Chem.* **15** 1782–8

[16] Rapoport L, Bilik Y, Feldman Y, Homyonfer M, Cohen S R and Tenne R 1997 Hollow nanoparticles of WS_2 as potential solid-state lubricants *Nature* **387** 791–3

[17] Redlich M, Katz A, Rapoport L, Wagner H D, Feldman Y and Tenne R 2008 Improved orthodontic stainless steel wires coated with inorganic fullerene-like nanoparticles of WS_2 impregnated in electroless nickel-phosphorous film *Dent. Mater.* **24** 1640–6

[18] Batra P 2018 Applications of nanoparticles in orthodontics *Dental Applications of Nanotechnology.* ed R Chaughule (Cham: Springer)

[19] Samorodnitzky-Naveh G R, Redlich M, Rapoport L, Feldman Y and Tenne R 2009 Inorganic fullerene-like tungsten disulfide nanocoating for friction reduction of nickel–titanium alloys *Nanomedicine* **4** 943–50

[20] Gavrilovic I 2014 White spot lesions in orthodontic patients: formation, prevention and treatment *Oral Hyg. Health* **2** 154

[21] Al-Anezi S A and Harradine N W T 2012 Quantifying plaque during orthodontic treatment: *Angle Orthod.* **82** 748–53

[22] Zachrisson S and Zachrisson B U 1972 Gingival condition associated with orthodontic treatment *Angle Orthod.* **42** 26–34

[23] Freitas A O A, Marquezan M, Nojima M da C G, Alviano D S and Maia L C 2014 The influence of orthodontic fixed appliances on the oral microbiota: a systematic review *Dental Press J. Orthod.* **19** 46–55

[24] Rosenbloom R G and Tinanoff N 1991 Salivary *Streptococcus mutans* levels in patients before, during, and after orthodontic treatment *Am. J. Orthod. Dentofacial Orthop.* **100** 35–7

[25] Campbell C H C T 2003 Alteração da microflora bucalempacientesportadores de aparelhoortodônticofixo *Ortodon.gaúch.* 98–109

[26] Borzabadi-Farahani A, Borzabadi E and Lynch E 2014 Nanoparticles in orthodontics, a review of antimicrobial and anti-caries applications *Acta Odontol. Scand.* **72** 413–17
[27] Govindankutty D 2015 Applications of nanotechnology in orthodontics and its future implications: a review *Int. J. Appl. Dental Sci.* **1** 166–71
[28] Allaker R P 2010 The use of nanoparticles to control oral biofilm formation *J. Dent. Res.* **89** 1175–86
[29] Cao B 2013 Preparation of an orthodontic bracket coated with an nitrogen-doped $TiO_{2-x}N_y$ thin film and examination of its antimicrobial performance *Dental Mater. J.* **32** 311–16
[30] Ahn S, Lee S, Kook J and Lim B 2009 Experimental antimicrobial orthodontic adhesives using nanofillers and silver nanoparticles *Dent. Mater.* **25** 206–13
[31] Poosti M, Ramazanzadeh B, Zebarjad M, Javadzadeh P, Naderinasab M and Shakeri M T 2013 Shear bond strength and antibacterial effects of orthodontic composite containing TiO_2 nanoparticles *Eur. J. Orthod.* **35** 676–79
[32] Elsaka S E, Hamouda I M and Swain M V 2011 Titanium dioxide nanoparticles addition to a conventional glass-ionomer restorative: influence on physical and antibacterial properties *J. Dent.* **39** 589–98
[33] Lin J, Zhu J and Gu X *et al* 2011 Effects of incorporation of nano-fluorapatite or nano-fluorohydroxyapatite on a resin-modified glass ionomer cement *Acta Biomater.* **7** 1346–53
[34] Enan E T and Hammad S M 2013 Microleakage under orthodontic bands cemented with nano-hydroxyapatite-modified glass ionomer *Angle Orthod.* **83** 981–6
[35] Cheng L, Zhang K, Weir M D, Melo M A S, Zhou X and Xu H H K 2015 Nanotechnology strategies for antibacterial and remineralizing composites and adhesives to tackle dental caries *Nanomedicine* **10** 627–41
[36] Cavalcanti A, Rosen L, Kretly L C, Rosenfeld M and Einav S December 2004 Nanorobotic challenges in biomedical applications, design and control *Proc. of the 2004 11th IEEE Int. Conf. on Electronics, Circuits and Systems, 2004. ICECS 2004* 447–50
[37] Arosarena O A 2007 Cleft lip and palate *Otolaryngol. Clin. North Am.* **40** 27–60
[38] Subtelny J D and Brodie A G 1954 An analysis of orthodontic expansion in unilateral cleft lip and cleft palate patients *Am. J. Orthod.* **40** 686–97
[39] Hong C, Song D and Lee D *et al* 2017 Reducing posttreatment relapse in cleft lip palatal expansion using an injectable estrogen-nanodiamond hydrogel *Proc. Natl. Acad. Sci. USA* **114** E7218–25
[40] Gurel H G, Memili B, Erkan M and Sukurica Y 2010 Long-term effects of rapid maxillary expansion followed by fixed appliances *Angle Orthodon.* **80** 5–9
[41] Raza H, Major P, Dederich D and El-Bialy T 2016 Effect of low-intensity pulsed ultrasound on orthodontically induced root resorption caused by torque: a prospective, double-blind, controlled clinical trial *Angle Orthod.* **86** 550–7
[42] El-Bialy T, Farouk K and Carlyle T D *et al* 2020 Effect of low intensity pulsed ultrasound (LIPUS) on tooth movement and root resorption: a prospective multi-center randomized controlled trial *J. Clin. Med.* **9** 804
[43] Lapatki B G and Paul O 2007 Smart brackets for 3D-force-moment measurements in orthodontic research and therapy—developmental status and prospects *J. Orofac. Orthop.* **68** 377–96
[44] Lapatki B G, Bartholomeyczik J, Ruther P, Jonas I E and Paul O 2007 Smart bracket for multi-dimensional force and moment measurement *J. Dent. Res.* **86** 73–8

IOP Publishing

Nanotechnology for Dentistry Applications

Ashutosh Kumar Shukla

Chapter 8

Nanotechnology in endodontics

Desigar Moodley and Amir I O Ibrahim

Endodontics is the branch of dental sciences that deals with the prevention, disease pathology and treatment of the root canal space and the periapical area infection. It also now encompasses regeneration of the pulp tissue. Endodontic diseases result mainly from invasion of various microbial species within the root canal system. The treatment of such diseases require the use of various chemical agents in an attempt to eradicate these endodontic pathogens. Furthermore, it requires one to fill (obturate) the root canal system three-dimensionally with a suitable filling material that prevents the recurrence of the disease. One of the main problems of these chemical agents is the development of microbial resistance. Furthermore, the presence of residual microbial species following endodontic treatment results in re-infection of the root canal space and the periapical area.

Pulp tissue regeneration in endodontics deals with repair and regeneration of injured tissue. Various techniques have been used to achieve this purpose.

In the era of nanotechnology, nanomaterials in endodontics provide a new scope of application in endodontics with their use ranging in antimicrobial agents, bioactive endodontic material and tissue regeneration.

8.1 Introduction

Endodontics is the branch of dentistry that involves the study of the normal morphology, physiology and pathology of the human dental pulp and periradicular tissues. It deals with the diagnosis, prevention and treatment of pulp and periradicular diseases. The role of microorganisms as a primary etiological factor in pulp diseases and apical periodontitis was recognized by Kakehashi *et al* (1965). They showed a direct correlation between the presence of microbial cells within the pulp tissue and the presence of an inflammatory process in the dental pulp and around the root apex if the microbial cells passed beyond the periapical foramen (Kakehashi *et al* 1965).

Furthermore, the ability of different microbial species to invade the dentinal tubules was observed in 1890 by Miller (Love 2004). This invasion plays an

doi:10.1088/978-0-7503-3671-0ch8

important role in the initiation of pulpal and periradicular diseases. Multiple pathways are available for the microorganisms to invade the root canal system. However, the crown of the tooth is the main portal of entry for infection of the root canal space (Tronstad and Sunde 2003). Microbial invasion can occur through carious lesions, faulty restorations, tooth wear and enamel and dentine cracks (Haapasalo *et al* 2003). Additionally, other portals of microbial entry to the pulp space may also occur through invasion of microorganisms from periodontal pockets and from the blood stream (Love 2004; Narayanan and Vaishnavi 2010). Microorganisms can invade the dentinal tubules to varying depths (Love 2004) and can extend into the periapical area causing periapical periodontitis (Kakehashi *et al* 1965; Nair 2004).

The presence of microorganisms in the dentinal tubules of the coronal part of the root canal system are mainly associated with pulpal infection, while those present in the radicular dentinal tubules are associated mainly with periradicular infection (Love 2004).

Endodontic infections are polymicrobial in nature (Sundqvist and Figdor 2003) and more than 400 microbial species have been identified from endodontic samples (Jhajharia *et al* 2015). *α-hemolytic streptococci*, *ϒ-streptococci*, and *enterococci* are the most commonly isolated bacteria from infected root canals (Farber and Seltzer 1988). However, other bacterial species are also observed in an infected root canal. In addition to that, other microorganisms were also observed in endodontic infections such as archaea and fungi (Siqueira 2002; Siqueira and Sen 2004; Waltimo *et al* 1997).

Root canal infections are classified into primary or secondary infections. Primary infection is usually associated with symptomatic or asymptomatic periradicular disease, while secondary infection is usually associated with persistent infection or failed root canal treated teeth (Siqueira 2002).

There are variations between the microbial species isolated from primary endodontic infection compared to those isolated from secondary endodontic infection. In primary infection, obligate anaerobic bacterial species are the most predominant bacterial species, while in secondary infection, the facultative anaerobic bacterial species are found in higher percentages (Gomes *et al* 2004). This transition between different bacterial species results from the changes in nutritional and environmental status inside the root canal system as the infection progresses. This will create a more anaerobic environment with depletion of nutrition, which offers a tough ecological niche for the surviving microorganisms (Sundqvist and Figdor 2003).

Secondary endodontic infection is caused by persistent pathogens within the root canal system or the periapical area following completed root canal treatment. These microorganisms can be present in the root canal system as a result of inadequate root canal disinfection, improper access to the root canal system, missed canal during treatment and leaking of microorganisms through defective restorations (Peciuliene *et al* 2008).

8.2 Endodontic infection

Initially, endodontic pathogens survive within the root canal system utilizing the remnants of pulp tissue as a nutrient media. As the infection advances, environmental factors such as the oxygen tension, pH, temperature and type of available nutrients in the root canal system change. As a result, the invading microorganisms will change their virulence factors in an attempt to accommodate the new environmental changes (Siqueira 2002).

Endodontic diseases are biofilm-induced diseases (Siqueira *et al* 2010). This form of microbial colonization inside the root canal was initially identified by Nair (1987). It appears as a surface-attached microbial community which is defined by Mohammadi *et al* as 'a sessile multicellular microbial community characterized by cells that are firmly attached to a surface and enmeshed in a self-produced matrix of extracellular polymeric substance' (Mohammadi *et al* 2013). However, certain microorganisms are also present in a planktonic state within the root canal system (Svensäter and Bergenholtz 2004; Usha 2010).

Microbial biofilm consists of approximately 10%–15% microbial content and 85%–90% extracellular matrix (Kokare *et al* 2009). The extracellular matrix is composed of a conglomeration of various types of hydrated biopolymers known as the extracellular polymeric substance (EPS) (Flemming and Wingender 2010).

Root canal biofilms can be categorized into three categories according to their location inside the root canal system; intra-canal biofilms, extra-radicular biofilms, periapical biofilms.

Biofilm formation generally occurs in phases in the presence of a microbial element, aqueous environment and solid surface (Dua *et al* 2012). The steps of biofilm formation include deposition of thin film as a conditioning for adhesion and colonization of planktonic microorganisms that float in the fluid media. Following microbial adhesion, the microorganisms start growing to form the biofilm. There can be detachment of some parts of the biofilm into the surrounding environment (Dua *et al* 2012).

Formation of endodontic biofilm starts gradually, from the coronal pulp chamber and migrates apically to the apical foramen (Svensäter and Bergenholtz 2004). Most of the microorganisms that contaminate the pulp originate from coronal carious lesions. As the carious lesion extends to the pulp chamber exposing the pulp tissue, the carious biofilm contaminates the coronal pulp tissue causing inflammation and an exudate which provides liquefied media for biofilm formation and a source of nutrition for the invading microorganisms (Siqueira *et al* 2010). This theory was supported by the similarity between the microbial community found in deep carious lesions and the one found in primary endodontic infection (Siqueira *et al* 2010).

Endodontic pathogens in a biofilm state within the root canal system show enhanced pathogenicity since the biofilm facilitates a different behaviour from those microorganisms found in a planktonic state, as it provides several virulence factors not present in the planktonic state (Flemming and Wingender 2010; Kolenbrander *et al* 2010; Siqueira and Rôças 2014; Tronstad and Sunde 2003).

Kishen (2010) listed the mechanisms by which endodontic pathogens can resist the commonly used root canal irrigants and medicaments. These mechanisms are usually associated with the extracellular polymeric matrix, rate of growth, availability of nutrients and ability to adopt a resistance phenotype.

The extracellular polymeric matrix provides a number of essential functions to the microbial community (Flemming and Wingender 2010). It facilitates adhesion of the biofilm to the surfaces and presents mechanical stability to the biofilm. It also has the ability to retain moisture to maintain a highly hydrated environment around the biofilm. Additionally, it accumulates extracellular enzymes around the biofilm which in turn allow the availability of nutrients around it. It can also serve as a source of nutrients in case of starvation. Moreover, it has a protective role by protecting the microbial cell from the host immune response and antimicrobial agents. Finally, it maintains the biofilm cell in close proximity to facilitate different microbial communication methods such as genetic exchange, communication between cells (quorum sensing) and pathogenic synergism (Flemming and Wingender 2010). All these factors can contribute to the high resistance of biofilm to the commonly used antimicrobial agents. Additionally, the extracellular polymeric matrix can facilitate a decrease in the penetration rate of the antimicrobial agents (Costerton *et al* 1999).

About 20% to 70% of microorganisms in biofilms demonstrate dissimilar gene expression patterns compared to those microorganisms found in a planktonic state. Thus, microbial biofilm can be more resistant to antimicrobial agents compared to equivalent microorganisms found in a planktonic state (Beloin *et al* 2004; Sauer *et al* 2002).

Another hypothesis for the low susceptibility of biofilm to antimicrobial agents is that the presence of slow-growing microbial cells in the biofilm is the result of nutrition limitation, making them more resistant to antimicrobial agents (Brown *et al* 1988).

8.3 Management of endodontic and periapical infection

The main goal in treating root canal infection is to eliminate the microbial contamination within the entire root canal system (Carrotte 2004). This goal is achieved by fully debriding the root canal system from any infectious microorganisms. The debridement is done through root canal cleaning and shaping in which all the necrotic tissue and the infected root canal dentine is removed from the root canal system (Hülsmann *et al* 2005). Thereafter, the root canal system is finally sealed 'obturated' three-dimensionally to prevent any reinfection (Torabinejad *et al* 2003).

As a result of the complex anatomy of the root canal system, eradication of endodontic pathogens should be done with the use of both mechanical and chemical root canal preparation (Siqueira 2001). Both methods are used simultaneously and cannot be considered separately, hence it is referred to as chemomechanical or biomechanical preparation (Hülsmann *et al* 2005).

The role of mechanical preparation in root canal disinfection is to mechanically remove part of the infected dentine and at the same time provide room for the antimicrobial irrigant solutions and intra-canal medicament to act against the remaining different pathogens and thus enhance the eradication of endodontic pathogens (Vaudt *et al* 2007). The techniques used for the mechanical preparation include manual and automated root canal preparation, sonic and ultrasonic preparation, use of laser, and a nickel titanium rotary system. Hülsmann *et al* (2005) listed the goals of mechanical preparation of the root canal system, namely: removal of the vital and necrotic tissue from the main root canal; creation of sufficient space for irrigation and medication; preservation of the integrity and location of the apical anatomy; avoidance of iatrogenic damage to the canal system and root structure; facilitation of the canal filling; avoidance of further irritation and/or infection of the peri-radicular tissue; and preservation of sound root dentine to allow long-term functioning of the tooth (Hülsmann *et al* 2005).

Although there is advancement in the technology of mechanical instrumentation as new instruments are introduced to the dental market, these instruments do not have the ability to clean the entire root canal system as a result of the complex anatomy of the root canal system and the presence of isthmus and lateral canals which are difficult to instrument (Del Fabbro *et al* 2018; Siqueira *et al* 2017).

Numerous root canal irrigants are currently available. Although their use showed some effectiveness against some pathogens, none of them have the ability to completely eradicate the persistent pathogens from the root canal system (Stuart *et al* 2006). In addition, mature microbial biofilm has the ability to resist the antibacterial action of commonly used root canal irrigants (Stojicic *et al* 2013).

Intra-canal medicaments are antimicrobial agents that are placed inside the root canal system in order to eradicate the remaining microorganisms that persist after mechanical instrumentation and irrigation (Mohammadi and Dummer 2011). Calcium hydroxide is the most commonly used intra-canal medicament, however its antimicrobial activity is only effective against some of the root canal pathogens (Ørstavik and Haapasalo 1990). In addition, the presence of factors such as dentine powder, dentine matrix and remnants of pulp tissue within the root canal system was shown to inhibit the antimicrobial properties of calcium hydroxide (Haapasalo *et al* 2000; Haapasalo *et al* 2007) as a result of the buffering action against its alkalinity (Portenier *et al* 2001). Furthermore, the development of resistant microbial species renders it difficult to disinfect the root canal system using commonly available root canal irrigants and intra-canal medicaments (Evans *et al* 2002; Wang *et al* 2012).

8.4 Regenerative endodontic

Regenerative endodontics is defined by Murray *et al* (2007) as a 'biologically based procedure designed to replace damaged structure including dentine and root structure as well as cells of the pulp-dentine-complex'. The scope of regenerative endodontics can be categorized into two sections:

Guided tissue regeneration:

The philosophy of guided tissue regeneration was first introduced by Östby in 1961. Its aim to develop new blood vessels by the aid of inducing blood clot from the periapical tissue and hence the name revascularization. Clinically, this is achieved by starting to disinfect the root canal system using intra-canal medicament such as calcium hydroxide for several weeks followed by placement of the triple antibiotic paste (mixture of Ciprofloxacin, Metronidazole and Minocycline) (Murray *et al* 2007).

Tissue engineering:

Tissue engineering was developed during 1993–2007. The philosophy of tissue engineering is based on using suitable stem cells induced by growth factors in the presence of a scaffold which controlled the development of the targeted tissue.

Several sources of stem cells are available for endodontic engineering. Such sources may include the dental pulp, periapical tissue and dental papilla. Numerous stem cells are available in the oral cavity and the craniofacial region such as the dental pulp, periapical tissue and dental papilla (He *et al* 2017).

To control and regulate the growth of the stem cells a growth factor is needed. Several growth factors are present such as platelet-derived growth factor, transforming growth factor-β, vascular endothelial growth factor (Kim *et al* 2012). The growth factors differentiate the stem cells into different cells phenotypes. The type of the differentiated cells is depend on the growth factor used, extra-cellular matrix and other conditions such as a proper scaffold.

The third component involved in tissue engineering is the use of a proper scaffold. The scaffold is a synthetic or natural biodegradable three-dimensional porous material designed to provide a definite position and support stem cells growth (Murray *et al* 2007). Several scaffold materials are used in endodontic tissue engineering such as blood clot, platelet-rich plasma and platelet rich fibrin (Ulusoy *et al* 2019) and most recently nanofibers of different synthetic and natural materials such as collagen (Gupte and Ma 2012).

8.5 Nanotechnology in endodontics

The concept of nanotechnology was first illustrated in 1960 by Richard P Feynman. Since then, the concept of nanotechnology was applied in numerous dental fields such as endodontics, dental materials periodontics and surgery. Nanotechnology is defined as a science that deals with the development of new materials with new properties and functions through controlling and restructuring of the materials on a nanometer scale 'less than 100 nm' and hence the name nanomaterials (Sanchez and Sobolev 2010).

Generally nanomaterials can be classified based on their morphology, dimension, state and chemical composition. Based on their dimension, nanomaterials can be classified into zero-, one-, two- or three-dimensional nanomaterials. Such classification includes nanoparticles and nanofibers as one-dimensional nanomaterials (Saleh 2020). The use of nanoparticles and nanofibers of various materials have been evaluated and used for potential uses in endodontics.

Dealing with a material at the nano-scale level results in a new characteristic feature of the material compared to its bulk state (Tiwari *et al* 2012). This motivates researchers to investigate new or existing materials at nano-scale level for a new application in endodontics. The application of nanomaterials in endodontics can be discussed from two points:

i. Application of nanoparticles in endodontics.
ii. Application of nanofibers in endodontics.

Different applications of nanoparticles and nanofibers are listed in table 8.1.

8.5.1 Application of nanoparticles in endodontics

Numerous nanoparticles have been investigated for potential use in endodontics. Raura *et al* (2020) classify the nanoparticles used in endodontics based on their

Table 8.1. Applications of nanotechnology in endodontics.

Applications of nanoparticles	Applications of nanofibers
As antimicrobial agents	As scaffold in regenerative endodontics
As a drug carrier	As part of root canal filling material
As active component of root canal filling material	
It has a role in regenerative endodontics	

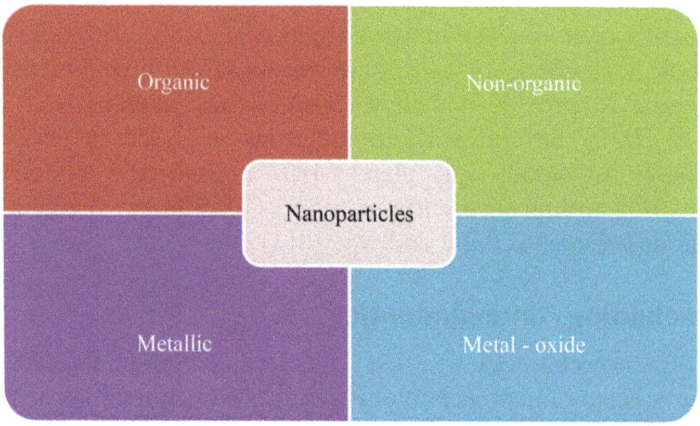

Figure 8.1. Diagram showing the classification of nanoparticles used in endodontics based on their chemical structure.

Table 8.2. Classification and types of nanoparticles used in endodontics based on their chemical structure.

Organic nanoparticles	Non-organic nanoparticles	Metallic nanoparticles	Metal—oxide nanoparticles
Graphene	Bioactive glass	Silver	Iron compound (FeO_x)
Chitosan	Mesoporous	Silica	Zirconia
Poly(lactic) co-glycolic acid	Hydroxyapatite		Titanium dioxide (TiO_2)
			Magnesium oxide (MgO) and calcium oxide (CaO)
			Copper oxide (CuO)

chemical structure into organic, non-organic, metallic and metal oxide nanoparticles (figure 8.1). Examples of the different types of nanoparticles based on their chemical structure are listed in table 8.2.

There are different types of nanoparticles and each has its unique properties. As a result, the application of nanoparticles in endodontics involves different applications based on the properties of the selected nanoparticles. Nanoparticles can be used as antimicrobial agent (Samiei *et al* 2016), drug carrier system (Arias-Moliz *et al* 2020), as active component of root canal filling material (Loyola-Rodríguez *et al* 2019) and in regenerative endodontics (Barreras *et al* 2016).

8.5.1.1 The use of nanoparticles as antimicrobial agents

The use of nanoparticles as antimicrobial agents has recently attracted considerable attention in pharmaceutical sciences as a result of their superior antibacterial properties (Seil and Webster 2012). Although most of the nanoparticles possess the same unique properties, which are their ability to activate multiple mechanisms to eradicate microbial cells and their low potentiality to produce microbial resistance, there may be slight differences in the antimicrobial mechanism among different types of nanoparticles.

The antimicrobial activity of nanoparticles is suggested to result from multiple mechanisms that occur concurrently (figure 8.2):

i. Electrostatic interaction between the negatively charged microbial cell wall and the positively charged nanoparticles. This attraction results in accumulation of a large number of nanoparticles in the microbial cell wall causing its lysis and penetration of the nanoparticles inside the microbial cell. Inside the microbal cell the nanoparticles bind to the mesosomes altering the respiration and division of the microbial cell and DNS replication (Pelgrift and Friedman 2013).
ii. Inside the microbial cell, nanoparticles release reactive oxygen species (ROS) causing oxidative stress inside the microbial cell, altering its

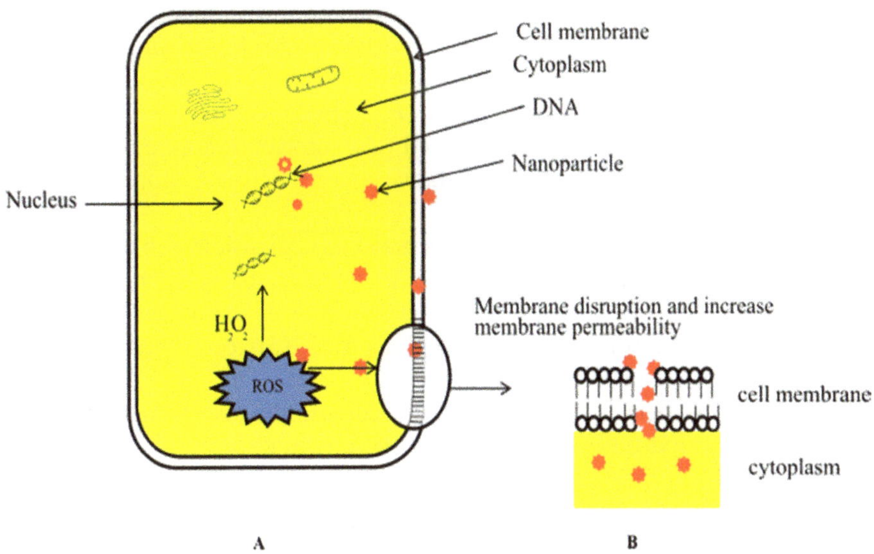

Figure 8.2. Diagram representing the antibacterial mechanisms of nanoparticles. (A) Toxicity through production of reactive-oxygen species (ROS). (B) Nanoparticles attach to bacterial cell membranes causing toxicity through cell damage. The diagram is copied from Ibrahim *et al* (2017) with permission of SciELO.

respiration and thus, reducing the production of ATP causing further disturbance of the microbial cell membrane (Valko *et al* 2007).

iii. The positive charged nanoparticles interact with the nucleic acid molecules affecting the chromosomal and DNA replication and thus inhibiting the signal transduction of the microbial cell (Giannousi *et al* 2014).
iv. Metallic nanoparticles affect the microbial cell homostasis by retarding its growth or even killing the microbial cells as a result of presence of excess metallic ions inside the cell (Wang *et al* 2016).
v. Nanoparticles also cause protein and enzyme dysfuction of the microbial cell by forming carbonyls (protein bound) inside the microbial cell that result in protein degradation, enzyme deactivation and alteration in the catalytic activity of the cell (Lynch and Dawson 2008).

In endodontics, one of the main objectives of root canal treatment is to disinfect the root canal system of any microbial contamination. The morphological and anatomical complexity of the root canal system and the ecological system of endodontic pathogens require an antimicrobial agent that has the ability to eradicate those pathogens while maintaining their activity inside the root canal system.

Several nanoparticles have been investigated for their potential use as an antimicrobial agent in endodontics. Ibrahim *et al* (2017) classify the antimicrobial nanoparticles into metallic or inorganic nanoparticles and non-organic bioactive glass nanoparticles. Raura *et al* (2020) give more detailed classification for the antimicrobial nanoparticles used in endodontics. They classify the antimicrobial nanoparticles into organic NP, non-organic NP, metallic NP and metal oxide NP.

Different nanoparticles have been evaluated for their antimicrobial efficacy against endodontic pathogens. Silver nanoparticles showed superior antimicrobial properties compared to 2.5% sodium hypochlorite against *E. faecalis* (Afkhami *et al* 2017). However, Rodrigues *et al* (2018) showed that 2.5% sodium hypochlorite showed better antibacterial effect against *E. faecalis* compared to silver nanoparticles. Furthermore, some studies showed that some endodontic pathogens such as *E. faecalis* are able to develop resistance against silver nanoparticles (Salas-Orozco *et al* 2019).

Clinically, the use of silver nanoparticles may cause discoloration (Moazami *et al* 2018) and a potential to produce cytotoxic effect, especially if used in large concentrations (Gomes-Filho *et al* 2010) thus limiting its use in endodontics. As a result of these drawbacks, Shrestha and Kishen (2016) recommend its use as a medicament rather than root canal irrigant.

Chitosan nanoparticle is one of the most commonly investigated non-organic nanoparticles in endodontics. Chitosan is a deacetylated form of chitin (Rinaudo 2006) (figure 8.3). It possesses antimicrobial properties against various Gram-positive as well as Gram-negative pathogens (Divya and Jisha 2018).

In endodontics, chitosan nanoparticles were evaluated for their possible use as a plausible endodontic antimicrobial agent. Chitosan nanoparticles showed significant antibacterial activity against *E. faecalis* when compared to sodium hypochlorite. Furthermore, chitosan nanoparticles showed greater tendency to remove the

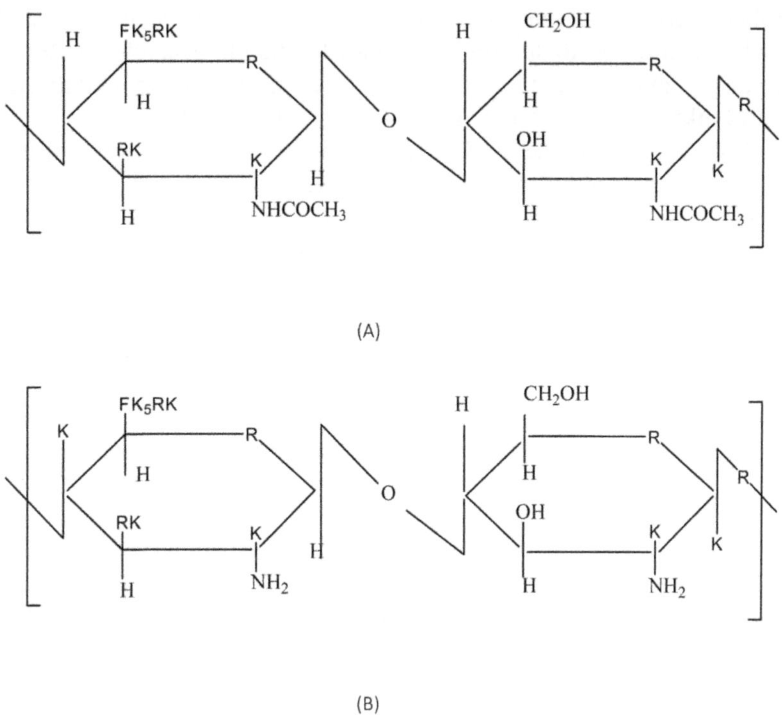

Figure 8.3. The chemical structure of (A) chitin, (B) chitosan in which the acetyl group was removed from chitin (chemical structure adapted from Ravi Kumar (2000)).

smear layer from the dentinal wall which may harbour some microbial species (del Carpio-Perochena *et al* 2015a). Chitosan nanoparticles also possess antibiofilm activity. Shrestha *et al* (2010) evaluated the effect of chitosan nanoparticles against *E. faecalis* biofilm. They showed that chitosan nanoparticles eradicated planktonic cells of *E. faecalis* with significant reduction in its biofilm state after 72 h contact. Unlike other endodontic antimicrobial agents such as calcium hydroxide and chlorhexidine which lose their antimicrobial efficacy when they contact the tissue inhibitory factors like dentine, dentine matrix and lipopolysaccharide (Haapasalo *et al* 2000; Haapasalo *et al* 2007), chitosan nanoparticles were able to maintain their antimicrobial activity in the presence of these factors. However, other factors such as the pulp remnants and serum albumin was found to inactivate the antimicrobial properties of chitosan nanoparticles (Shrestha and Kishen 2012).

Although chitosan is a biocompatible material (Dutta *et al* 2004), some investigators found that chitosan nanoparticles produced a cytotoxic effect against human dental pulp tissue (Alhomrany *et al* 2019).

Other nanoparticles such as magnesium oxide, zinc oxide and iron oxide nanoparticles have been also evaluated for their potential use as antimicrobial agents in endodontics. Iron oxide nanoparticles produced significant antibacterial activity compared to sodium hypochlorite and chlorhexidine, when tested against *E. faecalis* (Bukhari *et al* 2018). Magnesium oxide also showed superior antimicrobial activity against *staphylococcus aureus*, *E. faecalis* and *Candida albicans* when compared to sodium hypochlorite and chlorhexidine (Monzavi *et al* 2015). Similarly zinc oxide nanoparticles showed antibacterial activity against *E. faecalis* in planktonic and biofilm state (Shrestha *et al* 2010).

In conclusion, most of the studies that evaluated the antimicrobial efficacy of nanoparticles against endodontic pathogens evaluate their antimicrobial activity against limited number of pathogens such as *E. faecalis* and *C. albicans*. Further studies are needed to include multiple microbial species.

Furthermore, the use of nanoparticles as antimicrobial agents is aiming to utilize them as either root canal irrigants or intra-canal medicaments. To date some nanoparticles showed superior antimicrobial properties compared to the currently used root canal irrigants. However, the other requirements of root canal irrigants such as ability to dissolve organic tissue and removal of smear layer are still areas of further investigation. The same can be applied to intra-canal medicament.

8.5.1.2 The use of nanoparticles as drug carrier

Nanoparticles can be used as a drug carrier to deliver a specific chemical or biological agent inside the root canal system. This property of the nanoparticle can be achieved by altering or modifying it. Sanvicens and Marco (2008) and Ravindran *et al* (2013) showed that nanoparticles can be modified by either:

a. Altering their surface composition.
b. Altering their surface charge.
c. Altering the structure of the material while maintaining their original properties.

The process of nanoparticle's surface modification is known as functionalization. The newly functionalized nanoparticles can act as a drug carrier and enhance their biological properties such as antimicrobial activity (Shrestha and Kishen 2016).

In endodontics many nanoparticles have been functionalized for different applications. Shrestha *et al* (2014) functionalized chitosan nanoparticles with Rose Bengal (photosensitizer) to be used as photosensitizing agent. The functionalized chitosan nanoparticle could reduce the colony forming unit of *E. faecalis* and *S. mutans* in planktonic state. Furthermore, it completely eliminates a multispecies biofilm of *Streptococcus oralis*, *Prevotella intermedia* and *Actinomyces naeslundii* (Shrestha and Kishen 2014).

Pagonis *et al* (2010) loaded poly(lactic–co–glycolic acid) with methylene blue as photosensitizer and tested against planktonic cells of *E. faecalis*. The resulted functionalized nanoparticles significantly reduce the colony forming unit of *E. faecalis*.

Different types of nanomaterials such as liposomes, which are nano-vesicles with bilayered phospholipid membrane, have been evaluated for their potential use as a drug carrier to treat endodontic infection. Longo *et al* (2012) loaded liposomes with aluminum-chloride-phthalocyanine as photosensitizing agent and tested for their antimicrobial activity against cariogenic pathogens and for their cytotoxic effect toward dental pulp cells. The liposomes loaded aluminum-chloride-phthalocyanine resulted in 82% of the total carious microorganisms tested without reduction in the viable pulp cells.

Most of the studies presented in the literature regarding the use of nanoparticles as a drug delivery system evaluated functionalized nanoparticles with photosensitizing agent for photodynamic therapy. Further investigation is needed to evaluate the potential of using nanoparticles as a drug delivery system for various types of antimicrobial molecules in a different form.

8.5.1.3 The use of nanoparticles as part of root canal filling material

Following biomechanical preparation, the root canal system needs to be filled (obturated) with a suitable filling material. In general, root canal filling material consist of:

a. Core material (such as)
   i. Gutta percha.
   ii. Calcium silicate filling material such as MTA.
b. Root canal sealer

   It seals the gap between the gutta percha and the root canal surface and between the gutta percha itself. It includes:
   i. Zinc oxide eugenol base root canal sealers.
   ii. Calcium hydroxide base root canal sealers.
   iii. Glass ionomer base root canal sealers.
   iv. Resin based root canal sealers.
   v. Calcium silicate based root canal sealers.
   vi. Silicon based root canal sealers.

The objectives of filling the root canal system were listed by Tomson *et al* (2014). These objectives are aiming to:

a. Prevent the ingress of microorganisms into the root canal system coronally.
b. Prevent the ingress of periapical and periodontal fluids from entering the root canal system.
c. Act as a barrier to entomb any remnants of microbial species and prevent their multiplication by blocking the space available for that.

The term biomaterial was defined as ‘any matter, surface or construct that interacts with the biological system’ (Bhushan *et al* 2015). With the introduction of the concept of bioactive dental materials (dental material having biological effect) (Chan *et al* 2018), some nanoparticles were used to modify some of the property of endodontic materials and thus bioactivity of these materials can be achieved. The use of nanoparticles as an active component falls into the category of nano-biomaterial. This is mainly due to their ability to react with the surrounding tissue resulting in different reactions such as:

a. Antimicrobial activity.
b. Ability to adhere to the tooth structure such as bioactive glass nanoparticles.
c. Ability to repair the damaged tissue and regenerate the pulp tissue.

An example of such modification was demonstrated by Kishen *et al* (2008), in which zinc oxide nanoparticles and chitosan nanoparticles were loaded into zinc oxide based root canal sealer. The added nanoparticles improved the antimicrobial properties of the root canal sealer by continuous release on the antimicrobial nanoparticles. Furthermore, it resulted in reducing the ability of *E. faecalis* to adhere to the dentinal wall.

The effect of polymeric nanoparticles such as chitosan nanoparticles was evaluated on epoxy resin and calcium silicate based root canal sealers and tested for their antimicrobial efficacy. Chitosan nanoparticles were found to show antimicrobial properties of the sealers for up to four weeks (del Carpio-Perochena *et al* 2015b).

Similarly, quaternary ammonium polyethylene nanoparticles were added to AH plus and Pulp Canal Sealer EWT (resin based sealer) in two different concentrations (1% and 2%). The added nanoparticles improved the antimicrobial properties of the two sealers when tested against *E. faecalis* (Barros *et al* 2014).

A novel nanoparticle based root canal sealer was developed by (Baras *et al* 2019). They incorporated silver nanoparticles into dimethylaminohexadecyl methacrylate. The novel nanoparticle based root canal sealer showed reduction in colony forming units of microbial biofilm obtained from human saliva. The antimicrobial activity was greater when compared with AH plus sealer (resin sealer).

Silver nanoparticles were also added to MTA based root canal sealer and tested for their antimicrobial properties. The results showed that addition of silver nanoparticles increased the antimicrobial properties of MTA based root canal sealer against *E. faecalis* (Hemmanur and Nasim 2020).

The sealing ability of different core materials like gutta percha and resilon was tested when bioactive glass nanoparticles were added. Mohn *et al* (2010) showed that addition of bioactive glass results in hydroxyapatite formation on the surface of the two materials tested, thus improving their sealing ability.

Different nanoparticles were incorporated into gutta percha to increase its biological and mechanical properties (figure 8.4).

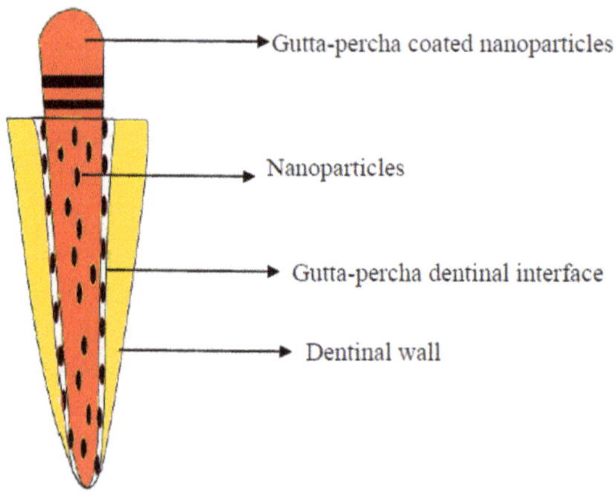

Figure 8.4. Diagram showing the gutta-percha coated nanoparticles adhere to the dentinal surface of the root canal system.

Chitosan and silver nanoparticles were also used to improve the antimicrobial properties of gutta-percha. They were used to coat the surface of gutta-percha and the antimicrobial properties were tested against *E. faecalis*. The results showed that chitosan and silver nanoparticles resulted in elimination of *E. faecalis* as a result of the oxidative stress on the bacterial cells membrane caused by the nanoparticles (Mohan *et al* 2020).

The effect of incorporating nanoparticles into calcium silicate based root canal filling material such as mineral trioxide aggregate (MTA) and Portland cement was also evaluated. Vazquez-Garcia *et al* 2016 incorporated silver nanoparticles into MTA and Portland cement. The aim was to reduce the solubility of the new material and increase the compressive strength and setting time of the material. Furthermore, silver nanoparticles increased the antimicrobial properties of both materials when tested against *E. faecalis*.

The cytotoxic effect of nanoparticles-coated gutta-percha is of biological concern, as one of the main properties of root canal filling material is its biocompatibility. Mozayeni *et al* (2017). They showed that, gutta-percha-coated with silver nanoparticles have an inflammatory response compared to uncoated gutta-percha. However, the inflammatory response reduces over time.

In conclusion, the use of nanoparticles as a part of endodontic filling material results in enhancement of the properties of different types of root canal filling materials biologically and mechanically.

8.5.1.4 The role of nanoparticles in regenerative endodontics

Regenerative endodontics deals with repairing and regeneration of damaged pulp tissue. The process of pulp regeneration involves disinfection of the root canal system. The use of nanoparticles as a antimicrobial agent improves disinfection of the root canal system prior to any regenerative procedure.

There are a limited number of studies that evaluate the role of nanoparticles in regenerative endodontics. Iron oxide nanoparticles were incorporated into calcium phosphate cement scaffold and tested for their effect on human dental stem cells attachment. The results showed that iron oxide nanoparticles improved the attachment of the stem cells into the scaffold and result in bone mineral synthesis (Xia *et al* 2018).

Titanium dioxide nanoparticles also improved dental pulp stem cells proliferation and morphology after seven days contact without eliciting cytotoxic effect. The results also showed that titanium dioxide assists in nucleation of collagen fibers that are necessary for hard tissue formation (Chuang *et al* 2021).

Mesoporous bioactive nanoparticles were also shown to induce odontoblast differentiation from rat dental stem cells and can act as an additive to the dental material to regenerate the damaged dentine (Lee *et al* 2016). The same effect of nanoparticles was demonstrated by Zhang *et al* (2015) where they showed that the incorporation of zinc bioglass nanoparticles into calcium phosphate cement promotes odontoblast differentiation.

Some nanoparticles based dental material also showed some degree of tissue regeneration capability. Tricalcium silicate nanoparticles containing cement (Biodentine™) showed an ability to differentiate odontoblast cells for dental tissue regeneration without eliciting cytotoxic effect (Jung *et al* 2020).

Table 8.3. Applications of nanofibers in the dental and medical field.

Applications of nanofibers in the medical field	Applications of nanofibers in the dental field
• As part of scaffold material for tissue regeneration and engineering • As wound dressing material • As drug carrier material • Biosensing	• As part of scaffold for pulp and periodontal tissue regeneration • As caries preventive agent • To reinforce the resin composite material • To modify the implant surface • For mucosal wound repair and healing • As drug delivery material

In conclusion, nanoparticles seem to promote cell differentiation, but the exact mechanism is not yet clear.

8.5.2 Application of nanofibers in endodontics

Nanofibers is one of the nanomaterials characterized by their large surface area and porosity compared to other nanomaterials, rendering them suitable for many applications. Nanofibers are usually synthesized from polymers either synthetic or natural, carbon-based nanomaterials, composite nanomaterials, semiconducting nanomaterials (Lim 2017). In the dental and medical field, nanofibers are involved in many applications, as listed in table 8.3.

8.5.2.1 Application of nanofibers in regenerative endodontics

One of the main dental applications of nanofibers is as a scaffold material for pulp regeneration. The rationale of using nanofibers as scaffold in regenerative endodontics results from their mimicking the connective tissue matrix that supports the growth of stem cells. Furthermore, the size of the nanofibers (50–500 nm) and their porous structure facilitate the growth of the regenerated tissue (Zein *et al* 2019). Different synthetic and natural nanofiber materials were evaluated for this purpose. Bottino *et al* (2015) synthesized nanofiber based scaffold using polydioxanone and halloysite nanotube and evaluated it for its cells proliferation support. The nanofiber material was found to support the proliferation and attachment of dental pulp derived cells while maintaining a high level of biocompatibility, giving the potential for using it as a bioactive scaffold material for regenerative endodontics.

Fibronectin loaded random and aligned nanofiber scaffold of polycaprolactone were tested against human apical papilla stem cells. The nanofiber scaffold showed high potential to promote the human apical papilla stem cells with different degrees based on the nanofiber used (Leite *et al* 2020).

Some nanofibers can be functionalized with bioactive molecules to produce additional effects. Functionalized nanofiber was evaluated by Bottino *et al* (2013) who synthesized polymer based polydioxanone nanofiber containing antibiotic (metronidazole and ciprofloxacin) and tested for their antimicrobial activity against *Prohyromonasgingivalis* and *Enterococcus faecalis* and cytocompatibility against human dental pulp stem cells. The nanofiber-antibiotic nanofiber scaffold showed antimicrobial activity against the tested pathogens with low cytotoxicity and thus a dual effect (antimicrobial and regenerative role) in regenerative endodontics. Similar results were obtained by Albuquerque *et al* (2017) when tested against microbial biofilm and the ability of the polydioxanone nanofiber containing antibiotics.

In conclusion, various materials are used to synthesize nanofiber. The main role of nanofibers in regenerative endodontics is mainly concentrated in its use as scaffold material. Their use also extended to involving some antimicrobial activity in addition to supporting the stem cell proliferation and differentiation.

8.6 Prospective of nanotechnology in endodontics

The role of microbiology in endodontic disease has been reported in the literature as one of its main etiological factors. The production and use of biologically active dental material that has the ability to eradicate the microbial contamination from the root canal system comprises one of the main factors that ensure the success of such treatment. Various antimicrobial root canal irrigants and intra-canal medicaments using different chemical agents and formulae have been developed to achieve this goal. However, the development of resistant microbial species renders the process of root canal disinfection more challenging and necessitates the development of new biologically active endodontic materials that have the ability to act as antimicrobial agents to overcome the shortcomings of the commonly used endodontic antimicrobial agents. At the same time they have the ability to provide biological activity such as repair of inflamed tissue or even regenerate damaged tissue.

The era of nanotechnology provides the advantages of synthesis of a material at the nano-scale level that possess new properties different from its bulk material. Furthermore, nanomaterials are shown to be highly biologically active giving advantage to their use as biologically active material.

In endodontics, nanomaterial in the form of nanoparticles as a prospective antimicrobial agent against endodontic pathogens has been evaluated extensively in different forms and was proven to be effective. The antimicrobial effectiveness has to be translated into clinical trials. Moreover, the ability of different nanoparticles to enhance tissue regeneration should be evaluated more extensively in terms of developing of nanoparticles that have the ability to induce different stem cells differentiation to specific cells. Furthermore, the production of new endodontic materials at the nano-scale level is an area that needs to be evaluated more.

The literature shows minimal use of nanofibers in endodontics, although some studies evaluated its potential to be used as a scaffold for pulp tissue regeneration. However, the applications of nanofibers in endodontics should be evaluated from a different perspective such as their ability to be incorporated into endodontic materials to enhance their biological activity.

References

Afkhami F, Akbari S and Chiniforush N 2017 Entrococcus faecalis elimination in root canals using silver nanoparticles, photodynamic therapy, diode laser, or laser-activated nanoparticles: an *in vitro* study *J. Endod.* **43** 279–82

Albuquerque M T, Nagata J and Bottino M C 2017 Antimicrobial efficacy of triple antibiotic-eluting polymer nanofibers against multispecies biofilm *J. Endod.* **43** S51–6

Alhomrany R, Zhang C and Chou L 2019 Cytotoxic effect of chitosan nanoparticles on normal human dental pulp cells *Nanosci. Nanotechnol.* **3** 1

Arias-Moliz M, Baca P, Solana C, Toledano M, Medina-Castillo A, Toledano-Osorio M and Osorio R 2021 Doxycycline-functionalized polymeric nanoparticles inhibit Enterococcus faecalis biofilm formation on dentine *Int. Endod. J.* **54** 413–26

Baras B H, Melo M A S, Sun J, Oates T W, Weir M D, Xie X, Bai Y and Xu H H 2019 Novel endodontic sealer with dual strategies of dimethylaminohexadecyl methacrylate and nanoparticles of silver to inhibit root canal biofilms *Dent. Mater.* **358** 1117–29

Barreras U S, Méndez F T, Martínez R E M, Valencia C S, Rodríguez P R M and Rodríguez J P L 2016 Chitosan nanoparticles enhance the antibacterial activity of chlorhexidine in collagen membranes used for periapical guided tissue regeneration *Mater. Sci. Eng.* C **58** 1182–7

Barros J, Silva M, Rodrigues M, Alves F, Lopes M, Pina-Vaz I and Siqueira J Jr 2014 Antibacterial, physicochemical and mechanical properties of endodontic sealers containing quaternary ammonium polyethylenimine nanoparticles *Int. Endod. J.* **47** 725–34

Beloin C, Valle J, Latour-Lambert P, Faure P, Kzreminski M, Balestrino D, Haagensen J A, Molin S, Prensier G and Arbeille B 2004 Global impact of mature biofilm lifestyle on *Escherichia coli* K-12 gene expression *Mol. Microbiol.* **51** 659–74

Bhushan M, Tyagi S, Nigam M, Choudhary A, Khurana N and Dwivedi V 2015 Bioactive materials: a short review *J. Orofac. Res.* **5** 138–41

Bottino M, Kamocki K, Yassen G, Platt J, Vail M, Ehrlich Y, Spolnik K and Gregory R 2013 Bioactive nanofibrous scaffolds for regenerative endodontics *J. Dent. Res.* **92** 963–9

Bottino M C, Yassen G H, Platt J A, Labban N, Windsor L J, Spolnik K J and Bressiani A H 2015 A novel three-dimensional scaffold for regenerative endodontics: materials and biological characterizations *J. Tissue Eng. Regen. Med.* **9** E116–123

Brown M R, Allison D G and Gilbert P 1988 Resistance of bacterial biofilms to antibiotics a growth-rate related effect? *J. Antimicrob. Chemother.* **22** 777–80

Bukhari S, Kim D, Liu Y, Karabucak B and Koo H 2018 Novel endodontic disinfection approach using catalytic nanoparticles *J. Endod.* **44** 806–12

Carrotte P 2004 Endodontics: Part 1 the modern concept of root canal treatment *Br. Dent. J.* **197** 181–3

Chan D C, Chung A K and Paranjpe A 2018 Antibacterial and bioactive dental restorative materials: do they really work? *Am. J. Dent.* **31**(Sp Is B) 3B–5B

Chuang Y-C, Chang C-C, Yang F, Simon M and Rafailovich M 2021 TiO_2 nanoparticles synergize with substrate mechanics to improve dental pulp stem cells proliferation and differentiation *Mater. Sci. Eng.* C **118** 111366

Costerton J W, Stewart P S and Greenberg E P 1999 Bacterial biofilms: a common cause of persistent infections *Science* **284** 1318–22

del Carpio-Perochena A, Bramante C M, Duarte M A H, de Moura M R, Aouada F A and Kishen A 2015a Chelating and antibacterial properties of chitosan nanoparticles on dentin *Restor. Dent. Endod.* **40** 195–201

del Carpio-Perochena A, Kishen A, Shrestha A and Bramante C M 2015b Antibacterial properties associated with chitosan nanoparticle treatment on root dentin and 2 types of endodontic sealers *J. Endod.* **41** 1353–8

Del Fabbro M, Afrashtehfar K I, Corbella S, El-Kabbaney A, Perondi I and Taschieri S 2018 *In vivo* and *in vitro* effectiveness of rotary nickel-titanium vs manual stainless steel instruments for root canal therapy: systematic review and meta-analysis *J. Evid. Based Dent. Pract.* **18** 59–69

Divya K and Jisha M S 2018 Chitosan nanoparticles preparation and applications *Environ. Chem. Lett.* **16** 101–12

Dua K, Singh G, Goel M and Sachdeva G 2012 Endodontic biofilm and its relevance in endodontic infection *Indian J. Dent. Sci.* **4** 110–14

Dutta P K, Dutta J and Tripathi V 2004 Chitin and chitosan: Chemistry, properties and applications *J. Sci. Indus. Res.* **63** 20–31

Evans M, Davies J K, Sundqvist G and Figdor D 2002 Mechanisms involved in the resistance of *Enterococcus faecalis* to calcium hydroxide *Int. Endod. J.* **35** 221–8

Farber P A and Seltzer S 1988 Endodontic microbiology. I. Etiology *J. Endod.* **14** 363–71

Feynman R P 1960 There's plenty of room at the bottom *Eng. Sci.* **23** 22–36

Flemming H C and Wingender J 2010 The biofilm matrix *Nat. Rev. Microbiol.* **8** 623–33

Giannousi K, Lafazanis K, Arvanitidis J, Pantazaki A and Dendrinou-Samara C 2014 Hydrothermal synthesis of copper based nanoparticles: antimicrobial screening and interaction with DNA *J. Inorg. Biochem.* **133** 24–32

Gomes-Filho J E, Silva F O, Watanabe S, Cintra L T A, Tendoro K V, Dalto L G, Pacanaro S V, Lodi C S and de Melo F F F 2010 Tissue reaction to silver nanoparticles dispersion as an alternative irrigating solution *J. Endod.* **36** 1698–702

Gomes B P F A, Pinheiro E T, Gadê-Neto C R, Sousa E L R, Ferraz C C R, Zaia A A, Teixeira F B and Souza-Filho F J 2004 Microbiological examination of infected dental root canals *Oral Microbiol. Immunol.* **19** 71–6

Gupte M and Ma P 2012 Nanofibrous scaffolds for dental and craniofacial applications *J. Dent. Res.* **91** 227–34

Haapasalo H K, Sirén E K, Waltimo T M T, Òrstavik D and Haapasalo M P P 2000 Inactivation of local root canal medicaments by dentine: an *in vitro* study *Int. Endod. J.* **33** 126–31

Haapasalo M, Qian W, Portenier I and Waltimo T 2007 Effects of dentin on the antimicrobial properties of endodontic medicaments *J. Endod.* **33** 917–25

Haapasalo M, Udnæs T and Endal U 2003 Persistent, recurrent, and acquired infection of the root canal system post-treatment *Endod. Topics* **6** 29–56

He L, Kim S G, Gong Q, Zhong J, Wang S, Zhou X, Ye L, Ling J and Mao J J 2017 Regenerative endodontics for adult patients *J. Endod.* **43** S57–64

Hemmanur S and Nasim I 2020 Antibacterial efficacy of nanoparticle-incorporated root canal sealer against common endodontic pathogens-an *in vitro* Study *J. Pharm. Res. Int.* 68–74

Hülsmann M, Peters O A and Dummer P M 2005 Mechanical preparation of root canals: shaping goals, techniques and means *Endod. Topics* **10** 30–76

Ibrahim A, Moodley D, Petrik L and Patel N 2017 Use of antibacterial nanoparticles in endodontics *S. Afr. Dent. J.* **72** 105–12

Jhajharia K, Parolia A, Vikram Shetty K and Mehta L K 2015 Biofilm in endodontics: a review *J. Int. Soc. Prev. Community Dent.* **5** 1–12

Jung Y, Yoon J-Y, Dev Patel K, Ma L, Lee H-H, Kim J, Lee J-H and Shin J 2020 Biological effects of tricalcium silicate nanoparticle-containing cement on stem cells from human exfoliated deciduous teeth *Nanomaterials* **10** 1373

Kakehashi S, Stanley H R and Fitzgerald R J 1965 The effects of surgical exposures of dental pulps in germ-free and conventional laboratory rats *Oral Surg. Oral Med. Oral Pathol. Oral Radiol.* **20** 340–9

Kim S G, Zhou J, Solomon C, Zheng Y, Suzuki T, Chen M, Song S, Jiang N, Cho S and Mao J J 2012 Effects of growth factors on dental stem/progenitor cells *Dent. Clin.* **56** 563–75

Kishen A 2010 Advanced therapeutic options for endodontic biofilms *Endod. Topics* **22** 99–123

Kishen A, Shi Z, Shrestha A and Neoh K G 2008 An investigation on the antibacterial and antibiofilm efficacy of cationic nanoparticulates for root canal disinfection *J. Endod.* **34** 1515–20

Kokare C, Chakraborty S, Khopade A and Mahadik K 2009 Biofilm: Importance and applications *Indian J. Biotechnol.* **8** 159–68

Kolenbrander P E, Palmer R J, Periasamy S and Jakubovics N S 2010 Oral multispecies biofilm development and the key role of cell–cell distance *Nat. Rev. Microbiol.* **8** 471–80

Lee J-H, Kang M-S, Mahapatra C and Kim H-W 2016 Effect of aminated mesoporous bioactive glass nanoparticles on the differentiation of dental pulp stem cells *PLoS One* **11** e0150727

Leite M L, Soares D G, Anovazzi G, Mendes Soares I P, Hebling J and de Souza Costa C A 2020 Development of fibronectin-loaded nanofiber scaffolds for guided pulp tissue regeneration *J. Biomed. Mater. Res.* B **109** 1244–58

Lim C T 2017 Nanofiber technology: current status and emerging developments *Prog. Polym. Sci.* **70** 1–17

Longo J P F, Leal S C, Simioni A R, Almeida-Santos M d F M, Tedesco A C and Azevedo R B 2012 Photodynamic therapy disinfection of carious tissue mediated by aluminum-chloride-phthalocyanine entrapped in cationic liposomes: an *in vitro* and clinical study *Lasers Med. Sci.* **27** 575–84

Love R M 2004 Invasion of dentinal tubules by root canal bacteria *Endod. Topics* **9** 52–65

Loyola-Rodríguez J P, Torres-Méndez F, Espinosa-Cristobal L F, García-Cortes J O, Loyola-Leyva A, González F J, Soto-Barreras U, Nieto-Aguilar R and Contreras-Palma G 2019 Antimicrobial activity of endodontic sealers and medications containing chitosan and silver nanoparticles against *Enterococcus faecalis J. Appl. Biomater. Funct. Mater.* **17** 2280800019851771

Lynch I and Dawson K A 2008 Protein-nanoparticle interactions *Nano Today* **3** 40–7

Moazami F, Sahebi S and Ahzan S 2018 Tooth discoloration induced by imidazolium based silver nanoparticles as an intracanal irrigant *J. Dent.* **19** 280

Mohammadi Z and Dummer P M H 2011 Properties and applications of calcium hydroxide in endodontics and dental traumatology *Int. Endod. J.* **44** 697–730

Mohammadi Z, Palazzi F, Giardino L and Shalavi S 2013 Microbial biofilms in endodontic infections: an update review *Biomed. J.* **36** 59–70

Mohan A, Dipallini S, Lata S, Mohanty S, Pradhan P, Patel P, Makkar H and Verma S 2020 Oxidative stress induced antimicrobial efficacy of chitosan and silver nanoparticles coated Gutta-percha for endodontic applications *Mater. Today Chem.* **17** 100299

Mohn D, Bruhin C, Luechinger N, Stark W J, Imfeld T and Zehnder M 2010 Composites made of flame-sprayed bioactive glass 45S5 and polymers: bioactivity and immediate sealing properties *Int. Endod. J.* **43** 1037–46

Monzavi A, Eshraghi S, Hashemian R and Momen-Heravi F 2015 *In vitro* and *ex vivo* antimicrobial efficacy of nano-MgO in the elimination of endodontic pathogens *Clin. Oral Investig.* **19** 349–56

Mozayeni M A, Dianat O, Tahvildari S, Mozayani M and Paymanpour P 2017 Subcutaneous reaction of rat tissues to nanosilver coated gutta-percha *Iran. Endod. J.* **12** 157

Murray P E, Garcia-Godoy F and Hargreaves K M 2007 Regenerative endodontics: a review of current status and a call for action *J. Endod.* **33** 377–90

Nair P 2004 Pathogenesis of apical periodontitis and the causes of endodontic failures *Crit. Rev. Oral Biol. Med.* **15** 348–81

Nair R P N 1987 Light and electron microscopic studies of root canal flora and periapical lesions *J. Endod.* **13** 29–39

Narayanan L L and Vaishnavi C 2010 Endodontic microbiology *J. Conserv. Dent.* **13** 233–9

Ørstavik D and Haapasalo M 1990 Disinfection by endodontic irrigants and dressings of experimentally infected dentinal tubules *Endod. Dent. Traumatol.* **6** 142–9

Östby B N 1961 The role of the blood clot in endodontic therapy an experimental histologic study *Acta Odontol. Scand.* **19** 323–53

Pagonis T C, Chen J, Fontana C R, Devalapally H, Ruggiero K, Song X, Foschi F, Dunham J, Skobe Z and Yamazaki H 2010 Nanoparticle-based endodontic antimicrobial photodynamic therapy *J. Endod.* **36** 322–8

Peciuliene V, Maneliene R, Balcikonyte E, Drukteinis S and Rutkunas V 2008 Microorganisms in root canal infections: a review *Stomatologija* **10** 4–9

Pelgrift R Y and Friedman A J 2013 Nanotechnology as a therapeutic tool to combat microbial resistance *Adv. Drug Delivery Rev.* **65** 1803–15

Portenier I, Haapasalo H, Rye A, Waltimo T, Ørstavik D and Haapasalo M 2001 Inactivation of root canal medicaments by dentine, hydroxylapatite and bovine serum albumin *Int. Endod. J.* **34** 184–88

Raura N, Garg A, Arora A and Roma M 2020 Nanoparticle technology and its implications in endodontics: a review *Biomater. Res.* **24** 1–8

Ravi Kumar M N V 2000 A review of chitin and chitosan applications *React. Funct. Polymers* **46** 1–27

Ravindran A, Chandran P and Khan S S 2013 Biofunctionalized silver nanoparticles: advances and prospects *Colloids Surf.* B **105** 342–52

Rinaudo M 2006 Chitin and chitosan: properties and applications *Prog. Polym. Sci.* **31** 603–32

Rodrigues C T, De Andrade F, De Vasconcelos L, Midena R, Pereira T, Kuga M, Duarte M and Bernardineli N 2018 Antibacterial properties of silver nanoparticles as a root canal irrigant against *Enterococcus faecalis* biofilm and infected dentinal tubules *Int. Endod. J.* **51** 901–11

Salas-Orozco M, Niño-Martínez N, Martínez-Castañón G-A, Méndez F T, Jasso M E C and Ruiz F 2019 Mechanisms of resistance to silver nanoparticles in endodontic bacteria: a literature review *J. Nanomater.* **2019** 7630316

Saleh T A 2020 Nanomaterials: Classification, properties, and environmental toxicities *Environ. Technol. Innov.* **20** 101067

Samiei M, Farjami A, Dizaj S M and Lotfipour F 2016 Nanoparticles for antimicrobial purposes in Endodontics: A systematic review of *in vitro* studies *Mater. Sci. Eng.* C **58** 1269–78

Sanchez F and Sobolev K 2010 Nanotechnology in concrete–a review *Constr. Build. Mater.* **24** 2060–71

Sanvicens N and Marco M P 2008 Multifunctional nanoparticles–properties and prospects for their use in human medicine *Trends Biotechnol.* **26** 425–33

Sauer K, Camper A K, Ehrlich G D, Costerton J W and Davies D G 2002 Pseudomonas aeruginosa displays multiple phenotypes during development as a biofilm *J. Bacteriol.* **184** 1140–54

Seil J T and Webster T J 2012 Antimicrobial applications of nanotechnology: methods and literature *Int. J. Nanomed.* **7** 2767–81

Shrestha A, Hamblin M R and Kishen A 2014 Photoactivated rose bengal functionalized chitosan nanoparticles produce antibacterial/biofilm activity and stabilize dentin-collagen *Nanomed. Nanotechnol. Biol. Med.* **10** 491–501

Shrestha A and Kishen A 2012 The effect of tissue inhibitors on the antibacterial activity of chitosan nanoparticles and photodynamic therapy *J. Endod.* **38** 1275–8

Shrestha A and Kishen A 2014 Antibiofilm efficacy of photosensitizer-functionalized bioactive nanoparticles on multispecies biofilm *J. Endod.* **40** 1604–10

Shrestha A and Kishen A 2016 Antibacterial nanoparticles in endodontics: a review *J. Endod.* **42** 1417–26

Shrestha A, Zhilong S, Gee N K and Kishen A 2010 Nanoparticulates for antibiofilm treatment and effect of aging on its antibacterial activity *J. Endod.* **36** 1030–35

Siqueira J, Perez A, Marceliano-Alves M, Provenzano J, Silva S, Pires F, Vieira G, Rocas I and Alves F 2017 What happens to unprepared root canal walls: a correlative analysis using micro-computed tomography and histology/scanning electron microscopy *Int. Endod. J,* **51** 501–8

Siqueira J F 2001 Strategies to treat infected root canals *J. Calif. Dental Assoc.* **29** 825–38

Siqueira J F 2002 Endodontic infections: Concepts, paradigms, and perspectives *Oral. Surg. Oral Med. Oral Pathol. Oral Radiol. Endod.* **94** 281–93

Siqueira J F and Rôças I N 2014 Present status and future directions in endodontic microbiology *Endod. Topics* **30** 3–22

Siqueira J F, Rôças I N and Ricucci D 2010 Biofilms in endodontic infection *Endod. Topics* **22** 33–49

Siqueira J F and Sen B H 2004 Fungi in endodontic infections *Oral. Surg. Oral Med. Oral Pathol. Oral Radiol. Endod.* **97** 632–41

Stojicic S, Shen Y and Haapasalo M 2013 Effect of the source of biofilm bacteria, level of biofilm maturation, and type of disinfecting agent on the susceptibility of biofilm bacteria to antibacterial agents *J. Endod.* **39** 473–7

Stuart C H, Schwartz S A, Beeson T J and Owatz C B 2006 *Enterococcus faecalis*: its role in root canal treatment failure and current concepts in retreatment *J. Endod.* **32** 93–8

Sundqvist G and Figdor D 2003 Life as an endodontic pathogen *Endod. Topics* **6** 3–28

Svensäter G and Bergenholtz G 2004 Biofilms in endodontic infections *Endod. Topics* **9** 27–36

Tiwari J N, Tiwari R N and Kim K S 2012 Zero-dimensional, one-dimensional, two-dimensional and three-dimensional nanostructured materials for advanced electrochemical energy devices *Prog. Mater Sci.* **57** 724–803

Tomson R M, Polycarpou N and Tomson P 2014 Contemporary obturation of the root canal system *Br. Dent. J.* **216** 315–22

Torabinejad M, Shabahang S, Aprecio R M and Kettering J D 2003 The antimicrobial effect of MTAD: an *in vitro* investigation *J. Endod.* **29** 400–3

Tronstad L and Sunde P T 2003 The evolving new understanding of endodontic infections *Endod. Topics* **6** 57–77

Ulusoy A T, Turedi I, Cimen M and Cehreli Z C 2019 Evaluation of blood clot, platelet-rich plasma, platelet-rich fibrin, and platelet pellet as scaffolds in regenerative endodontic treatment: a prospective randomized trial *J. Endod.* **45** 560–66

Usha H 2010 Biofilm in endodontics: new understanding to an old problem *Int. J. Contemp. Dent.* **1** 44–51

Valko M, Leibfritz D, Moncol J, Cronin M T, Mazur M and Telser J 2007 Free radicals and antioxidants in normal physiological functions and human disease *Int. J. Biochem. Cell Biol.* **39** 44–84

Vaudt J, Bitter K and Kielbassa A M 2007 Evaluation of rotary root canal instruments *in vitro*: a review *Endod. Pract. Today* **1** 189–203

Vazquez-Garcia F, Tanomaru-Filho M, Chávez-Andrade G M, Bosso-Martelo R, Basso-Bernardi M I and Guerreiro-Tanomaru J M 2016 Effect of silver nanoparticles on physicochemical and antibacterial properties of calcium silicate cements *Br. Dent. J.* **27** 508–14

Waltimo T M T, Sirén E K, Torkko H L K, Olsen I and Haapasalo M P P 1997 Fungi in therapy-resistant apical periodontitis *Int. Endod. J.* **30** 96–101

Wang D, Lin Z, Wang T, Yao Z, Qin M, Zheng S and Lu W 2016 Where does the toxicity of metal oxide nanoparticles come from: the nanoparticles, the ions, or a combination of both? *J. Hazard. Mater.* **308** 328–34

Wang Z, Shen Y and Haapasalo M 2012 Effectiveness of endodontic disinfecting solutions against young and old Enterococcus faecalis biofilms in dentin canals *J. Endod.* **38** 1376–9

Xia Y, Chen H, Zhang F, Wang L, Chen B, Reynolds M A, Ma J, Schneider A, Gu N and Xu H H 2018 Injectable calcium phosphate scaffold with iron oxide nanoparticles to enhance osteogenesis via dental pulp stem cells *Artif. Cells Nanomed. Biotechnol.* **46**(sup 1) 423–33

Zein N, Harmouch E, Lutz J-C, Fernandez De Grado G, Kuchler-Bopp S, Clauss F, Offner D, Hua G, Benkirane-Jessel N and Fioretti F 2019 Polymer-based instructive scaffolds for endodontic regeneration *Materials* **12** 2347

Zhang J, Park Y-D, Bae W-J, El-Fiqi A, Shin S-H, Lee E-J, Kim H-W and Kim E-C 2015 Effects of bioactive cements incorporating zinc-bioglass nanoparticles on odontogenic and angiogenic potential of human dental pulp cells *J. Biomater. Appl.* **29** 954–64

Chapter 9

Nanotechnology and prosthodontics: current trends and future

Ivy Fernandes Coutinho, Kathleen Manuela D'Souza, Meena Ajay Aras and Cora Abigail Coutinho

The pursuance of ideal restorative materials in prosthodontics has led to the exploration of nanomaterials which have immense potential after their scale was reduced from micron-size into nano-size by nanotechnology. The ongoing research in the realm of nano is due to the unique structure and properties of nanoparticles, thus gaining high impetus in their application. There are more atoms on the surface of the nanoparticle which present with unbound surfaces compared to the core of the nanoparticle. This gives them the potential to create new and strong bonds and makes them more reactive when compared to macro- or micro-particles which have more core than surface atoms and are easy to manipulate and use. Many materials used in prosthodontics like composites, metals, ceramics, resins, implants, dental adhesives, dental cements and maxilla-facial prosthesis have become much more potent in their properties such as modulus of elasticity, surface hardness, polymerization shrinkage, and filler loading after the addition of the nanomaterials.

This chapter gives the readership a detailed description of nanomaterials and their up-to-date applications as well as their future in prosthodontics.

9.1 Introduction

Prosthodontics is an important branch in dentistry dealing with dental and oro-facial defects. This branch provides rehabilitative treatment after tooth loss using inlays, onlays, crowns, fixed dental prostheses, removable dental prostheses and artificial prostheses for temporo-mandibular joint disorders and maxillofacial defects.

Nanomaterials is the buzzword in today's technology and it has captured significant attention in the field of oral health because of its unique structure and features and the similarity in the nanostructure of the tooth surface and its organic components [1].

doi:10.1088/978-0-7503-3671-0ch9

Since natural tooth is a biological nanomaterial composed of enamel, dentine and cementum with nanoscale particles, the field of nanotechnology has shown emerging nano materials with distinct biological properties like antifungal, antiviral and antimicrobial which can be used for applications in prosthodontics, restorative dentistry and implantology [1–3]. Nanoparticles in composites improve the mechanical properties of a restorative material and increase their bonding between the tooth and biomaterial. They also prevent crack propagation and improve the fracture toughness in dental ceramics [4, 5]. Although incorporation of nanoparticles in biomaterials is effective, *in vivo* results with long-term data are needed [6].

This chapter provides an in-depth study of application of nanomaterials like nanoresins, nanoceramics, nanometals and other nanomaterials in prosthodontics, and parts of this chapter have been extracted from the article 'Nanomaterials and their application in Prosthodontics: A Review' [7].

9.2 Poly (methylmethacrylate) (PMMA)

Poly (methylmethacrylate) (PMMA) resin is widely used as a denture base material. It is also routinely used in making orthodontic appliances. However, some drawbacks have been identified, namely surface porosities in PMMA, making the denture base prone to plaque accumulation and inadequate mechanical properties allowing frequent denture base fracture [8, 9].

Recent advances report incorporation of nanoparticles to achieve a nanocomposite possessing enhanced material properties [10]. Various nanoparticles such as ZrO_2, TiO_2, and carbon nanotubes have been used to improve the performance of PMMA, and the results showed that desired mechanical property enhancement can be achieved in those composites with small amounts of nanoparticles [11, 12].

Inclusion of carbon nanotubes into heat cure monomer has decreased polymerization shrinkage and has enhanced the mechanical properties. An improvement in the mechanical properties of the PMMA matrix material was noted when carbon fibers were replaced by carbon nanotubes and carbon nanofibrils, provided care was taken to allow uniform dispersion of the small diameter particles in the composite. Nanodiamonds (NDs) belonging to the nanocarbon family have proven advantageous over other metal/metal-oxide nanoparticles, exhibiting improved chemically stability and biocompatibility [13]. Addition of graphene, a carbon-based nanomaterial to autopoymerized PMMA helped to enhance mechanical and antibacterial properties [14].

Similarly, incorporation of metal-oxide nanoparticles like silver, platinum, titanium and iron have shown increase in flexural strength, antimicrobial properties, surface hydrophobicity, viscoelasticity, decrease in porosity and biomolecular adherence [15–18]. Incorporation of silver nanoparticles in acrylic resins exhibited an antimicrobial activity by releasing bioactive silver ions that interact with the bacterial cell membranes, however, a colour change was noted in concentrations above 80 ppm and cytotoxity in concentrations more than 40 ppm [17].

Addition of 0.4% TiO_2 nanoparticles into 3D printed PMMA denture base has shown significant antibacterial effects, especially against Candida species, and has also improved mechanical properties as well as thermal conductivity [19]. Increase in concentration of nanoparticles beyond the recommended limit can result in TiO_2 agglomeration and loose attachment of cluster formations, leading to a negative impact on the material properties. Silicon dioxide nanoparticles (SNPs) in low concentrations presented marked improvement in the mechanical properties and allowed uniform dispersion of nanoparticles, preventing SNPs agglomeration [19]. Arun Kumar *et al* reported increase in impact strength in PMMA incorporated with TiO_2 nanoparticles processed by microwave technique [20].

Zirconia nanoparticles have helped to improve mechanical strength and avoid premature fatigue and fracture failures in denture base acrylic resin [11, 21–23]. Addition of zirconium dioxide nanoparticles in heat cure PMMA has shown an increase in abrasion resistance, impact strength, fracture toughness and biocompatibility. On the other hand, a decrease in water sorption, solubility, and porosity was reported. However, the properties of the resulting nanocomposite are dependent on the concentration of the nanofillers [19]. Although the addition of zirconium dioxide nanoparticles provides good esthetic acceptability and minimal effect on the color of the final nanocomposite, it has been reported that the translucency decreases with the increase in content of nano zirconium oxide [11, 22]. Similarly, it was recorded that a concentration of zirconia within the range of 1.5wt.% to 5 wt.% should be followed to maximize the bond strength between heat-cure PMMA denture base impregnated with zirconia nanoparticles and acrylic teeth [24]. The results of the study by Ahmed *et al* using heat cure PMMA with 7% nano zirconium oxide has shown enhanced hardness levels, flexural strength and fracture toughness [23]. According to Gad *et al*, 2 wt% or 5 wt% nano zirconium oxide has increased the transverse strength of dentures repaired with auto polymerized resin [25]. They also suggested that cold-cured resin reinforced with 7.5 wt% zirconia nanoparticles had an inhibitory effect on adhesion of *candida albicans* to repaired PMMA dentures and interim removable prostheses [26]. Mohammed and Mudhaffar evaluated the addition of different percentages of zirconium oxide nanofillers (2 wt%, 3 wt% and 5 wt%) heat-cured PMMA. These nanofillers were modified by introducing reactive groups onto the filler surface following reaction with 3-tri (methoxysilyl) propylmethacrylate. It was noted that there was a significant increase in the abrasive wear resistance as well as the tensile and fatigue strength with 3 wt% and 5 wt% of nanofillers, compared to pure PMMA materials. Also, a decrease in water sorption, solubility and in porosity of heat-cure denture base resin was noted [11].

9.3 Composites

Composite resins are used widely as restorative materials as they have numerous benefits such as biocompatibility, esthetics and are easy to use. However, reduced longevity has been observed due to polymerization shrinkage, incomplete enveloping of the dentin matrix and biodegradation. The organic components of composite resins have been constant over a period of time. Addition of inorganic fillers has improved the properties, like increased wear strength, anti-bacterial properties, decreased polymerization shrinkage and good mechanical characteristics [27, 28].

Nanofillers of size 1–100 nm in the form of nanomers and nanoclusters have been incorporated into the resin matrix to produce nanocomposites [29].

Nanomers are monodispersed, non-aggregated and non-agglomerated particles of silica treated with 3-methacryloxy-propyltrimethoxysilane (MPTS). MPTS has helped in chemical bonding of the nanomer filler to the resin whilst curing [30]. They have good optical properties and dispersion rate, high polish and polish retention, superior hardness, flexural strength and good esthetics but showed poor rheological and handling properties [31].

Nanoclusters: These fillers range from 2–20 nm. They have been formed by lightly sintering nanomeric oxides to form clusters of a controlled particle size distribution. Nanoclusters of silica sol exclusively and mixed oxides of silica and zirconia have been synthesized [31]. They have shown the same advantages as nanomers with better rheological properties but the disadvantage of poor handling still persisted [31].

Pre-polymerized organic fillers when incorporated in nanomers are called nano-hybrid composites. They show improved esthetics and desirable rheological properties but reduced polish retention and surface gloss [31, 32].

Titanium dioxide nanoparticles treated with organosilaneallytriethoxysilane (ATES) have been used to improve microhardness and flexural strength of composites [33]. Alumina nanoparticles have shown increased hardness, strength and modulus of elasticity of the nanocomposites [34].

Composites containing Hydroxyapatite nanoparticles (HApNPs) that release calcium phosphate and calcium fluoride have shown remineralization of tooth structure [35, 36]. They have also maintained the level of calcium (Ca) and phosphate (P) ion release through recharge and release and hence have been called a 'smart' material whereby they have inhibited secondary caries [37–39].

Ormocers (Organically modified ceramics) consist of a polysiloxane backbone used for glass and ceramics. Iron oxide, titanium oxide and aluminium sulfosilicate pigments have been added for shade. These nanoceramic particles prevent micro-crack propagation [40]. Addition of ZrO_2 nanoparticles or functionalization with 4-Methylthiophenol to ormocers showed high translucency and no cytotoxic effects and proved to have a potential for 3D-printing of indirect restorations [41].

Nanocomposite denture teeth have homogenously distributed nanofillers and polymethyl methacrylate. They have shown high durability and polishability, increased shear strength, superior esthetics and higher abrasion resistance [42].

15 wt% (Zn-MSNs) zinc doped mesoporous silica nanoparticles exhibited improved mechanical properties (flexural strength, flexural modulus, compressive strength and hardness) and antibacterial properties with no compromise on the cytocompatibility of the composite resins, curing depth, degree of conversion and shrinkage [43].

Ca^{+2} doped mesoporous silica nanoparticles (MCS) in dental composites also improved mechanical properties of the resin and showed the ability to induce apatite-mineralization and inhibit bacterial growth. They also responded to wear and induced remineralization when the resin composites are worn [44].

Addition of silver nanoparticles to composites had an impact on the degree of conversion (DC) and mechanical properties. It was shown that the higher the silver nanoparticle concentration, the lower the DC. Among the polymerization methods used, the thermal polymerization resulted in the highest level of curing over photopolymerization. Hardness showed the least sensitivity to the DC, followed by the modulus, bending strength, and impact resistance [45].

Addition of bioactive glass in various concentrations of 5%, 10% and 15% and fixed concentration of 0.009% silver nanoparticles in resin based composites (RBC) having 70% fillers also showed decrease in the mechanical properties, like flexural strength, flexural modulus and hardness when compared with those of the commercial resin based composites. However, these composites could still be used in restoring class III, V lesions where stresses due to masticatory forces are less and for restoring deciduous teeth [46]. Nano-sized bioactive glass particles improved the alkalizing and hydroxyapatite-forming potential of composites with no negative effects on their DC and microhardness [47].

Graphene oxide zirconia (GO-ZrO_2) and hydroxyapatite (HA-ZrO_2) nanofillers in composites were shown to be well suited for restoration in load-bearing areas. The measured mechanical properties as well as materials' reliability were within or even above the mean values reported in the literature for materials of the same category [48]. The fracture toughness of Bis-GMA/TEGDMA dental resins improved about 35% by the inclusion of 5 wt% silica-loaded core–shell hybrid nanoparticles (polybutyl acrylate (PBA) rubbery core and polymethyl methacrylate [PMMA)/polystyrene (PS) shell] without sacrificing the flexural strength and flexural modulus [49].

Polybenzoxazine based composites reinforced with zirconia particles (60 wt % micro and 10 wt % nano-particles) could be used as a novel biomaterial owing to its excellent biocompatibility, good mechanical properties, appropriate viscosity and low rate of polymerization shrinkage [50].

Denture base composites surface modified with nano-silver loaded zirconium phosphate 6S-NP3 by using 3D printer printing exhibited increased flexural strength, flexural modulus and impact strength and good antibacterial properties against either *S. aureus* or *E. coli* [51].

Ag-BGN@MSN silver-doped bioactive glass/mesoporous silica nanoparticle effectively occluded the exposed dentinal tubules by forming a membrane-like layer. It presented acid-resistant stability, no cytotoxity, did not affect the microtensile bond strength between dentine and the adhesive system and also showed antibacterial effect [52].

The intermolecular crosslinks between polylactic acid (PLA) and Al_2O_3 nanoparticles Al_2O_3/PLA nanoscaffolds and also CaO/PLA nanoscaffolds significantly improved the mechanical properties like compressive strength, flexural strength and flexural modulus when these scaffolds were incorporated in dental nanocomposites [53, 54].

The composite containing brushite nanoparticles ($CaHPO_4{\cdot}2H_2O$) showed high mechanical properties after 28 days in water, maintaining fracture strength similar to the microhybrid composites used in severe occlusal load areas [55].

Nanocomposite sealants with synthetic strontium fluoride (SrF_2) nanoparticles, yttrium stabilized zirconia (YSZ) nanoparticles, and poly-ε-L-lysin (ε-PL) showed increased antibacterial activity against *Streptococcus mutans*, but caused mechanical weakening of the sealants that was partly compensated by incorporation of YSZ nanoparticles (up to 10 wt%). Based on physio-mechanical and antibacterial properties, the sealant composed of 5 wt% SrF_2, 5 wt% YSZ and 0.5 wt% ε-PL was selected as the optimal sealant as it showed good fluoride release and moderate cytotoxity [56].

9.4 Dental adhesives

Dental adhesives are materials that have been used to enhance adhesion and cohesion of two different substances or between a material and natural tooth structure. Polymerizable silane has been added to increase the cohesive strength. Discrete silane treated 5–7 nm size nanoparticles of silica and zirconia have been added to overcome the inconsistency in their performance due to settling of the filler particles during storage [57]. Zirconia has shown an added advantage of giving radiopacity to the adhesive. Nano adhesives have increased the bond strength to enamel and dentine, have longer shelf life, no separate etching required and have shown a durable marginal seal [57], for example, Adper Single Bond 2 and Adper Single Bond Plus (3M ESPE) [58].

Incorporation of silanized silica NPs into adhesives exhibited higher bond strength and mechanical properties with very small filler content of 0.2–0.5 wt% [59]. Nano-amorphous calcium phosphate has shown significant remineralization ability for up to three weeks without affecting the bond strength to dentine [60]. Addition of nitrogen-doped titanium dioxide nanoparticles to dental adhesive resins have shown strong antibacterial activity and hence are capable of preventing secondary caries [61]. Addition of 5/0.2 wt% of zinc oxide and copper (ZnO/Cu NP) to universal adhesive resulted in an increase in the antibacterial activity with no adverse effects on the biological, adhesive and mechanical properties [62]. Incorporation of silver nanoparticles to self-etch and etch-and-rinse systems has shown to reduce the degradation of resin–dentin bonds over a period of six months [63]. Poly (methacrylic acid) modified hybrid nanofillers in dentin adhesive resulted in enhanced shear bond strength to human dentin [64].

9.5 Nano-ceramics

Nano-ceramics have shown improved toughness, ductility, toughness and strength as compared to conventional ceramics [65]. Nano-glass ceramics have exhibited good translucency, excellent corrosion resistance, higher hardness and low modulus of elasticity when produced with the sol–gel method of a zirconia–silica system [66]. Nano-TiO_2 ceramics have shown higher toughness and hardness as compared to traditional TiO_2 ceramics [67]. Highly translucent zirconia materials have been introduced due to the development of ceramic materials to meet the patient preferences for all-ceramic restorations. The surface of highly translucent zirconia ($mZrO_2$) nanoparticles by dipping followed by sintering resulted in 1.3 times higher strength. 20% nanozirconia ceramics showed improved fracture toughness and hardness [68, 69].

According to results from an *in vitro* study, ceramics with up to 4% carbon nanotubes have significantly improved wear and mechanical properties [70]. Lava Ultimate Resin Nano Ceramic (RNC) blocks (3M ESPE) are innovative new CAD/CAM materials with superior esthetic results, durability and fracture resistance [71, 72]. Thermal coating of Y-TZP ceramic with hydroxy apatite nanoparticles (100 nm) enhanced their binding characteristics [73]. Feldspethic ceramic reinforced with titanium and silver nanoparticles improved fracture toughness [74]. Y-TZP enhanced with 5% TiO_2-nano tubes showed high shear bond strength values [75].

9.6 Tissue conditioners and soft liners

Tissue conditioners or soft liners are used to treat an inflamed and abused mucosa supporting a denture and allow them to return to normal. They are degradable with time and occasionally susceptible to microbial colonization. Thus, addition of silver nanoparticles in these materials has displayed anti-microbial properties against *S. mutans* and *S. aureus* at 0.1% and *C. Albicans* at 0.5% after 24 h incubation period [76].

Silver nanoparticles addition to soft silicone liner improved its antifungal properties but adversely affected its tensile bond strength to denture acrylic resin [77]. Tissue conditioners with Ag-zeolite (SZ) continuously released Ag ions making the tissue conditioner anti microbial [78].

The incorporation of 5 wt% Ag vanadate decorated on Ag nanoparticles ($AgVO_3$) to a soft denture liner was efficient in the control of *P. Aeruginosa*, *E. Faecalis*, and *C. Albicans*, whereas, 2.5 wt% and 10 wt% improved the adhesion properties between the liner and the denture base material [79]. 15 wt% ZnO nanoparticles incorporated into the tissue conditioner provided an antifungal effect up to two weeks without adverse effects on the penetration depth and tensile bond strength of the tissue conditioner and incorporation of ZnO–Ag nanoparticles into tissue conditioners inhibited bacterial proliferation [80, 81]. Addition of concentration of 5%–10% of chitosan nanopolymer completely inhibited the growth of all microorganisms [82]. Also, the addition of 5% concentration of Ag, ZnO, chitosan nanoparticles to tissue conditioners in complete dentures, reduced fungal and bacterial growth [83].
Solutions of chlorhexidine nanoparticles mixed with sodium triphosphate (TP), trimetaphosphate (TMP) or hexametaphosphate (HMP) provided antifungal property on silicone soft liners and obturators and chlorhexidine-HMP coating exhibited the most effective antifungal agent, thus enhancing the life of the prosthesis [84].

Incorporating licorice root extract-based fluorescent carbon nanoparticles FCNs, which are known for their antioxidant and antibacterial properties into acrylic based tissue conditioner had no adverse effects on their mechanical properties. A silicone-based soft denture liner, however, showed decreased tear strength and tensile bond strength but was within clinically acceptable limits [85].

9.7 Dental cements

Innovations and modifications by incorporating nanoparticles as fillers have enhanced the properties of cements. Glass ionomer cement with up to 5% w/w TiO_2 nanoparticles has shown increased fracture toughness, flexural strength, compressive strength and anti-bacterial activity [86]. Also, addition of 5% TiO_2 nanotubes has significantly increased the noncollagenous composition of extra cellular matrix which has improved the micro-hardness and fluoride release without influencing the surface roughness, hence it could be considered for restorations in higher stress bearing sites [87].

Nano light cure glass ionomer cement has hydroxyapatite/fluoro-alumino silicate technology and has been used for core build-up in prosthodontics owing to its significantly high compressive, tensile and biaxial flexural strength. It is commercially available as Ketac N100 light curing nano-ionomer (3M ESPE). Addition of nanosized apatite crystals has been shown to improve the bond strength to the tooth, and make the cement more stable and enhance fluoride release [88, 89].

Poly quaternary ammonium salt containing glass ionomer cement has shown significant antibacterial activity and high strength [90]. A new nanofilled resin modified glass ionomer with 65% nanofillers has been used to restore primary teeth and small cavities in permanent teeth [91]. Silver nanoparticles reinforced resin luting cements and restorative GIC exhibited long term inhibitory effect against *S. mutans* and favourable mechanical properties [92, 93].

Hybrid resin luting cements with less than 2.5% of 7 nm silica nanofillers have also shown improved properties [94]. TiO_2 nanotube reinforced self adhesive resin cement has been used in luting indirect dental restorations as it has exhibited superior physicochemical and biological properties compared to the unreinforced cements without cellular viability compromise [95].

Nanoparticle impregnated luting cements have increased the bond strength to enamel and dentine. Compressive and tensile strength of zinc polycarboxylate has been enhanced with addition of ZnO and MgO nanoparticles [96]. A novel crown cement containing nanosized amorphous calcium phosphate promoted remineralization of the tooth structure through release of Ca^{2+} and $PO4^{3-}$ ions [97].

9.8 Impression materials

Impression materials are available now with nanomaterials. Nanofillers in poly vinyl siloxane (PVS) have shown good flow, improved hydrophilic properties thus improving the wettability, and shown superior detail precision [98, 99]. These nanofilled silicone impression materials have shown a high degree of fluidity compared to the original viscosity. They have been designed to give a snap set with fewer errors caused by micro-movements. Addition of silver nanoparticles of 80–100 nm in a concentration of 0.5%–1% to irreversible hydrocolloid resulted in superior antimicrobial activity [100].

9.9 Tooth whitening agents

The majority of the dental bleaching agents are based on hydrogen peroxide or other reactive oxygen species such as carbamide peroxide which oxidizes discolorations, or chromophores, inside the teeth. Carbamide peroxide is an agent composed by hydrogen peroxide and urea. The most classic hydrogen peroxide-based whitening agents are effective, but can cause enamel demineralization, gingival irritation, or cytotoxicity and are excessively time-consuming. It is stated that the application of hydrogen peroxide just up to a maximum of 6% may be safe under well-defined conditions when applied following the relevant treatment guidelines.

To improve bleaching efficiency while avoiding the use of hydrogen peroxide above 6%, research aims to develop novel bleaching technologies [122]. Bleaching gels containing TiO_2 or TiO_2/Ag show less cytotoxicity than hydrogen peroxide-containing gel alone [123]. The clinical efficacy of the whitening gels was improved with incorporation of nanoparticles up to 5% of real carbamide peroxide. There was no *in situ* pulp damage by histological evaluation. The nanotechnology strategy of converting carbamide peroxide into polymeric nanoparticles revealed a new product with improved stability, a good approach for carbamide peroxide delivery [124].

The degradation of organic dyes via piezo-catalysis of $BaTiO_3$ (BTO) nanoparticles was done under ultrasonic vibration to simulate daily tooth brushing and teeth stained with black tea, blueberry juice, wine or a combination thereof could be notably whitened by the poled BTO turbid liquid after vibration for 3 h with remarkably less damage to both enamel and biological cells [125].

Addition of nanoparticles like calcium peroxide to the tooth whitening agents decreased the harmful effects and increased the whitening efficacy. These nanoparticles penetrate deeper into the tooth structure through micro- and nano-cracks and make longer surface contact and thus increase the efficiency of the bleaching agent [126]. Low concentration (6%) hydrogen peroxide bleaching gel catalyzed by titanium oxide nanoparticles and activated with a hybrid light (LED blue/laser infrared) showed good colour stable results at the end of one year when compared to the conventional 35% concentration of hydrogen peroxide [126].

Blue light activated nano-TiO_2@PDA (polydopamine) showed ideal whitening effect and exhibited no obvious damage on enamel surface, thus making it possible for an essentially highly effective and nondestructive tooth whitening agent [127].

9.10 Dental implants

Bacterial colonization and infection is documented to be the most common cause for dental implant failure. Thus, several micro- and nano-scale modifications, nanosurface coatings and functionalization have been suggested to enhance osseointegration in dental implants, by increasing the surface roughness of the dental implant, corrosion resistance, altering the surface energy or by varying the surface composition by adding different elements [101].

There is documented research that indicates that osteogenic cells respond to chemical and morphological surface nano-scale changes. The nanosurface reduces fibroblast proliferation, can increase the adsorption of proteins, stimulate the osteogenic cell differentiation, migration and proliferation, bone matrix formation, promoting osseointegration. In addition, controlled nanotopographies on electrochemically anodized titanium implants involving titania nanotubes (TNTs) have been researched to accommodate and elute potent antibiotic agents. This promotes desirable bioactivity between Ti implant and bone [102–104]. Moreover, nanorough Ti and nanostructured Ti showed better results compared to nanosmooth surfaces [105].

Literature suggests various subtracting and additive methods for creating nano-features on dental implants. All these techniques aim to increase the roughness, surface area and energy of the implant surface. The bioactive coatings can be applied either by dissolution and precipitation technique (ceramic coatings), by chemisorptions of active organic coating (self-assembly of monolayer), injection of high energy ions on the surface or injection of ions by kinetic energy (plasma spraying, sputtering, ion deposition), chemical methods (anodic oxidation, acid treatment, alkali treatment, sol–gel treatment, chemical vapor deposition) and nanoparticle deposition (sol–gel, discrete crystalline deposition, lithography and contact printing technique) [106, 107]. Among these, the most popular techniques include acid etching, plasma spraying, anodization, sand blasting as well as hydroxyapatite coating [108].

In a recent study, superhydrophobic and anti-biofilm coating on Ti dental implants was engineered by a one-step non-thermal technique using a low-pressure plasma technology. This coating was created to reduce polymicrobial adhesion (bacterial and fungal) and biofilm formation, increased surface roughness, enhanced corrosion resistance and improved pattern of protein adsorption, without affecting fibroblast growth and proliferation. Glow discharge plasma using Ar, O_2 and hexamethyldisiloxane gases (surface functionalizing agent) were used to create the superhydrophobic coating [109].

Etching procedures for dental implants can modify the roughness of dental implants at nanoscale level. These procedures, in combination with grit-blasting result in a nanostructured surface of dental implants [107].

During manufacturing of anodized titanium implants, the TiO_2 nanotubes are aligned at nanoscale level by modifying the anodic oxidation process. Various factors affect the nanotopography of the Ti surface during anodic oxidation, namely, the electric current of the electrochemical cell, temperature, the pH values of electrolyte solution, electrolyte, oxidation voltage and oxidation time. Electrochemical methods at different parameters of voltage, time and immersion solution determine the chemical composition, properties and dimensions (diameter and length) of the nanotubes. TiO_2 nanotube arrays are known to accelerate apatite formation by transformation of amorphous TiO_2 into the crystalline form. Huang *et al* reported that strontium (Sr) and silicon (Si) can be simultaneously incorporated into the titania nanotubes (TN) to enhance corrosion resistance and bone formation around these Sr–Si codoped TN-coated Ti implants. This was carried out by anodization and hydrothermal treatment [110].

Nanoparticles are ultrafine particles that exhibit a bone regenerative property. They can be applied as particle coatings to the implant surface to promote soft tissue integration. Hydroxyapatite nano-coated implants encourage a chemical bond with bone, act as an enriched source of calcium and phosphate and present with higher surface area and higher reactivity, producing a biomimetic coating on implants. Hence, they result in enhanced osseointegration, improved biological fixation and high implant stability. Contrary to the aforementioned advantages, some adverse effects such as instability because of nano-coating have also been recorded. Thus, the long-term stability of the HA/bone interface is still under investigation [108].

Surface functionalization with protein, peptides and bioactive compounds is an effective approach to design new multifunctional materials by changing composition, structure and/or morphology of surfaces without altering the bulk properties. The surface topographical changes as well as the modification in the physicochemical composition help to enhance the biomechanical anchorage and osseointegration of the implant at the histological level. The incorporation of bioactive molecules into the nanostructured surfaces has reduced early and late implant failures induced by the biofilm accumulation [107].

9.11 Maxillo-facial prosthesis

The most common reason for re-fabrication of silicone-based maxillofacial prostheses includes degradation of color and mechanical properties. Thus, there has been a need for materials that optimally match the patients' natural appearance, exhibit mechanical and physical properties that enhance the patients' function and are compatible with human tissue and skin. This has promoted the incorporation of nanofillers in organic polymer matrix to develop nanocomposites with superior mechanical properties such as high tear strength, low hardness and increased flexibility as well as improved biological and optical characteristics [111].

Han *et al* reported that incorporation of nanosized oxides of titanium, zinc, or cerium at concentrations of 2.0% and 2.5% improved the overall mechanical properties of the silicone A-2186 maxillofacial elastomer by offering strength to the flexible organic polymer matrix. These nano-oxides have higher surface energy than polymeric matrix and chemical reactivation of these particles promotes their interaction with the silicone elastomer, forming a three-dimensional network within the silicone polyethylene backbone [112].

Nanoparticles of rutile-titanium dioxide (TiO_2), fumed silica (SiO_2), silaned silica, zinc oxide (ZnO) have been successfully added to commercial silicone-based elastomer used for the fabrication of maxillofacial prostheses to alter material properties. These nanosized particles by their active function and strong interfacial interaction with the organic polymer also help to provide resistance to cracking and ageing caused by environmental stress [113, 114]. Shakir and Abdul-Ameer reported that incorporation of 0.25 wt% and 0.2wt% TiO_2 nanofiller into VST50F and Cosmesil M511 elastomers, respectively, resulted in improved tear strength, tensile strength, elongation percentage and hardness of the materials [115].

Incorporation of 0.2 wt% TiO_2 nanofiller and 0.25wt% intrinsic pigment in Cosmesil M511, an HTV silicone elastomer, and 0.25 wt% TiO_2 nanofiller and 0.25wt% intrinsic pigment in VST50F, an RTV silicone elastomer yielded significant improvement in the tear strength with slight increase in Shore A hardness [116]. Coating of silicone prosthetic materials with silver nanoparticles exhibited antifungal properties against candida albicans and had no toxic effect on the human dermal fibroblast cells [117].

A systematic review conducted by Barman *et al*, on the influence of filler particles on the mechanical properties of maxillofacial prosthetic silicone elastomers, reported that 1.5% $ZrSiO_4$, 3% SiO_2, 1.5% Y_2O_3, 2%–6% TiO_2, 2%–2.5% ZnO,

2%–2.5% CeO_2, 0.5% $TiSiO_4$ and 1% Ag-Zn zeolite can be used to reinforce these materials, and help them to better withstand mechanical degradation [118].

9.12 Alloys used in removable partial denture connectors

Titanium alloy and chrome cobalt used for skeletal prostheses in removable partial dentures and for fixed bases for crowns and bridges have poor corrosion resistance, affect tooth mobility and may cause inflammation of gingival. Hence, the use of titanium and zirconia nanoparticles have significantly reduced the number of adherent bacteria. Also, cobalt and cobalt oxide nanoparticles have been shown to have bactericidal properties [119–121].

9.13 Conclusion

Nanotechnology has a strong potential to revolutionize its applications in prosthodontics. Many properties like modulus of elasticity, surface hardness, polymerization shrinkage and filler loading of materials used in prosthodontics can be significantly enhanced after their scales were reduced from micro- to nano-size. In prosthodontics, nanotechnology built up its place by improving strength in resin and impression material, improving esthetics in composites and ceramics, providing a strong base for prostheses and enhancing osseointegration of implants.

References

[1] Zafar M S, Khurshid Z, Najeeb S, Zohaib S and Rehman I U 2017 Nanostructures for oral medicine, micro and nano technologies, *Therapeutic Applications of Nanotechnology in Dentistry* ed E Andronescu and A M Grumezescu (Amsterdam: Elsevier) pp 833–62

[2] Roy I, Stachowiak K and Bergey E J 2008 Non-viral gene transfection nanoparticles: function and applications in brain *Nanomedicine* **4** 89–97

[3] Li X, Feng Q, Cui R, Liu W, Sun L, Yu B, Fan Y, Feng Q, Cui F and Watari F 2013 The use of nanoscaled fibres or tubes to improve biocompatibility and bioactivity of biomedical materials *J. Nanomater.* **2013** 728130

[4] Salas M, Lucena C, Herrera L J, Yebra A, Bona A B and Pérez M M 2018 Translucency thresholds for dental materials *Dent. Mater.* **34** 1168–74

[5] Mirghani I, Mushtaq F, Allsop M J, Al-Saud L M, Tickhill N, Potter C and Manogue M 2018 Capturing differences in dental training using a virtual reality simulator *Eur. J. Dent. Educ.* **22** 67–71

[6] Burns J, McGoldrick N and Muir M 2018 Oral health literacy, oral health behaviours and dental outcomes *Evid. Based. Dent.* **19** 69–70

[7] Coutinho I F, Aras M A and D'souza K M 2018 Nanomaterials and their application in prosthodontics: a review *J. Res. Dent.* **6** 124–31

[8] Price C A 1994 A history of dental polymers *Aust. Prosthodont. J.* **8** 47–54

[9] Lessa F C R, Enoki C, Ito I Y, Faria G, Matsumoto M A N and Nelson- Filho P 2007 In-vivo evaluation of the bacterial contamination and disinfection of acrylic baseplates of removable orthodontic appliances *Am. J. Orthod. Dentofacial. Orthop.* **131** 705.e11

[10] Gad M M, Abualsaud R, Al-Thobity A M, Baba N Z and Al-Harbi F A 2020 Influence of addition of different nanoparticles on the surface properties of poly(methylmethacrylate) denture base material *J. Prosthodont.* **29** 422–28

[11] Mohammed B D S D and Mudhaffar B D S M 2012 Effect of modified zirconium oxide nano-fillers addition on some properties of heat cure acrylic denture base material *J. Bagh. Coll. Dent.* **24** 1–7

[12] Cooper C A, Ravich D, Lips D, Mayer J and Wagner H D 2002 Distribution and alignment of carbon nanotubes and nanofibrils in a polymer matrix *Compos. Sci. Technol.* **62** 1105–112

[13] Fouda S M, Gad M M, Ellakany P, Al-Thobity A M, Al-Harbi F A, Virtanen J I and Raustia A 2019 The effect of nanodiamonds on candida albicans adhesion and surface characteristics of pmma denture base material—an in-vitro study *J. Appl. Oral. Sci* **27** e20180779

[14] Bacali C *et al* 2020 Flexural strength, biocompatibility, and antimicrobial activity of a polymethyl methacrylate denture resin enhanced with graphene and silver nanoparticles *Clin. Oral. Investig.* **24** 2713–25

[15] Acosta- Torres L S, Lopez- Marin L M, Nunez- Anita R E, Hernandez-Padron G and Castano V M 2011 Biocompatible metal oxide nanoparticles: nanotechnology improvement of conventional prosthetic acrylic resins *J. Nanomater.* **2011** 941561

[16] Acosta- Torres L S, Mendieta I, Nunez- Anita R E and Castano V M 2012 Cytocompatible antifungal acrylic resin containing silver nanoparticles for dentures *Int. J. Nanomed.* **7** 4777–86

[17] Monteiro D R, Gorup L F, Takamiya A S, de Camargo E R, Filho A C and Barbosa D B 2012 Silver distribution and release from an antimicrobial denture base resin containing silver colloidal nanoparticles *J. Prosthodont.* **21** 7–15

[18] Totu E E, Nechifor A C, Nechifor G, Aboul- Enein H Y and Cristache C M 2017 Polymethymethacrylate with TiO_2 nanoparticles inclusion for stereographic complete denture—the future in dental care for elderly edentulous patients? *J. Dent.* **59** 68–77

[19] Gad M M, Fouda S M, Al-Harbi F A, Näpänkangas R and Raustia A 2017 PMMA denture base material enhancement: a review of fiber, filler, and nanofiller addition *Int. J. Nanomedicine.* **12** 3801–12

[20] Arun Kumar C, Ravi Kumar C, Vamshikiran K, Deepthi G, Naveen Kumar G and Akhilesh M 2019 Evaluation of impact strength of dental acrylic resins by incorporation of TiO_2 nanoparticles using two different processing techniques *J. Contemp. Dent. Pract.* **20** 1184–89

[21] Zidan S, Silikas N, Alhotan A, Haider J and Yates J 2019 Investigating the mechanical properties of ZrO_2-impregnated PMMA nanocomposite for denture-based applications *Materials (Basel)* **12** 1344

[22] Gad M M, Abualsaud R, Rahoma A, Al- Thobity A M, Al Abidi K S and Akhtar S 2018 Effect of zirconium oxide nanoparticles addition on the optical properties of polymethyl methacrylate denture base material *Int. J. Nanomedicine.* **13** 283–92

[23] Ahmed M A and Ebrahim M I 2014 Effect of zirconium oxide nanofillers addition on the flexural strength, fracture toughness and hardness of heat polymerized acrylic resin World *J. Nano. Sci. Eng.* 4 50–7

[24] Zidan S, Silikas N, Haider J, Alhotan A, Jahantigh J and Yates J 2020 Assessing tensile bond strength between denture teeth and nano-zirconia impregnated PMMA denture base *Int. J. Nanomed.* 15 9611–25

[25] Gad M, ArRajaie A S, Abdel-Halim M S and Rahoma A 2016 The reinforcement effect of nano-zirconia on the transverse strength of repaired acrylic denture base *Int. J. Dent.* **2016** 7094056

[26] Gad M M, Al-Thobity A M, Shahin S Y, Alsaqer B T and Ali A A 2017 inhibitory effect of zirconium oxide nanoparticles on candida albicans adhesion to repaired polymethyl methacrylate denture bases and interim removable prostheses: a new approach for denture stomatitis prevention *Int. J. Nanomed.* **12** 5409–19

[27] Opdam N J M, Collres K, Hickel R, Bayne S C, Loomans B A, Cenci M S and Wilson N H F 2018 Clinical studies in restorative dentistry: new directions and new demands *Dent. Mater.* **34** 1–12

[28] Donovan T E, Marzola R, Murphy K R, Cagna D R, Eichmiller F, McKee J R and Troeltzsch M 2018 Annual review of selected scientific literature: a report of the committee on scientific investigation of the American Academy of Restorative Dentistry *J. Prosthet. Dent.* **120** 816–78

[29] Chen M H 2010 Update on dental composites *J. Dent. Res.* **89** 549–60

[30] Mitra S B, Wu D and Holmes B N 2003 An application of nanotechnology in advanced dental materials *J. Am. Dent. Assoc.* 134 1382–90

[31] Pal K S, Ranganath L M, Gaikwad A V, Sarapur S and Jain S K 2015 Nanoparticles in prosthodontics—boon or bane *IJOCR.* **3** 32–9

[32] Swift E J 2005 Ask the experts-nanocomposites *J. Esthet. Restor. Dent.* **17** 3–4

[33] Xia Y, Zhang F, Xie H and Gu N 2008 Nanoparticle—reinforced resin based dental composites *J. Dent.* **36** 450–5

[34] Al-Haik M, Hanson C, Luhrs C, Tehrani M and Miltenberger S 2009 Mechanical performanance of dental fillers based on alumina nanoparticles *Proc. SEM. Ann Conf.* (Albuquerque, NM: Society for Experimental Mechanics Inc.) 1992–99

[35] Vollenweider M, Brunner T J, Knecht S, Grass R N, Zenhder N, Imfeld T and Stark W J 2007 Remineralisation of human dentine using ultrafine bioactive glass particles *Acta Biomater.* **3** 936–43

[36] Xu H H K, Weir M D and Sun L 2007 Nanocomposites with Ca and PO_4 release, effects of reinforcement, di-calcium phosphate particle size and silanization *Dent. Mater.* **23** 1482–91

[37] Xie X, Wang L, Xing D, Arola D D, Weir M D, Bai Y and Xu H H K 2016 Protein-repellant and antibacterial functions of a calcium phosphate rechargeable nanocomposite *J. Dent.* **52** 15–22

[38] Jardim R N, Rocha A A, Rossi A M, Neves A A, Portela M B, Lopes R T, Dos Santos T M P, Xing Y and da Silva E M 2020 Fabrication and characterization of remineralizing dental composites containing hydroxyapatite nanoparticles *J. Mech. Behav. Biomed. Mater.* **109** 103817

[39] Li L, Pan H, Tao J, Xu X, Mao C, GuX and Tang R 2008 Repair of enamel by using hydroxyapatite nanoparticles as the building blocks *J. Mater. Chem.* **18** 4079–84

[40] Malathi S, Sujatha V and Mahalaxmi S 2014 Nanotechnology: an asset to dentistry!!! *Int. J. Comm. Dent.* **5** 27–31

[41] Kolb C, Gumpert K, Wolter W and Sextl G 2020 Highly translucent dental resin composites through refractive index adaption using zirconium dioxide nanoparticles and organic functionalization *Dent. Mater.* **36** 1332–42

[42] Chandki R, Kala M, Kumar K N, Brigit B, Banthia P and Banthia R 2012 Nanodentistry—exploring the beauty of the miniature *J. Clin. Exp. Dent.* **4** e119–124

[43] Bai X, Lin C, Wang Y, Ma J, Wang X, Yao X and Tang B 2020 Preparation of Zn doped mesoporous silica nanoparticles (Zn-Msns) for the improvement of mechanical and antibacterial properties of dental resin composites *Dent. Mater.* **36** 794–807

[44] Zhang Y, Huang C and Chang J 2018 Ca-doped mesoporous SiO_2/dental resin composites with enhanced mechanical properties, bioactivity and antibacterial properties *J. Mater. Chem. B.* **6** 477–86

[45] Barszczewska-Rybarek I and Chladek G 2018 Studies on the curing efficiency and mechanical properties of Bis-GMA and TEGDMA nanocomposites containing silver nanoparticles *Int. J. Mol. Sci.* **19** 3937

[46] Hanif A and Ghani F 2020 Mechanical properties of an experimental resin based composite containing silver nanoparticles and bioactive glass *Pak. J. Med. Sci.* **36** 776–81

[47] Odermatt R, Par M, Mohn D, Wiedemeier D B, Attin T and Taubock T 2020 Bioactivity and physico-chemical properties of dental composites functionalized with nano- vs. micro-sized bioactive glass *J. Clin. Med.* **9** 772

[48] Ilie N, Sarosi C, Rosu M C and Moldovan M 2021 Synthesis and characterization of graphene oxide-zirconia (GO-ZrO_2) and hydroxyapatite-zirconia (HA-ZrO_2) nano-fillers for resin-based composites for load-bearing applications *J. Dent.* **105** 103557

[49] Roshanali M, Nodehi A and Atai M 2020 Synthesis and characterization of core–shell nanoparticles and their application in dental resins *Mech. Behav. Biomed. Mater.* **110** 103926

[50] Lotfi L, Javadpour J and Naimi-Jamal M R 2018 Biological and nano-indentation properties of polybenzoxazine-based composites reinforced with zirconia particles as a novel biomaterial *Biomed. Mater. Eng.* **29** 369–87

[51] Liao W, Zheng S, Chen S, Zhao L, Huang X, Huang L and Kang S 2020 Surface silanization and grafting reaction of nano-silver loaded zirconium phosphate and properties strengthen in 3D-printable dental base composites *J. Mech. Behav. Biomed. Mater.* **110** 103864

[52] Jung J H, kim D H, Yoo K H, Yoon S Y, kim Y, Bae M K, Chung J, Ko c.c., Kwon Y H and Kim Y L 2019 Dentin sealing and antibacterial effects of silver-doped bioactive glass/mesoporous silica nanocomposite: an in-vitro study *Clin. Oral. Investig.* **23** 253–66

[53] Ranjbar M, Noudeh G D, Hashemipour M A and Mohamedzad I 2019 A systematic study and effect of PLA/Al_2O_3 nanoscaffolds as dental resins: mechanochemicalproperties *Artif. Cells. Nanomed. Biotechnol.* **47** 201–9

[54] Ranjbar M, Pardakhty A, Tahmipour B and Mohamadzadeh I 2019 Novel Cao/polylactic acid nanoscaffold as dental resin nanocomposites and the investigation of physicochemical properties *Luminescence* **34** 360–67

[55] Rodriguesa M C, Chiaria M D S, Alaniaa Y, Natalea L C, Arana- Chaveza V E, Mierb M M, Fadalb V S, Vichic F M, Hewerd L R and Bragaa R R 2018 Ion-releasing dental restorative composites containing functionalized brushite nanoparticlesfor improved mechanical strength *Dent. Mater.* **34** 746–55

[56] Hesaraki S, Karimi M and Nezafati N 2020 The synergistic effects of SrF_2 nanoparticles, YSZ nanoparticles, and poly-ε-L-lysin on physicomechanical, ion release, and antibacterial-cellular behavior of the flowable dental composites *Mater. Sci. Eng. C. Mater. Bio. Appl.* **109** 110592

[57] Silikas N, Masouras K, Satterthwaite J and Watts D C 2007 Effects of nanofills in adhesives and aesthetic properties of dental resin composites *Int. J. Nano. Biomater.* **1** 116–27

[58] Abiodun-Solanke I M F, Ajayi D M and Arigbede A O 2014 Nanotechnology and its applications in dentistry *Ann. Med. Health. Sci. Res.* **4** S171–77

[59] Azad E, Atai M, Zandi M, Shokrollahi P and Solhi L 2018 Structure—properties relationships in dental adhesives: effect of initiator, matrix monomer structure, and nano-filler incorporation *Dent. Mater.* **34** 1263–70

[60] Zhang L, Weir M D, Hack G, Fouad A F and Xu H H 2015 Rechargeable dental adhesive with calcium phosphate nanoparticles for long term ion release *J. Dent.* **43** 1587–95

[61] Florez F L, Hiers R D, Larson P, Johnson M, O'Rear E, Rondinone A J and Khajotia S S 2018 Antibacterial dental adhesive resins containing nitrogen-doped titanium dioxide nanoparticles *Mater. Sci. Eng. C. Mater. Biol. Appl.* **939** 31–43

[62] Gutiérrez M F *et al* 2019 Biological, mechanical and adhesive properties of universal adhesives containing zinc and copper nanoparticles *J. Dent.* **8** 245–55

[63] Jowkar Z, Shafiei F, Asadmanesh E and Koohpeima F 2019 Influence of silver nanoparticles on resin-dentin bond strength durability in a self-etch and an etch-and-rinse adhesive system *Restor. Dent. Endod.* **44** e13

[64] Solhi L, Atai M, Nodehi A and Imani M 2020 Poly (methacrylic acid) modified spherical and platelet hybrid nanoparticles as reinforcing fillers for dentin bonding systems: synthesis and properties *J. Mech. Behav. Biomed. Mater.* **109** 103840

[65] Wang W, Liao S, Liu M, Zhao Q and Fu Y 2015 Recent applications of nanomaterials in prosthodontics *J. Nanomater.* **2015** 408643

[66] Persson C, Unosson E, Ajaxon I, Engstrand J, Engqvist H and Xia W 2012 Nano grain sized zirconia-silica glass ceramics for dental applications *J. Eur.Ceram. Soc.* **32** 4105–10

[67] Raj V and Sirajudeen M M 2014 Formation and surface characterization of nanostructured Al_2O_3–Tio_2 coatings *Bull. Mater. Sci.* **37** 1141–48

[68] Fujii Y, Okada M, Taketa H and Matsumoto T 2021 Coating method for smooth and reinforced surface layer of highly translucent zirconia *Dent. Mater. J.* **40** 532–37

[69] Li C H, Hou Y L, Liu Z R and Ding Y C 2011 Investigation in temperature field of nano-zirconia ceramics precision grinding *Int. J. Abras. Technol.* **4** 77–89

[70] An J W, You D H and Lim D S 2003 Tribological properties of hot-pressed alumina-CNT composites *Wear* **255** 677–81

[71] Chen C, Trinidade F Z, de Jager N, Kleverlaan C J and Feilzer A J 2014 The fracture resistance of a CAD/CAM resin nano ceramic (RNC) and a CAD ceramic at different thickness *Dent. Mater.* **30** 954–62

[72] Koller M, Arnetzl G V, Holly L and Arnetzl G 2012 Lava ultimate resin nanoceramic for CAD/CAM: customization case study *Int. J. Comput. Dent* **15** 159–64

[73] Azari A, Jamnani S N, Yazdani A, Atri F, Rasaie V and Yazdi A F 2017 Deposition of crystalline hydroxyapatite nanoparticles on Y-TZP ceramic: a potential solution to enhance bonding characteristics of Y-TZP ceramics *J. Dent.* **14** 62

[74] Karthikeyan V, Chander N G, Reddy J R and Muthukumar B 2019 Effectsof incorporation of silver and titanium nanoparticles on feldspathic ceramic toughness *J. Dent. Res. Dent. Clin. Dent. Prospects.* **13** 98–102

[75] Magalhães A P, Ramos-Tonello C M, Galli M Z, Gomes O P, Pacheco L E, Fortulan C A, Lisboa-Filho P N, Furuse A Y and Borges A F 2020 Yttria-stabilized tetragonal zirconia polycrystal/resin luting agent bond strength: influence of titanium dioxide nanotubes addition in both materials *J. Prosthodont. Res.* **64** 408–16

[76] Nam K Y 2011 In vitro antimicrobial effect of the tissue conditioner containing silver nanoparticles *J. Adv. Prosthodont.* **3** 20–4

[77] Habibzadeh S, Omidvaran A, Eskandarion S and Shamshiri A R 2020 Effect of incorporation of silver nanoparticles on the tensile bond strength of a long term soft denture liner *Eur. J. Dent.* **14** 268–73

[78] Samaranayake Y H and Samaranayake L P 2001 Experimental oral candidiasis in animal models *Clin. Microbiol. Rev.* **14** 398–429

[79] Krevel S, Oliveira V C, Bachmann L, Alves O L and Dos Reis A 2019 Influence of $AgVO_3$ Incorporation on antimicrobial properties, hardness, roughness and adhesion of a soft denture liner *Sci. Rep.* **9** 11889

[80] Homsiang W, Komankhantikul K, Arksornnukit M and Takahashi H 2021 Effect of zinc oxide nanoparticles incorporated into tissue conditioner on antifungal, physical, and mechanical properties *Dent. Mater. J.* **40** 481–86

[81] Mousavi S A, Ghotaslou R, Akbarzadeh A, Azima N, Aeinfar A and Khorramdel A 2019 Evaluation of antibacterial and antifungal properties of a tissue conditioner used in complete dentures after incorporation of Zno–Ag nanoparticles *J. Dent. Res. Dent. Clin. Dent. Prospects.* **13** 11–18

[82] Mousavi S A, Ghotaslou R, Kordi S, Azin, Khoramdel A, Aeenfar A, Kahjough S T and Akbarzadeh A 2018 Antibacterial and antifungal effects of chitosan nanoparticles on tissue conditioners of complete dentures *Int. J. Biol. Macromol.* **118** 881–5

[83] Mousavi S A, Ghotaslou R, Khorramdel A, Akbarzadeh A and Aeinfar A 2020 Antibacterial and antifungal impacts of combined silver, zinc oxide, and chitosan nano-particles within tissue conditioners of complete dentures *in vitro* IR *J. Med. Sci.* **189** 1343–50

[84] Garner S J, Nobbs A H, McNally L M and Barbour M E 2015 An antifungal coating for dental silicones composed of chlorhexidine nanoparticles *J. Dent.* **43** 362–72

[85] Ataol A S, Ergun G, Cekic-Nagas I, Alas M O and Genc R 2020 The effects of adding fluorescent carbon nanoparticles on variousmechanical properties of denture liners *Dent. Mater. J.* Dec **24** Online ahead of print

[86] Elsaka S E, Hamouda I M and Swain M V 2011 TiO_2 nanoparticles addition to conventional glass ionomer restoratives: influence on physical and antibacterial properties *J. Dent.* **34** 589–98

[87] Cibim D D *et al* 2017 Novel nanotechnology of TiO_2 improves physical-chemical and biological properties of glass ionomer cement *Int. J. Biomater.* **2017** 7123919

[88] Najeeb S, Khurshid Z, Zafar M S, Khan A S, Zohaib S, Martí J M N, Sauro S, Matinlinna J P and Rehman I U 2016 Modifications in glass ionomer cements: nano-sized fillers and bioactive nanoceramics *Int. J. Mol. Sci.* **17** 1134

[89] Moshavernia A, Ansari S, Movasaghi Z, Billington R W, Darr J A and Rehman I U 2008 Modification of conventional glass ionomer cements with n-vinylpyrrolidone containing polyacids, nano-hydroxy and fluoroapatite to improve the mechanical properties *Dent. Mater.* **24** 1381–90

[90] Xie D, Weng Y, Guo X, Hhao J, Gregory R L and Zheng C 2011 Preparation and evaluation of a novel glass-ionomer cement with antibacterial functions *Dent. Mater.* **27** 487–96

[91] Baheti M J and Toshniwal N G 2014 Nanotechnology: a boon to dentistry *J. Dent. Sci. Oral. Rehab.* **5** 78–88

[92] Dola S, Kumar T V, SubbaRao P S, Ravikumar R V S S N and Gopala Krishna A 2017 Green synthesis and characterization of Ag nanoparticles from mangifera indica leaves for dental restoration and antibacterial application *Prog. Biomater.* **6** 57–66

[93] Yoshida K, Tanagawa M, Matsumoto S, Yamada T and Atsuta M 1999 Antibacterial activity of resin composites with silver containing materials *Euro. J. Oral. Sci.* **107** 290–96

[94] Habekost L V, Camacho G B, Lima G S, Ogliari F A, Cubas G B and Moraes R R 2012 Nanoparticle loading level and properties of experimental hybrid resin luting agents *J. Prosthodont.* **21** 540–45

[95] Ramos-Tonello C M, Lisboa-Filho P N, Arruda L B, Tokuhara C K, Oliviera R C, Furuse A Y, Rubo J H and Borges A F S 2017 Titanium dioxide nanotubes addition to self-adhesive resin cement: effect on physical and biological properties *Dent. Mater.* **33** 866–75
[96] Karimi M A, Roozbahani S H, Reza A, Hatefi A, Mashhadizadeh M H, Behjatmanesh-Ardakani R, Mohammad M A, Kargar H and Zebarjad S 2011 Synthesis and characterization of nanoparticles and nanocomposites of ZnO and MgO by sonochemical method and their application for zinc polycarboxylate dental cement preparation *Int. Nano. Lett.* **1** 43–51
[97] AlSahafi R, Balhaddad A A, Mitwalli H, Ibrahim M S, Melo M A S, Oates T W, Xu H H K and Weir M D 2020 Novel crown cement containing antibacterial monomer and calcium phosphate nanoparticles *Nanomaterials (Basel)* **10** 2001
[98] Padmasree S, Vaishnavi R and Sridharan R 2018 OSC45: comparative evaluation of physical properties of nanoparticle incorporated addition silicone with other elastomeric impression materials *J. Indian. Prosthodont. Soc.* **18** S27–8
[99] Jhaveri H M and Balaji P R 2005 Nanotechnology: the future of dentistry *J. Indian. Prosthodont. Soc.* **5** 15–17
[100] Ginjupalli K, Shaw T, Tellapragada C, Alla R, Gupta L and Perampalli N U 2018 Does the size matter? Evaluation of effect of incorporation of silver nanoparticles of varying particle size on the antimicrobial activity and properties of irreversible hydrocolloid impression material *Dent. Mater.* **34** e158–65
[101] Matos G R M 2021 Surface roughness of dental implant and osseointegration *J. Maxillofac. Oral. Surg.* **20** 1–4
[102] Chopra D, Gulati K and Ivanovski S 2021 Understanding and optimizing the antibacterial functions of anodized nano-engineered titanium implants *Acta. Biomater.* **127** 80–1
[103] Yeo I L 2019 Modifications of dental implant surfaces at the microand nano-level for enhanced osseointegration *Materials (Basel)* 13 89
[104] Yao C, Slamovich E B and Webster T J 2008 Enhanced osteoblast functions on anodized titanium with nanotube-like structures *J. Biomed. Mater. Res.* A **85** 157–66
[105] Puckett S, Pareta R and Webster T J 2008 Nano rough micron patterned titanium for directing osteoblast morphology and adhesion *Int. J. Nanomed.* **3** 229–41
[106] Praveena C, Chaughule R S and Satyanarayana K V 2021 Advances in dental implantology using nanomaterials and allied technology applications *Nanotechnology in Implant Dentistry* ed R S Chaughule and R Dashaputra (Cham: Springer) pp 9–17
[107] Souza J C M, Sordi M B, Kanazawa M, Ravindran S, Henriques B, Silva F S, Aparicio C and Cooper L F 2019 Nano-scale modification of titanium implant surfaces to enhance osseointegration *Acta. Biomater.* **94** 112–31
[108] Yazdani J, Ahmadian E, Sharifi S, Shahi S and Maleki Dizaj S 2018 A short view on nanohydroxyapatite as coating of dental implants *Biomed. Pharmacother.* **105** 553–7
[109] Souza J G S *et al* 2020 Targeting pathogenic biofilms: newly developed superhydrophobic coating favors a host-compatible microbial profile on the titanium surface *ACS Appl. Mater. Interfaces.* **12** 10118–29
[110] Huang Y, Shen X, Qiao H, Yang H, Zhang X, Liu Y and Yang H 2018 Biofunctional Sr- and Si-loaded titania nanotube coating of Ti surfaces by anodization-hydrothermal process *Int. J. Nanomed.* **13** 633–40
[111] Zayed S M, Alshimy A M and Fahmy A E 2014 Effect of surface treated silicon dioxide nanoparticles on some mechanical properties of maxillofacial silicone elastomer *Int. J. Biomater.* **2014** 750398

[112] Han Y, Kiat-amnuay S, Powers J M and Zhao Y 2008 Effect of nanooxide concentration on the mechanical properties of a maxillofacial silicone elastomer *J. Prosthet. Dent.* **100** 465–73
[113] Akay C, Cevik P, Karakis D and Sevim H 2018 In vitro cytotoxicity of maxillofacial silicone elastomers: effect of nano-particles *J. Prosthodont.* **27** 584–87
[114] Abdalqadir M, Faraj S and Azhdar B 2021 An evaluation of a technique to improve the mechanical properties of maxillofacial silicone elastomers with zinc oxide nanoparticles *J. Prosthet. Dent.* in press
[115] Shakir D A and Abdul-Ameer F M 2018 Effect of nano-titanium oxide addition on some mechanical properties of silicone elastomers for maxillofacial prostheses *J. Taibah. Univ. Med. Sci.* **13** 281–90 Erratum in: *J.Taibah. Univ. Med. Sci.***13** 407
[116] Abdul-Ameer F M 2020 Impact of a mixture of nanofiller and intrinsic pigment on tear strength and hardness of two types of maxillofacial silicone elastomers *Dent. Res. J. (Isfahan)* **17** 251–57
[117] Meran Z, Besinis A, De Peralta T and Handy R D 2018 Antifungal properties and biocompatibility of silver nanoparticle coatings on silicone maxillofacial prostheses *Vitro J. Biomed. Mater. Res. B. Appl. Biomater.* **106** 1038–51
[118] Barman A, Rashid F, Farook T H, Jamayet N B, Dudley J, Yhaya M F B and Alam M K 2020 The influence of filler particles on the mechanical properties of maxillofacial prosthetic silicone elastomers: a systematic review and meta-analysis *Polymers (Basel)* **12** 1536
[119] Arafa K A O 2016 Comparing the effects of titanium alloy and chrome cobalt in removable partial denture connectors on tooth mobility, bone loss and tissue reaction *Saudi. J. Dent. Res.* **7** 112–7
[120] Moradpoor H, Safaei M, Rezaei F, Golshah A, Jamshidy L, Hatam R and Abdullah R S 2019 Optimisation of cobalt oxide nanoparticles synthesis as bactericidal agents *Open. Access. Maced. J. Med. Sci.* **7** 2757–62
[121] Iravani S and Varma R S 2020 Sustainable synthesis of cobalt and cobalt oxide nanoparticles and their catalytic and biomedical applications *Green. Chem.* **22** 2643–61
[122] Joiner A 2006 The bleaching of teeth: a review of the literature *J. Dent.* **34** 412–19
[123] Kurzmann C, Verheyen J, Coto M, Kumar R V, Divitini G, Shokoohi-Tabrizi H A, Verheyen P, De Moor R J G, Moritz A and Agis H 2019 *In vitro* evaluation of experimental light activated gels for tooth bleaching *Photochem. Photobiol. Sci.* **18** 1009–19
[124] Lima F V, Mendes C, Zanetti-Ramos B G, Nandi J K, Cardoso S G, Bernardon J K and Silva M A S 2019 Carbamide peroxide nanoparticles for dental whitening application: characterization, stability and *in vivo/in situ* evaluation *Colloids. Surf.* B **179** 326–33
[125] Wang Y, Wen X, Jia Y, Huang M, Wang F, Zhang X, Bai Y, Yuan G and Wang Y 2020 Piezo-catalysis for nondestructive tooth whitening *Nat. Commun.* **11** 1328
[126] Vildosola P, Bottner J, Avalos F, Godoy I, Martin J and Fernandez E 2017 Teeth bleaching with low concentrations of hydrogen peroxide (6%) and catalyzed by LED blue (450 ± 10 nm) and laser infrared (808 ± 10 nm) light for in-office treatment: randomized clinical trial 1-year follow-up *J. Esthet. Restor. Dent.* **29** 339–45
[127] Zhang F, Wu C, Zhou Z, Wang J, Bao W, Dong L, Zhang Z, Ye Jing, Liao L and Wang X 2018 Blue-light-activated nano-TiO_2@PDA for highly effective and nondestructive tooth whitening *ACS. Biomater. Sci. Eng.* **4** 3072–77

IOP Publishing

Nanotechnology for Dentistry Applications

Ashutosh Kumar Shukla

Chapter 10

Nanotechnology in cancer—theragnostics

Divya Uppala, Neeharika Mortha, Anupama Bonam and B Alekhya

Nanotechnology as a rapidly evolving domain has changed the very foundation of cancer diagnosis, treatment and prevention. Since everything is made up of atoms and molecular matter, nanotechnology has provided a way to rearrange and restructure matter on the atomic scale thus allowing us to reach down to the very roots of any problem. This emerging field is a multidiscipline science, essential for electronic engineering, drug-delivery, optical computer tomography photo-thermal therapy, imaging, sensing, catalysis etc. Nanotechnology had improved cancer detection and diagnosis in the inceptive stage. The present chapter gives information on the use of nanotechnology to solve the enigma in the field of oncology.

10.1 Introduction

Nano: it is one-billionth of a meter. The concept of nanotechnology was first initiated by Richard P Feynman in 1959. The term nanotechnology by itself was defined by Norio Taniguchi of the Tokyo Science University in a scientific paper termed 'Nanotechnology' in the year 1974. It was defined as **the processing of, separation, consolidation and deformation of materials by one atom or one molecule** [1]. More than 1000 nanostructures have been reported, many with potential medical applications [2]. A plethora of nanostructures have been described according to the size, morphology shapes including solid nanoparticles (e.g., spheres, rods, triangles, cubes), nanoshells (within inner and outer cores), nanocages and various physical and chemical properties, with varying compositions (e.g., gold, iron oxide, carbon, dielectric materials, molecular, liposomal) and nanowires, nanotubes, branched dendrimers, and polymeric and organic lipid nanoparticles [2].

Nanoparticles range from 1 to 100 nm, whereas nanomaterials have unique materials with size ranging up to 100 nm. Nanomaterials can be broadly classified as **naturally occurring and artificially occurring**. Naturally occurring biological substances from viruses to materials which make up the bone are all-inclusive under the spectrum of nanoparticles. Artificial nanoparticles range from carbon, metal, dendrimers and composites. The carbon-based ones are mostly in the form of

doi:10.1088/978-0-7503-3671-0ch10

hollow tubes, spheres, ellipsoids and tubes [3]. Basically, the one dimensional, two dimensional and three dimensional ones are called sheets or nanowires [4]. The cylindrical ones are called nanotubes. The metal-based ones include quantum dots, nanogold, nanosilver and metal oxides, such as titanium dioxide [5].

Nanoparticles can be broadly divided according to the varying compositions. The compositions range from gold, carbon, dielectric materials, iron oxide, molecular or liposomal. Nanoparticles can be evenly divided according to the numerous shapes available. They can be spheres, rods, triangles and cubes. Among others, nanoshells are mentioned, they are thin dielectric spherical structures with an outer gold covering. These are widely used in medical imaging and cancer diagnosis. The features of the drug delivery structures are mentioned below.

10.2 Liposomes

Though the terms nanoparticles and liposomes are used interchangeably. They mostly deal with the controlled and systematic delivery of medicine to the target area. The major advantage of this system is its low toxicity. Specific combinations of various lipid and lipid bilayers are used to process this system. Because of its similarity to the naturally occurring liposomes, the bioavailability of such structures is immense and they are used as a major mode of drug delivery systems.

10.3 Polymer micelles

Polymer micelles are again structures which provide an efficient mode of delivery of cancer-targeted drugs which are poorly water-soluble. A micelle can be described as an aggregate of an electrically charged molecule. Thus, based on such composition polymer micelles can be categorized as amphophilic micelles, polyion complex micelles and micelles derived from various metal complexes

Dendrimers: These are especially branched, nano-sized, artificial macromolecules which are highly defined with functional groups. They have also been termed as cascade molecules [6]. Dendrimers generally increase in diameter in subsequent generations, which basically provides for effective drug delivery mechanisms.

Polyacid nanoparticles: Polyelectrolyte copolymers are mixed with oppositely charged species forming electrostatic self-assembly of stable 'supermicellar' aggregates. The charge distribution on these polyelectrolytes controls their conformation and interactions resulting in colloidal complexes. The interactions among the colloids, and the polymeric matrices influence the features of nanoparticles, their optical properties, and the long-term stability of the colloid dispersions [7].

10.4 Advantages of using nanotechnology in the diagnosis/treatment of cancer

The mainstay of this type of technology is that it is multifunctional and different approaches can be integrated into this towards a holistic treatment of cancer [8]. The imaging and treatment modalities go hand in hand in this type of technology integrated into cancer treatment modality.

10.5 Properties of nanoparticles/nanomaterials

10.5.1 Properties of nanoparticles

Size: The unique properties of nanoparticles are because of their small size and large surface area, as such they act as a proper bridge between the main bulk materials and the smaller atoms.

Their small size, which runs to billionths of a cm, are being proven to be of high value in the area of medical imaging, especially for angiogenesis imaging.

Melting point: As the size reaches 100 nm, the melting point of the particles also reduces.

Optical properties: The size of such nanoparticles also defy laws such as the quantum effect. According to this law, as the size of the nanoparticle is too small, it doesn't follow the quantum laws as these laws are obeyed by electrons and nanoparticles have their sizes less than an electron.

In quantum size effect, as the size of the particle decreases, the quantum bandwidth increases, in turn, increasing the frequency which would not be present in bulk material. For example, as the size of gold decreases, the colour of the particle changes from yellow to red. Hence, nanoparticles follow this principle and contain high energy capacity.

10.5.1.1 Plasmon resonance property

Application of SPR in gold nanotechnology in head and neck cancer: The other eminent optical property of nanoparticles is the SPR or surface **plasmon resonance** property of nanomaterials. When materials are stimulated by light, a specific set of electrons starts to resonate coherently producing a resonance (figure 10.1) In this, the accurate localization of the point on the nanoparticle where resonance has occurred can be found. The resonance at which the electrons get excited on a nanomaterial and then calibrated is called the SPR. This property is used in various imaging modalities as this principle provides better resolution, which results in better visualization during the various phases of malignancy. Generally, gold nanoparticles are coated with antibody, and the shift in the resonance is observed, which is interpreted in colour. This property of nanoparticles is also used in calculating the ligand and analyte molecule bonding.

Kah *et al* have shown that gold nanoparticles can provide an optical contrast to discriminate between cancerous and normal cells and their conjugation with antibodies also allows them to map the expression of relevant biomarkers for molecular imaging [9].

Nanoparticles seem to increase Raman scattering of saliva samples of oral cancer patients [10].

10.5.1.1.1 Biocompatibility

Inorganic nanoparticles, as already discussed above, have unique properties in electronic, photothermal, catalytic, and optical areas [2]. Of special mention are the plasmonic gold nanoparticles because of the simplicity in their fabrication, multifunctional nature, facile surface chemistry, biodistribution properties, and relatively

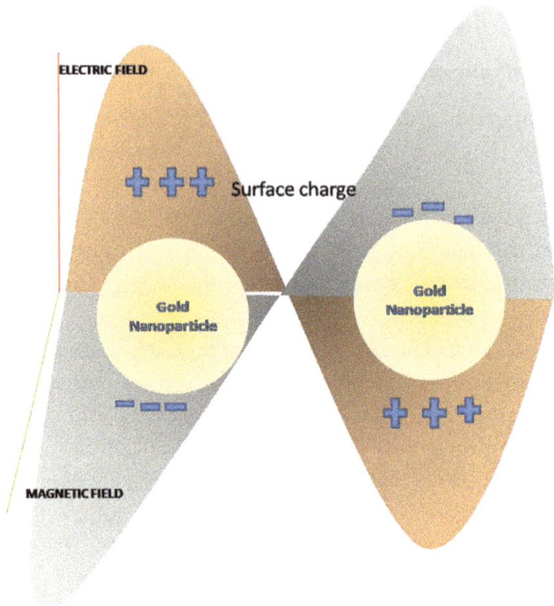

Figure 10.1. Schematic representation of surface plasmon oscillation of gold nanoparticles under electromagnetic field. When materials are stimulated by light, a specific set of electrons starts to resonate coherently producing a resonance.

low toxicity [2]. Further, these particles greatly enhance the signals which in turn tends to lower the background noise of the cells and tissues.

10.6 Facile surface chemistry

To the surface of nanoparticles a variety of materials have been attached to escape macrophage capture, refine targeting ability, circulation and life span which include: polyethylene glycol (PEG) ligand, folate receptor ligands and also conjugation of hydrophilic surfaces, thereby enhancing the delivery system [11, 12]. Also, to impart a binding surface for specific cells, secondary coating molecules are attached to minimize non-specific binding on other tissues [13].

10.7 Toxicity

For any therapeutic or diagnostic agent being administered, its safety and non-toxicity becomes the crucial aspect. The limitation of such particles is toxicity, chances of embolization and the risk of uptake by phagocytic cells. Nanoparticles containing selenium, arsenic, cadmium, lead or some of their complexes are known to cause risks concerning carcinogenicity, immunotoxicity, reproductive risks in vertebrates [14]. Therefore, nanoparticles should have an ideal half-life span in the blood and effective elimination from the body along with properties of selective binding and minimal non-specific binding. With longer retention of nanoparticles in the body the likelihood of potential toxicity is increased. [15]

10.8 Application

Nanoparticles can be used in various theragnostic areas. With surface modifications of nanoparticles, their ability of covalent/noncovalent binding properties can be improved. For example, the covalent conjugation of gold NPs stabilizes the conjugates for the purpose of imaging. It facilitates in non-covalent attachment of other molecules through specific binding affinity or electrostatic and hydrophobic interactions resulting in modified gold nanoparticles that are suitable for easy targeting. Also, super-paramagnetic iron nanoparticles or SPIONs are iron-oxide nanoparticles which are coated with biocompatible materials such as dextran and are further conjugated with a cancer-targeting ligand for targeting cancer. Hence, such ligand-bound particles are visualized using MRI, and eventually drain into the lymph nodes.

Quantum dots or QDs are frequently used in imaging, and when compared with other fluorescent particles, these exhibit sharper images and enhanced visualization with photostability. One-dimensional nanostructures like **nanotubes** have also shown to have great potential in medical imaging as they consist of high optical absorption in the near-infrared and photoluminescence and strong resonance Raman scattering.

Nanotechnology in diagnostic imaging in oral cancers (figure 10.2).

10.8.1 Magnetic resonance imaging

The frequently used contrast agents which have been complexed with diethyltri-amine-Penta acetic acid (Gd-DTPA) or tetra azocycloodo—decane 1,4,7,10—tetra-acetic acid (Gd- DOTA) are used [16]. Though these contrast agents distribute throughout the entire body and provide good contrast, they are not specifically found in any organ as the contrast is flushed out due to blood circulation. In addition, the half-lives of such agents are very short, ranging from 1–1.5 h leading to less usable time period for various diagnostic applications [17]. In contrast, such materials have been used by coating specific nanoparticles on their surface. The most frequently used nanoparticles are a superparamagnetic iron oxide (SPIO), and ultra-small super magnetic iron oxide (USPIO), being used for liver and spleen associated disorders as negative contrast agents [18].

Nanoparticles are also used in oral cancers with amazing results. Asifkhan *et al* reported using a combination of chitosan and magnetic polyglycolide to create an MRI contrast and a treatment modality, in its use in oral cancers [10]. They have been reported to have no adverse reactions in this study.

10.8.1.1 Optical computer tomography

OCT is another non-invasive, real-time measure between normal and abnormal. It works in a similar principle to that of ultrasound but works with infrared light hence gives a good resolution up to 10 micrometre, which remains higher than the other mainstream non-invasive diagnostic techniques [19]. But the major dis-advantage remains the fact that the contrast offered is still low, especially to diagnose between normal and neoplastic. Meanwhile, Kim *et al*, have used gold-coated nanoparticles to enhance the contrast to up to 150% of the images of oral dysplasia in hamster models [20].

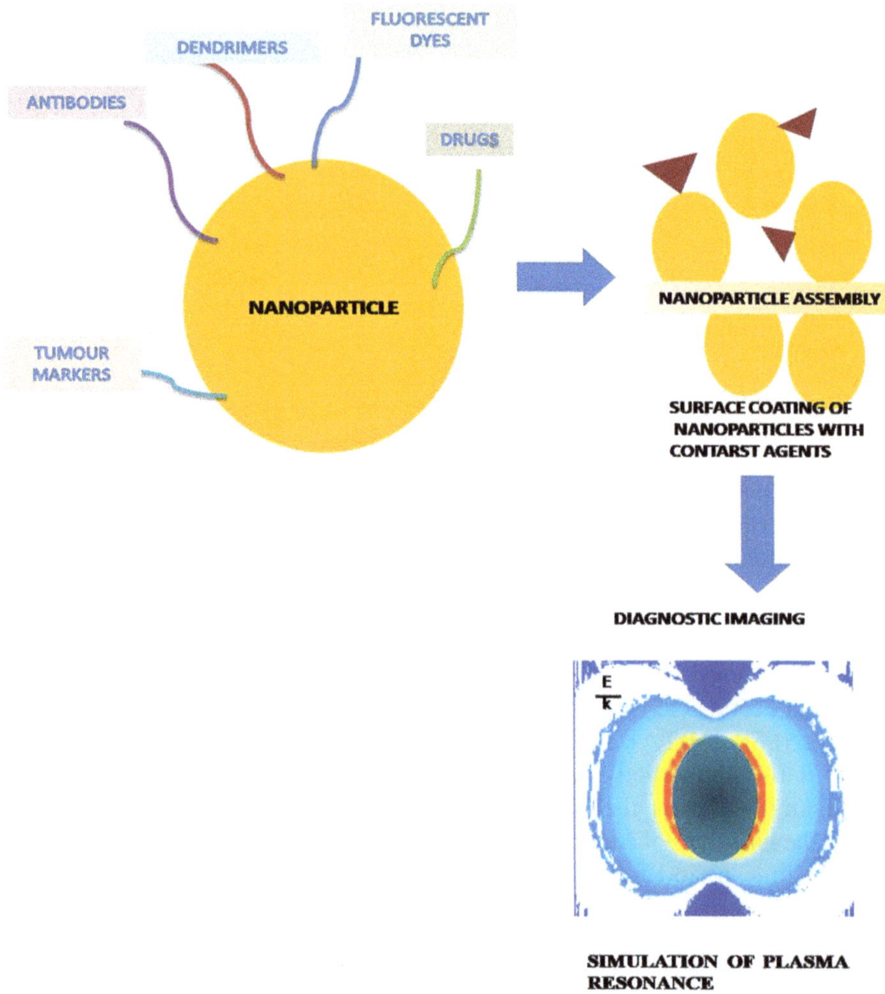

Figure 10.2. Gold nanoparticles are coated with antibody, and the shift in the resonance is observed, which is interpreted in colour. This property of nanoparticles is also used in calculating the ligand and analyte molecule bonding.

Gold nanoparticles have also been used in the concept of photoacoustic imaging, especially in the areas of brain, breast and prostate cancer diagnosis. They have provided increased and better contrast imaging.

10.8.1.2 Diffusion reflection imaging

Diffusion reflection imaging is a process where the light remnant after absorbing and transmitting gets reflected, and this property is utilized to visualize an image. The image is smooth when the reflecting object is smooth and sharp, for example, a mirror; if the reflecting object is slightly bumpy, then the reflected image would not be clear. This property of the reflected light is used to visualize normal, oral potentially malignant disorder and oral cancer. Gold nanorods or GNRs have proven to differentiate between the surgical margins with high specificity and sensitivity of 97% and 87%, respectively, with the help of diffusion reflection imaging process which by itself has a huge margin of accuracy.

10.8.1.3 Quantum dot imaging

Quantum dots are unique nanomaterials with properties intermediate between bulk semiconductors and discrete molecules [21]. Their size ranges from 2 to 10 nanometers and they possess special characters in producing specific colours depending on the size of the particle. As discussed in the above section, quantum dots are basically classified as core type, core alloyed type and alloyed type quantum dots.

Quantum dots work on the Bohr principle where it states that electrons occupy certain and fixed positions in orbit. Therefore, the colour of the emitted light can be altered by just changing the size of the quantum dot.

Studies have used quantum dots with 800 nm emission spectrum when conjugated with EGFR antibodies [10]. This technique has been proven to have potential in the treatment of oral squamous cell carcinoma.

10.9 Sensitive biomarker detection

Compared to the routine biomarkers available nowadays like the ELISA, western blot and polymerase chain reaction, nanotechnology has proven to have a higher detection sensitivity rate which can be useful for detecting minute quantities in tissue samples or other body fluids.

Routine biomarkers such as EGFR, TNF Alfa, IL 6, VEGF and the likes of these have been used for cancer detection [10]. Nano based biotechnology has been successfully used in further detection of lower or minute quantities of such markers using MALDI-TOF-MS.

Besides this method, a gold protein chip method also was used for TNF alfa detection in oral cancer [9]. Such studies have given a positive take on the usefulness of nanoparticles in cancer diagnosis. Further, oral cancer detection rates could be markedly improved if saliva was used as an alternative method to improve cancer detection rates, as shown in table 10.1.

Drug Delivery: Drug delivery can be broadly classified into passive targeting and active targeting of cancer cells. Passive targeting can include drug delivery through (a) vascularity, (b) tumor microenvironment, (c) local drug application [22].

Active targeting includes (a) carbohydrate targeted, mostly lectin carbohydrate binding, (b) receptor targeted, (c) antibody targeted [16].

10.10 Why an alternative is required?

The main reasons why cancer therapeutics now requires an alternative platform for drug delivery can be attributed to a myriad of causes. The most fundamental question is of disorderedly pharmacokinetics which ultimately leads to increased toxicity, and the other reason is owing to the repeated IV routes which eventually become problematic. It has been seen that the traditional mode of drug delivery has always come with a price tag: the healthy cell being killed as well as the cancerous cell. Thus, various alternatives are of keen interest to scientists.

Table 10.1. Nanoparticles, properties and their therapeutic applications.

Nanoparticles	Properties	Theragnostics
Liposomes	Specific combinations of various lipid and lipid bilayers similar to liposomes	Systematic delivery of medicine to the target area
Polymer micelles	Aggregate of an electrically charged molecule	Efficient mode of delivery of cancer-targeted drugs which are poorly water-soluble
Dendrimer	Artificial macromolecules which are highly defined with functional groups	Basically provides for effective drug delivery mechanisms
Nano gold particles	Small gold particles with a diameter of 1 to 100 nm which once dispersed in water are known as colloidal gold	Drug-delivery, OCT photo-thermal therapy, imaging, sensing, catalysis
Super paramagnetic iron nanoparticles (SPIONs)	Iron-oxide nanoparticles which are coated with biocompatible materials such as dextran	Targeting cancer
Quantum dots	Tiny semiconductor particles a few nanometres in size	Laser, medical imaging
Nanotubes	A tubular molecule composed of a large number of carbon atoms	Imaging

10.11 Nanoparticles in oral pathology

Oral pathology from a nanoscale perspective will improve our knowledge in particular diseases such as dental caries, tooth hypersensitivity and oral cancer [23]. As the size of nanoparticles is comparable to the sizes and shapes of normal subcellular structures, it can thus allow us to understand and demonstrate their interactions. For example, DNA complexes with carbon nanotubes. Nanoparticles are considered to potentially penetrate non-phagocytic cells which then may access subcellular compartments that are normally inaccessible to micron-sized particles.

Nanoparticles may invade targets that were once unreachable like the brain through endocytic pathways, by evading phagocytosis, or a combination of both, some nanoparticles may either enter cells through diffusion or with their unique energy-independent mechanisms [24].

Studies showed the histopathologic observation of early single-walled carbon nanotubes apparently interacting with chromatin in an abnormal mitotic figure. Further investigations demonstrated single-walled carbon nanotubes interactions with chromatin, centrosomes and microtubules of the mitotic spindles [25]. Individual malignant cells can be located using specific antibodies against them. Application of a novel set of lipid-coated, quantum dots in targeting the individual malignant cell surfaces with specific biomarkers has revolutionized early detection of malignancies [26].

The knowledge of the molecular pathways in disease development and progression provides researchers abundant scope in using possible molecular targets to generate new diagnostic strategies. This includes profiling of tumour biomarkers obtained from single tumour sections by conjugating nanoparticles to antibodies [27]. The use of antibody-conjugated nanoparticles, quantum dots conjugated to peptides to localize and demonstrate the cell surface proteins like c-erbB2, epidermal growth factor receptor and CA-125 have been used. Nanoparticles are used for *in vitro* detection of tumor cells, by concentrating and protecting a marker from degradation. For effective proteomic profiling using multiplexed testing with arrays, micro- and nanocantilevers are highly sensitive, compatible along with their advantage for microfluidic integration [28].

Magnetic nanoparticles are promising in the detection of bacteria at <100 cfu ml^{-1} concentration because of their easy entry into cells in comparison to magnetic beads. Goldman *et al* developed a multiplex immunoassay for the simultaneous and sensitive detection of cholera toxin, ricin, shiga-like toxin 1, and staphylococcal enterotoxin B using the antibodies conjugated to QDs [29].

10.12 Future directions

In view of the significant research dedicated to this field, it may be expected that humans will greatly benefit from nanotechnology in the very near future, especially in tumor therapy. Currently, its application to routine clinical practice will require a multidisciplinary approach guided by clinical, ethical, and social perceptions, the mainstay being the development of a multifunctional nanoscale-sized platform with the highest targeting efficacy.

The engineered nanomaterials can be used to identify malignant cells to deliver the drugs at the affected sites. Nanotechnology may enormously contribute to *in vitro* and *in vivo* applications of nanoparticles in the field of nanomedicine [30]. The targeted drug delivery of nanomedicines without toxicity will have greater importance for theranostic systems.

10.13 Conclusion

This review aimed to highlight the various types and efficiency, and the progress of nanotechnology towards cancer detection with a special focus on oral cancer. This review hopes to cover the recent developments of nanotechnology and the strides made so far in the field of cancer detection and treatment.

References

[1] Taniguchi N 1974 Proc. Int. Conf. Prod. Eng., Part II, Japan Society of Precision Engineering, Tokyo *J. Jpn. Soc. Precis. Eng.* **40** 516–21

[2] El-Sayed I 2010 Nanotechnology in head and neck cancer: the race is on *Curr. Oncol. Rep.* **12** 121–8

[3] Karatutlu A, Barhoum A and Sapelkin A 2018 Liquid-phase synthesis of nanoparticles and nanostructured materials *Micro and Nano Technologies, Emerging Applications of Nanoparticles and Architecture Nanostructures* (Amsterdam: Elsevier) pp 1–28

[4] de Souza Rastelli A N, Carreira E T, Dias H B and Hamblin M R 2016 Nanobiomaterials in dentistry *Nanobiomaterials in Dentistry* (Boston, MA: William Andrew Publishing) pp 1–25

[5] AZoNano 2020 Classification of nanomaterials, the four main types of intentionally produced nanomaterials https://www.azonano.com/article.aspx?ArticleID=1872

[6] Abbasi E, Aval S, Akbarzadeh A, Milani M, Nasrabadi H, Joo S, Hanifehpour Y, Nejati-Koshki K and Pashaei-Asl R 2014 Dendrimers: synthesis, applications, and properties *Nanoscale Res. Lett.* **9** 247

[7] Mayer A, Johnson R, Hausner S and Mark J 1999 Colloidal silver nanoparticles protected by water-solublenonionic polymers and 'soft' polyacids *J. Macromol. Sci., Part* A **36** 1427–41

[8] Zhao C-Y, Cheng R, Yang Z and Tian Z-M 2018 Nanotechnology for cancer therapy based on chemotherapy *Molecules* **23** 826

[9] Poonia M, Ramalingam K, Goyal S and Sidhu S K 2017 Nanotechnology in oral cancer: a comprehensive review *J. Oral Maxillofac. Pathol.* **21** 407–14

[10] Chen X J, Zhang X Q, Liu Q, Zhang J and Zhou G 2018 Nanotechnology: a promising method for oral cancer detection and diagnosis *J. Nanobiotechnol.* **16** 52

[11] Ghosh P, Han G, De M, Kim C and Rotello V 2008 Gold nanoparticles in delivery applications. *Adv. Drug Delivery Rev.* **60** 1307–15

[12] Cho K, Wang X, Nie S, Chen Z and Shin D 2008 Therapeutic nanoparticles for drug delivery in cancer *Clin. Cancer Res.* **14** 1310–16

[13] Choi H and Frangioni J 2010 Nanoparticles for biomedical imaging: fundamentals of clinical translation *Mol. Imag.* **9** 291–310

[14] Nie S, Xing Y, Kim G and Simons J 2007 Nanotechnology applications in cancer *Annu. Rev. Biomed. Eng.* **9** 257–88

[15] Frangioni J 2008 New technologies for human cancer imaging *J. Clin. Oncol.* **26** 4012–21

[16] Klaassen C D 2001 *Casarett and Doull's Toxicology: The Basic Science of Poisons* (New York: McGraw-Hill)

[17] Cheng W, Ping Y, Zhang Y, Chuang K H and Liu Y 2013 Magnetic resonance imaging (MRI) contrast agents for tumor diagnosis *J. Healthc. Eng.* **4** 23–45

[18] Bennett K M, Jo J, Cabral H, Bakalova R and Aoki I 2014 MR imaging techniques for nano-pathophysiology and theranostics *Adv. Drug. Deliv. Rev.* **74** 75–94

[19] Brigger I, Dubernet C and Couvreur P 2002 Nanoparticles in cancer therapy and diagnosis *Adv. Drug Deliv. Rev.* **54** 631–51

[20] Pande P, Shrestha S, Park J, Gimenez-Conti I, Brandon J and Applegate B E *et al* 2016 Automated analysis of multimodal fluorescence lifetime imaging and optical coherence tomography data for the diagnosis of oral cancer in the hamster cheek pouch model *Biomed. Opt. Express* **7** 2000–15

[21] Kim C S, Wilder-Smith P, Ahn Y C, Liaw L H, Chen Z and Kwon Y J 2009 Enhanced detection of early-stage oral cancer *in vivo* by optical coherence tomography using multi-modal delivery of gold nanoparticles *J. Biomed. Opt.* **14** 034008

[22] Sharma A, Jain N and RashmiSareen 2013 Nanocarriers for diagnosis and targeting of breast cancer *BioMed Res. Int.* **2013** 960821

[23] Tomsia A P, Launey M E, Lee J S, Mankani M H, Wegst Ulrike G K and Saiz E *et al* 2011 Nanotechnology approaches for better dental implants *Int. Oral Maxillofac. Implants.* **26** 25–44

[24] Zheng M *et al* 2003 Structure-based carbon nanotube sorting by sequence-dependent DNA assembly *Science* **302** 1545–8

[25] Shvedova A A *et al* 2008 Inhalation vs. aspiration of single-walled carbon nanotubes in C57BL/6 mice: inflammation, fibrosis, oxidative stress, and mutagenesis *Physiol. Lung Cell. Mol. Physiol.* **295** L552–65

[26] Sargent L M *et al* 2012 Single-walled carbon nanotube-induced mitotic disruption *Mutat. Res.* **745** 28–37

[27] Akerman M E, Chan W C, Laakkonen P, Bhatia S N and Ruoslahti E 2002 Nanocrystal targeting *in vivo Proc. Natl Acad. Sci. USA* **99** 12617–21

[28] Paciotti G F, Myer L and Weinreich D *et al* 2004 Colloidal gold: a novel nanoparticle vector for tumor directed drug delivery *Drug Deliv.* **11** 169–83

[29] Goldman E R, Clapp A R, Anderson G P, Uyeda H T and Mauro J M *et al* 2004 Multiplexed toxin analysis using four colors of quantum dot fluororeagents *Anal. Chem.* **76** 684–8

[30] Ahamed M, AlSalhi M S and Siddiqui M K J 2010 Silver nanoparticle applications and human health *Clin. Chim. Acta* **411** 1841–8

IOP Publishing

Nanotechnology for Dentistry Applications

Ashutosh Kumar Shukla

Chapter 11

Possible effects/future scope of using nano materials in oral dentistry

Eman Osman

These days, nanotechnology has been coordinated with distinctive zones of science because it gives different critical ways to meet logical and therapeutic issues. Nanoparticles (NPs), not at all like other biomaterials, show natural properties and can be utilized in novel applications in therapeutic dentistry, prosthetic dentistry, endodontics, implantology, oral cancers, and periodontology. Although the advancement and application of nanodentistry is of significant interest, information with respect to the conceivable harmfulness of such materials must be analyzed with greater attention and precision, and possible advantages must be weighed against the risks in order to discover security flaws.

NPs have enormous potential on account of their antimicrobial, antiviral, and antifungal properties. The security evaluation forms in place for solutions and therapeutic devices remain strong, and while individual harmfulness tests may need to be tweaked to function with engineered nanomaterials (ENMs), the overall security strategy is sound. However, the financial opportunities offered by nanomaterials must be balanced against a number of realistic problems, practical and societal issues, as well as human health and safety considerations. Nanomaterials, in particular, may have key harmful features that are currently unknown, which are related to their one-of-a-kind physico-chemical properties, which might pose a threat to a wide range of employees and consumers exposed during the overall lifecycle of nanomaterials.

In any event, the same characteristics that make NPs harmful are their preferences: tiny size, surface features, quantum state, mobility, aggregation, transformation, and the generation of free radicals. As a result, because to the practically limitless realms of application, NP research is currently one of the most widely studied branches of science, and hence administrative and security concerns must be reviewed and addressed in oral care products.

doi:10.1088/978-0-7503-3671-0ch11

11.1 Introduction

Nanotechnology has been integrated into various fields of research in recent years because it provides unique solutions to logical and remedial issues. In recent years, there has been an incredible increase in interest in the field of nanoscience. There is a growing belief that nanotechnology in medicine and dentistry will lead to significant advancements in the diagnosis, treatment, and prevention of disease. The growing interest in nanotechnology's potential therapeutic applications is fueling the growth of a new field termed nanomedicine. Nanomedicine must overcome the obstacles to its use, making significant progress in our understanding of the pathophysiologic basis of disease, introducing more modern symptomatic apertures, and offering more effective therapies and preventative characteristics. Atomic innovation is destined to become the cornerstone of all pharmaceutical and dental innovation in the 21st century [1].

Nano is derived from the Greek word for shadow and is usually coupled with an object to form phrases like nanometer, nanotechnology, and nanorobot. A nanometer is one billionth of a meter (10^{-9}). Using NPs, biotechnology, counting tissue creation, and eventually dental nanorobotics, nanodentistry will make it possible to support total oral health [2]. Local anesthetic, dentition renaturalization, changeless itch therapy, entire orthodontic realignments in a single office visit, covalently reinforced diamondized finish, and uninterrupted oral wellness upkeep employing mechanics are some of the most recent possible treatment openings in dentistry [3–5].

NPs could provide a previously unknown means of treating and preventing dental problems [6]. Because of their huge surface area and high charge density, NPs are able to bind more strongly with the negatively charged surface of bacterial cells, resulting in improved antibacterial movement. NPs, unlike other biomaterials, have recognizable natural qualities and can be used in unique applications in dentistry, prosthetics, endodontics, implantology, oral malignancies, and periodontology. The antibacterial, antiviral, and antifungal capabilities of NPs give them tremendous promise. NP consolidation prevents biofilm formation over the composite, preventing micro-leakage and secondary caries [7, 8]. These NPs improve the mechanical properties of restorative materials and the overall binding between dentin and biomaterials, affecting bond strength. Adhesive frameworks with NPs can be used in orthodontic treatments to prevent white spot lesions [9].

Furthermore, NPs coupled with polymers or coated onto biomaterial surfaces were discovered to have strong antibacterial activities in the oral cavity [10]. Because of their broad-spectrum bactericidal capabilities, metal and organic NPs have been linked in a few areas of dentistry [11]. Here below are a few expressions to be considered when managing with nano issues.

11.2 Nanotechnology

This is the molecular-scale engineering of function systems. Nanotechnology is emerging as a fascinating topic that is rapidly advancing and has resulted in massive advances in health and dentistry. Nanomaterial-based design can mimic a few mechanical and fundamental features of native tissue, which can help with biointegration [12]. For dental rebuilding (restoration), a variety of manufactured NPs such as hydroxyapatite, bioglass, titanium, zirconia, and silver NPs have been proposed [13].

11.3 Nanomedicine

Robert A Frietas Jr first proposed the notion of nanomedicine in 1993, and it was defined as follows: 'Nanomedicine is the maintenance and improvement of human health utilizing molecular instruments and molecular knowledge of the human body' [14]. Its efforts have expanded to include anything from medication delivery to tissue scaffolding using nanoscale particles [15].

11.4 Nanodiagnostics

This is the term utilized for the application of nanobiotechnology in atomic diagnosis, which is imperative for creating personalized cancer therapy. Doctors will be able to offer patients comprehensive health exams quickly and routinely as diagnosis breakthroughs advance [16]. Oral cancer detection and therapy are aided by nanotechnology. Nanotechnology finds tumor cell biomarkers, allowing for earlier detection and increased test affectability.

Metallic NPs, such as gold nanoparticles (AuNPs), are being investigated as a way to circumvent the limitations of imaging and chemical-based diagnostics. Kah *et al* demonstrated the use of AuNPs in combination with Raman spectroscopy to assess chemical changes in oral cancer patients' saliva. To discriminate between dangerous and nonmalignant keratinocyte cells, Huang *et al* coupled gold nanorods with anti-epidermal growth factor. This technique is extremely effective at detecting cell alterations [17]. The probability of a false-negative outcome would be almost nil. NPs have an incredibly high surface area due to the small size of their atoms, which allows diverse function groups to attach to them, potentially tying them to tumor cells [18, 19].

11.5 Nanodentistry

The term and field of nanodentistry were coined in the year 2000. Dentistry began to advance in the field of nanotechnology as nanomedicine evolved. Nanotechnology is expected to have an impact on diagnosis, materials, restorative dentistry, and surgery. Nanorobotics, nanodiagnosis, nanomaterials, nanosurgery, and nanodrugs, among other energising untapped disciplines, will have a big impact on clinical dentistry in the not-too-distant future. Science fiction will become reality in the not-too-distant future [20].

According to the definition of nanodentistry, it is 'the science and technology of diagnosing and preventing oral and dental problems, treating pain, conserving, and improving dental health with nanotechnology or nanostructured material.'

Using nanomaterials, biotechnology, counting tissue engineering, and eventually dental nanorobotics, nanodentistry will make it possible to support total oral health. Local anaesthetic, dentition renaturalization, permanent hypersensitivity cure, complete orthodontic realignments in a single office visit, covalently bonded diamondized enamel, and ongoing oral health support are some of the untapped potential therapeutic opportunities in dentistry [21–25].

Based on application, use and technology there are four approaches of nanotechnology in dentistry, as follows [22]:

1. **Top-down approach:** Attempts to make smaller devices by combining larger ones in order to achieve precision in construction and assembly. These solid-state materials can also be used to make NEMS (nanoelectromechanical frameworks) devices, which are used in cancer treatments [23].
2. **Bottom-up approach:** seeks to assemble smaller components into more complex congregations with incredibly strong covalent bonds [24].
3. **Functional approach:** seeks to build components that perform a certain purpose regardless of how they can be combined.
4. **Biomimetic approach:** seeks to use biomolecules in nanotechnology applications.

The basic two ways, top-down approach and bottom-up approach, are the most widely employed of the four. Both can be summarised in figure 11.1 [25].

11.6 Nanorobotics

These are artificially manufactured spider-like body objects made of the chemically inert shape of a carbon component and able to defuse unimpeded within the human body and connected with particular cells at the atomic level by themselves.

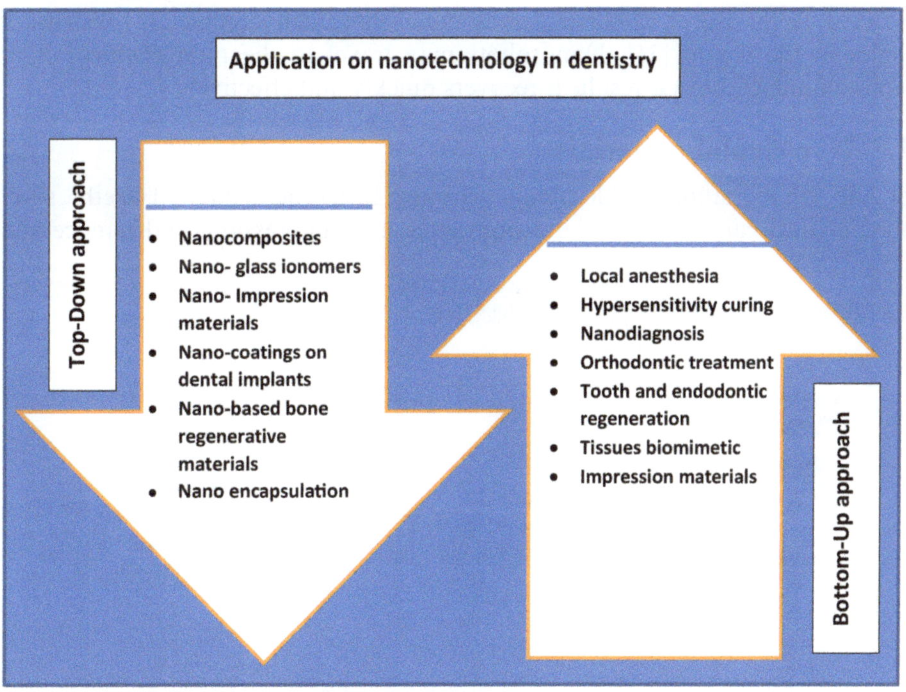

Figure 11.1. The top-down and bottom-up nanodentistry schemes are depicted in this diagram.

Their breadth is 0.5–2 micron, as part of the measurement of 1–100 nm [26, 27]. The most common nanorobotics functions in dentistry are:

- induce local anaesthesia;
- avoid discomfort of the patient;
- dentrifices (dentrofrobots);
- major tooth repair.

Nanorobots can distinguish between different cell types by examining their surface antigens. Sensors, actuators, control, power, communications, and interfacial signals are used to build nanorobots across spatial scales and between organic/inorganic as well as biotic/abiotic systems.

When the nanorobots have fulfilled their mission, they can be recovered by allowing them to excrete through normal human excretory pathways. Dynamic scavenger frameworks can also eject them [28].

Nanotechnology can be used in several ways in dentistry, as illustrated in figure 11.2.

NPs can be connected in orthodontics to create quick and easy tooth fixing, pivoting, and vertical repositioning solutions in minutes to hours. However, the reality of such a speculative idea is still a long way off. Sims suggested that brackets may be replaced with nanorobots that are adjusted to manage bone and periodontal tendon responsiveness in order to achieve tooth formation [29, 30].

Dental hypersensitivity is one area where nanotechnology might unquestionably benefit dentistry. Dentinal tubules, which have double the breadth of non-sensitive teeth and eight times the surface thickness of dentinal tubules, are gathered in easily impacted teeth. Reconstructive dental nanorobots can completely obstruct the tubules in the dentin [31]. Nanorobots may travel to the pulp chamber in less than a minute [32]. As a result, it recovers quickly and effectively.

11.7 Nanodental materials

For the rehabilitation of decaying, carious, lost, and shattered teeth, several nanomaterials can be used. Nanocomposites, nanoimpression, and nanoceramics

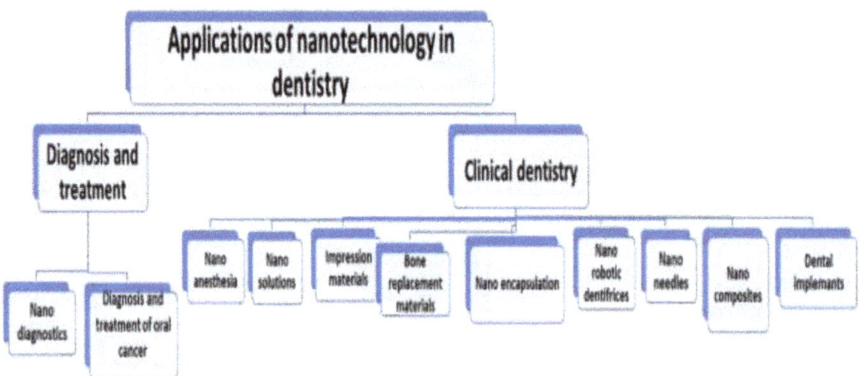

Figure 11.2. Different applications of nanotechnology in dentistry.

have all been introduced into clinical dentistry as a result of later advances in nanomaterials. Composite materials should have two key characteristics: strength and aesthetics. The size of the filler particle has an impact on each of these qualities. The aesthetic properties of filler particles have progressed from traditional filler particles to microfilled particles, but not the strength properties. Nanofillers were created in an attempt to create an underused substance with both of these traits, especially mechanical strength and aesthetic and polishing capabilities have been presented [33].

Polymethyl methacrylate (PMMA) has been used for practically all detachable dental prosthesis due to its superior features such as excellent optical qualities, biocompatibility, ease of handling, and reparability. However, PMMA has a number of drawbacks, including low strength, low break resistance, and microbial adherence. To compensate for these flaws, nanostructuring materials like TiO_2 and Fe_2O_3 are mixed into PMMA. PMMA has improved its modulus and strength after being nanostructured, and it may maintain or improve its ductility and aesthetic characteristics. Furthermore, PMMA has the potential to have antibacterial characteristics [34].

Because of their great strength, acceptable color, and low thermal and electrical conductivity, ceramics have been utilized in prosthodontics to construct dental crowns and dentures. Their great brittleness, however, may be a source of concern. The use of nanozirconia ceramic provides the advantages of increased hardness and toughness. Nano ceramic grains have exceptional fracture toughness, translucency, and erosion resistance.

Furthermore, nanocomposite prosthetic teeth are a recent advancement in this sector. Nanoinorganic fillers are dispersed throughout the composite lattice. These are more durable as well as quite safe to use because to their prominent hue and resistance to abrasion [35].

One of the most common causes of implant failure is insufficient bone arrangement around the implant. Surfaces of inserts must be modified to incorporate nanoscale topography and/or coatings for better and faster osseointegration of implants in order to achieve appropriate bone growth. The development of nanostructured ceramic implants that combine inertness and mechanical responsiveness to a dental embed combination. When compared to alumina, nano zirconia-alumina materials have comparable hardness values and are not vulnerable to hydrothermal instability [36].

A good impression material should accurately record the points of interest in a patient's mouth. To advance the characteristics of polyvinylsiloxanes, nanofillers are added to them. These materials have improved flow, improved hydrophilic characteristics, improved point-by-point precision, and fewer voids at the edges [37].

11.8 Nanosurgery

Nanoneedles and nanotweezers are also being developed, which will make cell surgery a reality in the not-too-distant future [38]. Suture needles made of nanosized stainless steel crystals are being developed and could be used to deliver entrance points to cells [39].

11.9 Nanodrugs

Periodontal infection is treated using NPs that contain triclosan. In the near future, hollow spheres, core–shell structures, nanotubules, and nanocomposite could be used as periodontal drug delivery frameworks. Aresin, a periodontal pocket treatment with tetracycline, can be used to treat periodontal pockets [38]. Nanomaterials, according to Friedman, speed up the wound healing process. Burn wounds in mice were treated with curcumin NPs, which resulted in faster wound healing [39].

11.10 Cosmetics nanomaterials in dentistry

Teeth whitening, cleaning glues for enamel surface, dental implant coatings, dental filling, antisensitivity agents, and the anticipation of care are some of the most common domains of NPs in dentistry [40]. Although we have many ideas for nanodentistry, most of them are not feasible due to various problems such as engineering challenges, biological challenges, societal challenges, and so on. It's really difficult to precisely position and collect the atomic size part.

Cosmetics, which rely heavily on NPs, are one segment that customers are particularly interested in. Cosmetics is defined as any substance intended to be in contact with external parts of the human body, including mucous layers and buccal depth teeth, with the primary or sole purpose of changing their appearance, cleaning them, flavouring them, adjusting body odours, maintaining, or protecting them in a healthy state. Toothpaste is a cosmetic hygiene item that must be used on a regular basis. The most significant segment of the oral care market is toothpaste. The toothpaste market was valued at USD 26.1 billion in 2018, and it is expected to grow to USD 37.0 billion by 2024 [41].

Mouthwash is anticipated to be a productive development in future with the developing acknowledgment of biofilm disorganization, and its expanding use by customers for oral quality of life. The combination of flavoured mouthwash and components identified as distinctive should increase demand. Periodontal infection, interdental death, halitosis, and gingivitis, as well as the administration of more experience in human prosthesis, should all contribute to the demand for curable and non-medicated mouthwash.

The European Commission published a database of all NPs used in restorative items available on the market in 2018. The catalogue listed NPs used as additives, UV filters, and colourants, as well as the conditions of their introduction and the types of cosmetics they were used in [42]. Three NPs were identified as colorants, four as UV-filters, and 22 as 'other' capacities, including oral hygiene product items, among the 29 nanomaterials listed in this catalogue. These various nanomaterials have a significant impact on consumer prophylaxis, caries prevention, and periodontitis prevention, as well as consumer quality of life [43]. Fabricating biofriendly nanomaterials is also a challenge since nanomaterials can be pyrogenic. Ethics, public acceptance, and human regulation are all key issues that need to be addressed [27].

11.11 Safety concerns and implications of nanotechnology

Nanotechnology has a place in every sector, and it has the potential to make our lives easier and faster. However, it is limited in terms of our safety concerns. Nanomaterials have a large surface area-to-volume ratio, which causes particles to accumulate on the surface. Nanomaterials are thus more responsive and have a higher rate of absorption through the skin, lungs, and digestive tract. As a result, following continuous use, they will collect in various organs and are transmitted to other organs via blood. They have the potential to irritate the alveoli in the lungs, resulting in cell damage. DNA, RNA, and other intracellular components can also be reacted to by these tiny particles. These microscopic particles can react to DNA, RNA, and other intracellular components, causing alterations [29]. Nanotechnology is a relatively new field that operates in an unfamiliar area. The benefits are not, however, entirely grasped, or the probable dangers and safety problems. In time, clinical use will decide the genuine possibilities and dangers of nanotechnology in medication and dentistry [44].

To validate the different implications of nanotechnology, it should be addressed from different concerns, the most important of which are listed below:

11.11.1 Ethical implications

Any dental or medical nanoproduct undergoes extensive preclinical *in vitro* testing after the investigation and advancement stage to investigate its mechanical, toxicological, and immunological features. Several organizations, including the **United States Environmental Protection Agency and the National Institute of Occupational Safety and Health**, have proposed guidelines for assessing the risks of nanomaterials [45, 46], creating a multidisciplinary administrative structure to analyze and govern nanotechnology, as well as resolve ethical concerns that fall under the four categories of: (i) metaphysics; (ii) equity; (iii) privacy; and (iv) security, which could be a long-term legislative challenge [47]. Despite the fact that animal studies provide a useful insight of what to expect when starting a stage I trial, true adverse responses have been observed when human participants were subjected to a nanomedicine dose 500 times lower than the recorded detrimental restraint in animal studies [40]. As a result, subjects must obtain the following:

- Each clinical trial must specify the level of risk associated with novel materials and information, as well as safety data sheets.
- Identify and record any negative side effects early on.
- Look for abnormalities in data handling, and ensure the safety and well-being of test subjects.

The unpredictability of nanomaterials puts dental professionals in a moral bind when faced with a wide range of materials to choose from, some of which have a long track record of clinical use, such as hybrid or microfilled composite resins, and others, such as nanofilled composite resins, which are intriguing in concept and backed by short-term clinical studies. The traditional ethical decision-making framework, based on utilitarianism, is unable to keep up with the rapid pace and uncertain future of nanotechnology. As a result, a deeper understanding of the science is necessary, as well as risk/benefit analysis and ethical considerations throughout the improvement process. This leads to the proposal of the anticipatory ethics and administration concept, which is designed to recognize and address ethics and societal suggestions through moral investigation models while the innovation is still in its early stages, allowing it to be easily adjusted and guided to a morally worthy outcome [48, 49].

11.11.2 Society implications

Because society is the buyer, the financier, and the originator of approaches and choices, the public's attitude toward nanotechnology is critical to its success or failure; in other words, society is the judge and jury. Typically guided by morality, ethics, and values, which have recently been more accepting of modern beginning advancements as the perceived benefits outweigh the perceived dangers [50]. However, while nanotechnology is currently coordinating in areas that directly affect the public, such as vitality supply, health care and diagnostics, broadcast communications, and contamination control, this has raised concerns that these advancements may cost the public thousands of jobs in exchange for a more machinery-dependent framework [51]. Different activities were put in place to bridge the gap between society and nanoscience in an effort to solve societal concerns. According to the National Nanotechnology Initiative study, success in innovation would necessitate a previously untapped age of prepared individuals with advanced operational and administrative skills [52].

'Portable manufacturing systems would be able to make nearly any physical product from information for low cost, thereby supplying for our physical requirements at almost no cost,' said technologist Ray Kurzweil in 2003 [50]. This necessitates immediate involvement with the public to resolve concerns and raise understanding about existing and future nanotechnology applications in order to gain and maintain public acceptance.

11.11.3 Health implications

The **US government and state agencies** have been given a four-stage system to examine and assess the severity of any health risk. The system stages follow the setup [54]:

i. problem distinguishing proof;
ii. dose-response appraisal;
iii. exposure evaluation; and
iv. risk characterization.

Nanomaterials' effects are entirely size dependent, which means that nontoxic 100 nm measured particles appear to change significantly into poisonous components as their size decreases to 1 nm, and vice versa. Non-toxic nanomaterials may break down or clump, resulting in poisonous NPs. This peculiarity of how our bodies react to nanomaterials is due to how our immune systems react to the nanoproduct, as research has revealed. NPs may react in an unexpected different way in a cell culture than in a living organism, according to studies.

NPs have been found to be inhaled and to pass cell layers, allowing them to reach the liver, lymph nodes, spleen, and bone marrow [55]. Although assertions of nanotoxic effects on inward breath have been clearly stated, the obtained data require sufficient logical evidence to support or refute these claims [56]. Because there are so many aspects that influence toxic quality, it's difficult to generalize the health concerns associated with nanomaterial introduction; each current nanomaterial must be examined separately, and all material features must be taken into account.

Concerns about worker and customer safety are growing as the use of nanomaterials grows around the world. To address these concerns, the **Swedish Karolinska Institute** conducted a study in which human lung epithelial cells were exposed to several NPs [57]. According to the findings, which were published in 2008, iron oxide NPs caused minor DNA damage and were non-toxic. NPs of zinc oxide were slightly worse. Titanium dioxide only caused DNA damage, while carbon nanotubes caused DNA damage at the molecular level. Copper oxide was found to be the most egregious offender, and the analysts identified it as the only nanomaterial that posed a clear health risk [58]. The **National Institute for Occupational Safety and Health** is looking into how NPs interact with the body's systems, as well as how professionals may be exposed to nano-sized particles during the manufacture or mechanical application of nanomaterials.

11.12 Potential effects of the interaction of nanomaterials with the human body

11.12.1 Possible interactions with gingiva [59]

Materials containing NPs might be in close contact with gingiva, thus possible antagonistic effects of disintegrated NPs from the particles' surfaces should be considered. Examinations surveying the association of NPs with gingiva that incorporate teeth or dental additions are restricted. Possible unfavorable leverages have to be inspected concerning convenience and agnostic result. Regular materials used in dentistry might unfairly affect both delicate tissue and additionally encompassing bone. Exemplary instances of such a circumstance is as per the following.

11.12.1.1 Fillings made of amalgam

These contain common metals which may actuate color removal of neighboring gingiva by oxidative products, commonly known as 'amalgam tattoo'. The orthodontic wires or brackets may cause such bluish-black discoloration [60, 61]. To date, no information is accessible with respect to the effect of metallic and nonmetallic nanomaterials actuating mucosal discolorations, nor is there adequate information on response of associated gingiva to other nanomaterials used in dental rebuilding, inserts or prosthetic gadgets.

11.12.1.2 AgNPs

Medina *et al* conducted research work in the lab that revealed exposure of fibroblasts of gingiva to Ag and iron NPs results in extended cell oxidative stress, aggravation, and apoptosis [62]. Specific concerns were taken into consideration due to these interactions in the field of nanomaterial technology for dental caring, where fluoride is the most common active component in the dental-caring items. Moreover, internalization of NPs and intracellular particles discharge were remarked in the study of Jiravova *et al* causing direct DNA harm in fibroblast cell lines within [63]. Furthermore, Gaillet and Rouanet claimed that AgNPs in toothpaste and other products can aggravate the gastrointestinal tract [64]. Weight loss, blood organic chemistry disturbances, and liver enzyme malfunction were the most hazardous effects seen. In any event, as toothpaste is intended for topical use rather than systemic use, antagonistic effects are unlikely to be observed in such small doses.

11.12.1.3 AuNPs

Toxicology is yet unknown when it comes to the negative effects of AuNPs. However, Jia and colleagues determined in one study that AuNPs are hazardous when used in biological systems because of the concentration range [65]. *In vitro* research revealed that when AuNPs enter bacterium cells, they trigger the production of reactive oxygen species (ROS). This oxidative stress causes unneeded cytotoxicity, which damages DNA, causes cell death through apoptosis and necrosis, and causes cell-cycle arrest.

Gunduz and colleagues recently published an article on the application of refined vascular endothelial cells to non-lethal AuNPs. They showed that after reaching the maximum amount of intracellular NPs, there was constant fatigue but no cell death. Endoplasmic reticulum stretch was reduced as a result of this intake. *In vitro*, NPs take-up is efficiently controlled by cells, according to this paper, as is the long-term effect of NPs exposure [66]. Inhalation of AuNPs by Sprague–Dawley rats for 6 h per day or 5 days per week for 90 days revealed that dose-related alterations were observed within the lungs [67].Because it is well known that *in vitro* and *in vivo* outcomes can differ, additional *in vivo* research is required to reach more reliable conclusions.

11.12.1.4 ZnO NPs

These are typically found in oral care products like toothpaste and mouthwash as antibacterial agents. In addition, the anti-inflammatory activity of ZnO NPs in response to pathogens is an intriguing property for fighting gingivitis or caries [68].

Because of their antibacterial properties, ZnO NPs in toothpaste and mouthwash aid in the fight against gingivitis. Toothpaste containing ZnO has also been shown to have a favorable impact on the dentin by reducing demineralization in an extracted tooth. Zn has also been added to bioactive glass toothpaste because of its antibacterial and anti-gingivitis properties. Increasing the Zn content in toothpaste appears to provide a far greater antibacterial and anti-gingivitis effect [69].

Enamel discoloration is one issue associated with using ZnO NP mouthwash. According to Eslami and colleagues, extracted teeth submerged in ZnO NPs showed the most significant color change, followed by CuO NPs, Ag NPs, TiO_2 NPs, and chlorhexidine, in that order. ZnO NPs are a biologically safe substance that does not cause toxicity in human cells. ZnO NPs are more hazardous to bacteria than their micron counterparts [70].

11.12.1.5 TiO_2 NPs

Later studies show that the utilization of TiO_2 NPs can have harmful impacts [71]. Furthermore, Geraets and colleagues demonstrated that TiO_2 NPs can accumulate in tissues in rats after repeated oral administration [72]. Using explanatory methodologies, Rompelberg and colleagues concluded that toothpaste (in young children only) is the item that contributes the most to TiO_2 consumption, but they did not reach a judgement about the risk to general health [73]. In another study, Baranowska-Wójcik and colleagues found that TiO_2 NPs in modest but consistent doses can affect the intestinal mucosa, the heart, the brain, and other inner organs, potentially increasing the risk of developing systemic diseases or cancers [74].

However, decreasing the toxic impact of metallic NPs on gingiva fibroblasts by employing either capping agents (such as lipoic corrosive or polyethylene glycol) or alloying silver with gold by laser ablation in liquid may be the solution. The downside of such changes is that they may reduce the needed AgNPs antibacterial characteristics of committed for 'targeted on tissue/organ' applications [75].

11.12.2 Possible interactions with alveolar bone

The expected harmful impact of NPs on bone ought to be assessed with respect to biomaterials utilized for root canal dental implants (RCTs) or dental implants. As previously stated, there are nanomaterials intended to interact with bone in order to advance its recovery, such as nanoscaffolds, nano hydroxyapatite (n-Hap), etc. Whether NPs or their by-products apply a deleterious impact on bone turnover osseointegration and embed outcome remain unsolved, particularly for metallic nanomaterials being utilized as an antibacterial additional component to embed surfaces. Added noble nanometals such as Ag, Au and Pt have been postulated as possible antimicrobials that seem to diminish the risk of periimplant infection. In spite of demonstrated antibacterial action, concern has been raised because of the close contact that exists between such materials, approximately possibly destructive impacts related with the aggregation of the NPs into bone and encompassing soft tissue [76].

To date there have been few research works available, generally in labs, that assess the possible antagonistic impact on bone due to direct exposure. A few of them propose, in spite of high antibacterial characteristics, that they could initiate cytotoxic impact to chosen cells and tissues [77]. From recent work by Zielinska *et al*, it appeared that inducible nitric oxide synthase, inducible nitric oxide synthase protein, and increased nitrogen oxide production are linked to apoptosis and necrosis in osteoblast cell lines where they are concentration- and time-dependent of silver NPs [63].

The poisonous effect of titanium resulted from silver NPs on ordinary human osteoblasts was emphasized by Pokrowiecki *et al*, and was attributed to the 'burst release' of Ag particles from the embed surfaces [77]. Furthermore, after incubation with AgNPs, Pauksch *et al* discovered a time- and dose-dependent impairment in cell reasonability and cell multiplication of bone-forming cells [78]. Geng *et al* published a new study that demonstrated the influence of NPs on osseointegration. The researchers used multimodal correlative imaging approaches and quantitative histological considerations to measure bone ingrowth into the implanted titanium and titanium–silver porous scaffolds in a rat demonstration experiment.

Sulfidation of Ag^+ inferred from AgNPs in neighboring freshly formed bone, the researchers concluded, could be a component that reduces cytotoxicity while also reducing antibacterial activity [79]. Furthermore, the type, size, and exposure period of free NPs affect the bone reaction. According to Aruoja *et al*, nano-titanium particles had the lowest harmful effect of all metallic NPs, and the mechanism of cytotoxicity was connected to cell entanglement rather than ions released or NP accumulations [80]. This is very encouraging since it suggests that when custom-made appropriately, Ti-nanostructured implants can have excellent osseointegration while also having an antimicrobial impact. Bone resorption surrounding teeth or dental inserts re-established with nanomaterials, gingival color fading, disruptions, and, most importantly, genotoxicity and mutagenicity are all potential local antagonistic effects of NPs in the oral cavity that have to be studied properly.

11.12.3 Potential interactions of NPs with oral cells and tissues

Because the release of NPs of various sizes in the oral cavity increases the danger, the influence of nanomaterials on local (tooth, gums, mucosa) and general health (GI tract, essential organs) should be examined. Excessive particles discharge from the surface of the NPs often follow the 'burst release,' and their coordinated contact with the cell membrane, changing its porousness and triggering death through oxidative damage, were the first suggested mechanisms of cytotoxicity [81]. A few particles have tampered with proteins and/or are unable to penetrate the cell layer of mammals and reach their interior. The oxidation state and soundness of other ionic shapes of components such as Cu, Fe, Mn, and Zn, which are critical cofactors of proteins, are assigned to their intracellular accumulation and directed by metal particles protein transporters, which play a key role in maintaining homeostasis [82]. Furthermore, hazardous particles may be absorbed by these channels—the DMt1 Fe^{2+} transporter has been reported to carry silver and cadmium particles [83].

Another technicality of toxic ability was hypothesized in which NPs penetrate the cell through endocytosis and then, following ionization, launch ions that influence the cells from within. This technicality is referred to as the 'Trojan-horse' impact and became determined within the work of Park *et al* [84]. The authors defined that SNPs entered the cells through phagocytosis, however, they had been now no longer been determined in lifeless cells, and AgNPs had been launched from the broken cells to the lifestyle medium. According to research by Hsiao *et al* [80], disintegration of intracellular AgNPs became associated ordinarily with H_2O_2 and reactive oxygen species (ROS), and Ag^+ was found to have potential for thiol proteins and peptides, causing toxicity. The consequences mentioned above, proved that the 'Trojan horse' may very probably be one of the toxicity techniques of NPs for human cells. It is extremely likely that those mechanisms of the cells with ions arise: firstly, inside the cell throughout the first several minutes after exposure and through burst launch; secondly, within the Trojan-horse mechanism within the cell [85].

Aside from the toxicological effect of the released ions, which results in ROS activation, NPs can be internalized in one-of-a-kind cell compartments, including the nucleus, resulting in DNA damage. Internalization and accumulation of NPs appear to be non-homogeneous and are linked to NP structure, stabilizers, particle size, surface ligands, biodistribution, and retention period, as well as cell type (phagocytic or nonphagocytic) [86]. Direct interactions of NPs with membrane of the cell no longer take place in the living organisms because the system is now correctly set up. However, protein coronas (PCs) are formed by the creation of NPs protein complexes. NPs released into body fluids interact with their organic constituents, which are usually proteins, producing complexes of proteins linked to the surface of NPs, which can also additionally display strikingly one-of-a-kind biodistribution, clearance, activity, and toxicity.

PC generation is a dynamic, time-based process that is linked to individual NPs and environmental conditions [87]. A PC is formed within seconds of NPs being exposed to the organic fluid of the oral cavity space, and within the first minutes, the so-called 'gentle corona' is sculpted by proteins with high mobility. The vast majority of PC investigations used fluids and human plasma, but there are no statistics on PC formation in saliva.

Proteins will be adsorbed on particles having small sizes more than larger ones, according to the most recent study in this area. Proteins will stretch to evolve on particle surfaces since NPs are larger than proteins. Smaller NPs will no longer be necessary for protein structural changes. Nanomaterial absorption, intracellular distribution, cytotoxicity, irritation, and oxidative stress could all be affected by the PC shape [88]. This impact could additionally fluctuate among each wholesome and unhealthy individual. Because it is the outside shell that the cell can perceive, the PC may also cause increase or decrease the poisonous effect. As a result, it appears to be more important in determining an immune reaction than the particle's inherent material qualities [89].

11.12.4 NPs penetration through the oral and stomach mucosa

When different NPs come into contact with the biological milieu of the mouth cavity and the gastrointestinal tract, they should disintegrate and release potentially toxic particles. Corbo *et al* described such a 'path' of NPs through a living creature as 'from synthetic identification to physiological reaction through biological identity.' [90]. As an illustration, orally overseen NPs with particular properties (engineered character) will begin with shape PCs in saliva (organic identity) and after that interface with other fluids (new natural character) of the gastrointestinal tract (GIT) when swallowed, where a really low pH environment and chemicals will modify the properties and bioavailability of NPs (physiological response). When ingested, altered NPs will be exchanged to other sections of the GIT and, inevitably, to the circulatory system. There's right now no data accessible that can uncover how the different NPs will associate with the various components of the GIT. Orally controlled and held NPs have been found within the lungs, bone marrow, brain, skin, eyes, bladder, stomach, teeth, tongue, salivary organs, thyroid, and parathyroid, according to Gaillet and Rouanet [91]. Oral obstructions are the foremost critical points of entry since verbal mucosa has a 4- to 4000-fold greater penetrability than skin in specific areas.

Mucosa can be partitioned into three sorts:

- masticatory;
- specialized tongue mucosa; and
- lining (buccal mucosa 'BM').

The last mentioned is the most penetrable and, along with the rest of the GIT, should be attended to in terms of NPs-related potential toxicological effects [92]. Despite the fact that BM is a commonly used sample for drug administration and pharmaceutical considerations, its connection with NPs is unknown. Typically, this is owing to a broader understanding of NPs within the oral cavity, which cannot be replicated *in vitro*. The ability of saliva particles to clump together is one of the most important factors in preventing remote NPs from entering the body. Although protein corona organization is one of these processes, there is still limited information about it within the mouth cavity. SNPs may increase their size up to 36% after molding a PC in saliva according to Walczak *et al*, and with their entrance via subsequent regions of the GIT, may experience noteworthy changes [93].

It was established in a research study by Cheng *et al* [89] that in some situations, PC arrangement reduces the attachment onto the cell membrane. It is possible that saliva's agglomerating ability plays a role in reducing the harmfulness of free NPs in the oral cavity. NPs or their organic complexes must be able to penetrate past the mucus layer and the epithelial region of the mucosal border to enter the body.

Roblegg *et al* discovered that the BM layer at the epithelium's site can effectively block the entry of bigger polystyrene NPs with a negative charge [94]. In any case, other exploratory studies have found that small NPs have a positive or neutral charge and due to the lack of adhesive intuitive with mucin, hydrophilic surfaces have a high likelihood of penetration into the mucus layer, and such particles may enhance the chance of aggregation within the body [95].

Konstantinova *et al* did another study on BM penetrability, evaluating nano-TiO_2 entry by re-making ordinary human BM *in vitro*. It was discovered that the depth of infiltration is related to the morphology of the NPs, with circular nano-TiO_2 reaching deeper portions of the epithelium than spindle-shaped nano-TiO_2. In any event, this behavior was observed at higher concentrations within the first few hours. The thickness of the epithelial layer was maintained all through the consider, but desquamation was seen, particularly when BM was exposed to round nano-TiO_2 for longer periods of time and at higher doses, suggesting that continuous interaction of BM with NPs can result in desquamation in negative impacts on human mucosa [93].

Teubl *et al* conducted another study to analyze nano-TiO_2 penetration of BM, and their findings revealed that cellular take-up of NPs in BM is forcefully affected by NPs critical size. Indeed, agglomerated NPs have discernible reactivity and may be absorbed by cells within 10 min of exposure [96]. The researchers also discovered that exposing nano-TiO_2 to light for a short period of time caused the buccal cells to produce ROS. They properly pointed out that risk assessments should encompass the mouth cavity due to the interaction of NPs with yet-to-be-discovered cellular organelles.

Because BM has a 14-day turnover, NPs are expected to reside in the BM for up to 14 days before being moved to the GIT. When NPs are swallowed, they travel they go from the esophagus to the stomach, where they come into contact with gastric fluid, altering their physico-chemical properties once more [97]. The size, surface charge, agglomeration/aggregation state, and surface of NPs can be influenced by gastric pH, buffer capacity, stomach associated substances, proteins, osmolality.

The size of NPs regulated influences their aggregation through several organs. Wang *et al* confirmed this notion by examining the gathering (collection) of nano zinc oxide with diameters of 20 nm and 120 nm in rodents. 20 nm NPs caused injury in the spleen, liver, and pancreatic. ZnO with a diameter of 120 nanometers has an obsessive effect on the stomach mucosa, liver, heart, and spleen [98]. Ebabe Elle *et al* observed increased cholesterolemia, low-density lipoprotein-cholesterol, and liver in mice after oral administration of 20 nm AgNPs for 81 days, showing hepatic toxicity [99]. When nano-anatase TiO_2 was compared to its bulk size, Liu *et al* found that it caused liver damage, ionic substance, and nutrition admissions. The accumulation of ROS, the amount of lipid peroxidation, and the decline of hepatic antioxidative frameworks were considered in mice treated with nano-TiO_2 [100].

11.13 Conclusion

Nanodentistry may be a very promising apparatus in revolutionizing oral health care and esthetics. In any case, the novel prospects are related with discussions about biocompatibility and harmfulness, social dangers, and moral questions included, particularly when mass production is expected. The advancement of nanotechnology has significantly improved the dentistry market in recent years by introducing a modern era of materials that give far superior prophylactic applications and the results of treatment. Nanodentistry remains an unfamiliar field, however, because NPs are normally connected with cells in a distinctive way from little atoms. There is to date a lack of knowledge concerning the potential negative effects of the supplied NPs on 'nontargeted' tissues in the mouth cavity and throughout the GIT.

Furthermore, the available evidence on nanotoxicity is contradictory, as test methods vary from article to study, making comparison difficult. Because of their incredibly small molecule size, NPs can easily enter the body and aggregate in organs, causing harmful side effects. Furthermore, the factors that influence the environment of oral decline and the chemical constitution of saliva are connected to both closely related (periodontal illness, cleanliness measures, use of tobacco, different tooth caries, parafunctions, and so on) and common variables (diabetes, autoimmunological diseases, salivary organ infections, and so on) in each individual.

Table 11.1. Some NPs used in dentistry properties and their toxicity [101, 102].

Name of the material	Purpose of use	Advantage of this material	Toxicity/side effect known
Carbon nanotubes	Teeth filling, a coating of the teeth surface.	Large surface area allows active agents to reach live cells and attaches well to tooth surfaces, dentin surfaces, and cementum surfaces.	Carbon nanotube reactivity is influenced by their structure, size, surface, and purity. By crossing membrane barriers, nanotubes can cause inflammatory and fibrotic reactions in rare cases.
Graphene	Teeth coating, suitable for implantation, biofilm reduction.	Treatment of bacterial biofilms, cost effectiveness, fracture resistance, low density, produces a homogenous crystal structure.	The toxicity of graphene is determined by its structure, size, and oxidative state. Metallic contaminants can be introduced into post-synthesis processing processes, causing a range of toxicity reactions.
Hydroxy apatite (HAp)	Reduces dental hypersensitivity, also acts as cavity filler, retards auxiliary demineralization, repair of enamel surface.	HAp particles that are nanoscale can easily be incorporated into tooth tubules. Similar to teeth and bone in composition, biocompatible, adsorbed to enamel of the teeth, protects teeth by forming a coating of artificial enamel around the tooth, and corrects periodontal flaws.	They have the ability to bind to proteins and form protein particle complexes, which are then destroyed by macrophages in tissues. Blood carries these particles to the lungs, spleen, and liver, where they dissipate. NP toxicity can influence the inflammatory response, signaling system, and oxidative stress.
Zirconia	Reduces bacterial adhesion to the tooth surface, provides protection against dental carries, effective polishing agent.	Low cytotoxicity, good biocompatibility, and excellent fracture resistance are all characteristics of this material, which has mechanical qualities and a hue similar to that of a tooth.	After exposure to zirconium oxide NPs, some short- and long-term concerns have been observed, including considerable DNA damage in human T-cells, induction of apoptosis, and reduction of cell proliferation in human mesothelioma

(Continued)

Table 11.1. (*Continued*)

Name of the material	Purpose of use	Advantage of this material	Toxicity/side effect known
			and mouse fibroblast cell lines. It was also discovered to cause cellular oxidative stress, which resulted in cell death. According to studies, these NPs can interrupt the cell cycle as well as overcome numerous physiological barriers, causing negative effects.
Silica	Dental filling agent, tooth polishing, prevents dental caries, an antibacterial agent, to treat dental hypersensitivity.	They are biocompatible, have a minimal toxic effect, are low in density, have a high adsorption capacity, and are, most importantly, cost effective. When used as a polishing agent, the roughness of the teeth surface is reduced.	Toxic effects are determined by the entrance route and physiochemical characteristics. Recent research has found that silica NPs, like crystalline particles, can cause silicosis and lung cancer. Cytotoxicity is caused by SiNPs. Furthermore, depending on the size and dose, it might cause oxidative stress and mediate apoptosis. SiNPs have also been shown to have genotoxic (DNA damage, regulation of genes involved in apoptosis and autophagy) and immunotoxic effects in several investigations.
Titania	Mainly dental implants.	Surface modification has a longer-term effect on dental implants, providing additional benefits such as reduced bacterial adherence and increased hardness.	The toxicity of NPs is higher than that of fine particles. It enters the body via inhalation. Workers in TiO_2 manufacturing plants develop cancer (revealed in epidemiologic studies).

			By penetrating the blood–brain barrier in the cortex and hippocampus, TiO_2 NPs can accumulate inside the brain. Microglia activation, formation of ROS, and activation of signaling pathways all play a part in cell death and inflammation when exposed to TiO_2.
Silver	Antimicrobial agent, dental restorative material, dental prosthetics, dental implants.	Because of their tiny size, they can easily penetrate bacterial membranes, reducing bacterial colonization and improving dental health. They are biocompatible, with minimal toxicity and antibacterial activity that lasts for a long time.	Toxicity is caused by AgNPs. Chronic silver exposure can cause argyria. The toxic effects of AgNPs are caused by the formation of ROS. The toxicity is caused by both silver ions and AgNPs. AgNPs play a role in oxidative stress and genotoxicity, as well as lysosomal AcP activation, actin disruption, hemocyte phagocytosis stimulation, and Na-K-ATPase inhibition.
Aluminum oxide	Has been used to increase the strength of dental porcelains for more than four decades.	Endodontic posts, orthodontic brackets, dental implants, crowns and bridges, and ceramic abutments have all been made with alumina.	The amounts of aluminum in the tissues are not disclosed. Hepatic superoxide dismutase inhibition at day 3 post-exposure, which was substantially gone by day 14 post-exposure.

Until the present, no studies on the cytotoxicity of NPs on humans have been carried out. A variety of factors determine the harmful effect of NPs, including dose, type, particle size, distribution, duration of activity, interaction with other components, and so on. As a result, comparing the harmful effect (toxicity) of different NPs is difficult due to barriers and differences in methods, as well as complicated organic chemistry of the mouth cavity that cannot be replicated *in vitro*. All of these variables could have confounding effects and be a major source of discrepancy across published experiments.

11.14 Future perspectives

The following difficulties should be considered in future research investigations when it comes to the neighborhood impacts of nanomaterials within oral decline:

- In terms of safety and health, nanotechnology should 'go green.'
- In this case, an expert meeting should be called to carefully weigh the benefits of green nanotechnology against the potential costs to society, particularly in terms of environmental, public, and occupational health. This method of thinking will maximise the benefits to the environment and society, as well as the health benefits and cost savings.
- To properly assess the risks, logical research, inventive, legislation, and workforce endeavors must be used to characterize the adverse impact of NPs. This would provide useful information for developing defensive measures for comprehensive hazard management for both the general public and occupationally exposed people.
- Better guidance to experts on nano-consolidated goods in terms of quiet security and word-related health. The nano-specific naming on the numerous distinct dental care items could be advanced to clearly separate the nano ingredient for transparency with patients. As a result, it will be clear whether an item has an ENM and what the proposed mode of action or benefit of the novel item is to the consumer. More in-depth investigation work is required in the last two items.
- The impacts of NPs having covers on osseointegration and tooth structures, for example: enamel, dentin, and cementum, as well as the effect of NPs on the oral mucosa and gingiva.
- The problems with protein crowns organization in saliva appear to be a missed link between experiments in labs (*in vitro*) and those performed on living organisms (*in vivo*) considerations, which can provide insight on the nanodentistry disadvantages previously noted.

List of abbreviations

NPs:	nanoparticles
NMTs:	nanomaterials
ENMs:	engineered nanomaterials
n-HAP:	nano hydroxyapatite
RCT:	root canal dental implant

GIT: gastrointestinal tract (digestive system)
DMT1: Divalent metal (ion) transporter
ROS: reactive oxygen species
PC: protein corona
BM: buccal mucosa
TiO_2: titanium dioxide
ZnO: zinc oxide
CuO: copper oxide
AgNPs: silver nanoparticles
AuNPs: gold nanoparticles

References

[1] Mallanagouda P, Dhoom Singh M and Sowjanya G 2008 Future impact of nanotechnology on medicine and dentistry *J. Ind. Soc. Periodontol* **12** 34–40

[2] Reshma R 2021 Review article nanotechnology in interdisciplinary dentistry *Indian J. Med. Sci.* **73** 226–9

[3] Freitas R Jr 2005 Nanotechnology: nanomedicine and nanosurgery *Int. J. Surg.* **3** 243–46

[4] Chen H F, Clarkson B H, Sunk and Mansfield J F 2005 Self-assembly of synthetic hydroxyaptite nanorods into enamel prism like structure *J. Colloid Interf. Sci.* **188** pp 97–108

[4] Maryam M 2013 Review future of dentistry, nanodentistry, ozone therapy and tissue engineering *J. Develop. Biol. Tissue Eng.* **5** 1–6

[5] Cao W and Zhang Y 2018 Novel resin-based dental material with anti-biofilm activity and improved mechanical property by incorporating hydrophilic cationic copolymer functionalized nanodiamond *J. Mater. Sci. Mater. Med.* **29** 162

[6] Amini H, Shenkin J D and Chi D L 2019 Dental public health issues in pediatric dentistry *Pediatr. Dent.* **617** 159–68

[7] Thomas N, Blake S, Morris C and Moles D R 2018 Autism and primary care dentistry: parents' experiences of taking children with autism or working diagnosis of autism for dental examinations *Int. J. Paediatr. Dent.* **28** 226

[8] Loke K F, Mohammad M F, Armita F M and Mohadeseh S *et al* 2020 Review article: applications of nanomaterials in diverse dentistry regimes *RSC Adv.* **10** 15430–60

[9] Saafan A and Zaazou M H 2018 Assessment of photodynamic therapy and nanoparticles effects on caries models *Open Access Maced. J. Med. Sci.* **6** 1289–95

[10] Magalhaes A P and Moreira F C 2016 Silver nanoparticles in resin luting cements: antibacterial and physiochemical properties *J. Clin. Exp. Dent.* **8** 415–22

[11] Bhuvaneswarri J, Nazish Alam M, Chandrasekaran S C and Sathya M S 2013 , Future impact of nanotechnology in dentistry—a review Int. J. Nanotechnol. Appl. *(IJNA)* **3** 15–20

[12] Abiodun-Solanke I M F, Ajayi D M and Arigbede A O 2014 Nanotechnology and its application in dentistry *Ann. Med. Health Sci. Res.* **4** 171–77

[13] Robert A F 2005 Progress in nanomedicine and medical nanorobotics https://researchgate.net/publication/241612132

[14] Shaeesta K B 2014 Current applications of nanotechnology in dentistry: a review *Gen. Dent.* **62** 72–7

[15] Neetha J S, Swati P and David K 2013 Review article, nanorobots: future in dentistry *Saudi Dent. J.* **25** 49–52

[16] Banu V 2012 Applications of nanomedicine in oral cancer *Oral Health Dent. Manag.* **11** 62–68

[17] Nikalje A P 2015 Nanotechnology—its applications in medicine *Med. Chem.* **5** 81–89
[18] Hatziantoniou S and Demetzos C 2006 An introduction to nanotechnology in health care *Pharmakeftiki* **19** 86–88
[19] Himanshu A, Varun K, Shashank U and Pooja T 2015 Nanodentistry: is just a fiction or future *J. Oral Biol. Craniofac. Res.* **5** 207–11
[20] Freitas R Jr 2005 Nanotechnology, nanomedicine and nanosurgery *Int. J. Surg.* **3** 243–46
[21] Sneha S, Sashi R A and Vidya S 2013 Review article: nanodentistry *Indian J. Sci. Res.* **4** 233–38
[22] Rybachunk A V and Chelkman I S 2009 Nanotechnology and nanoparticles in dentistry *Pharmacol. Pharm.* **1** 18–21
[23] Rita C, Kala M, Kiran B, Priyank B and Ruchi B 2012 Review article: 'Nanodentistry': exploring the beauty of miniature *J. Clin. Exp. Dent.* **4** 119–24
[24] Zohaib K, Muhammad Z, Saad Q, Sana S, Mustafa N and Ammar A R 2015 Advances in nanotechnology for restorative dentistry *Mater. Eng.* **8** 717–31
[25] Shellart W C and Oesterle I J 1999 Uprighting molars without extrusion *JADA* **130** 38
[26] Bumb S S, Bhaskar D J and Punia H 2013 Nanorobots and challenges faced by nanodentistry *Guident.* pp 67–69
[27] Neeta S and Swati P 2013 Nanorobots: future in dentistry *Saudi Dental J.* **25** 49–52
[28] Upadhyay Y 2013 Current state and future perspectives of nanotechnology in dentistry *Int. Organ. Sci. Res. J. Pharm.* **3** 68–71
[30] Hernandez J D, Perez M A and Valenzuela M C 2011 Use of new technologies in dentistry *Rev. Odontol. Mex.* **15** 158–63
[31] Fakruddin M, Hossain Z and Afroz H 2012 Prospects and applications of nanobiotechnology: a medical perspective *J. Nanobiotechnol.* **10** 1–8
[32] Rao K V and Kumar J S 2013 Nanotechnology in dentistry *Kerala Dent. J.* **36** 56–59
[33] Lainovic T, Vilotic M, Blazic L, Kakas D, Markovic D and Ivanisevic A 2013 Determination of surface roughness and topography of dental resin-based nanocomposites using AFM analysis *Bosn. J. Basic. Med. Sci.* **13** 35–43
[34] Wang W, Liao S, Zhu Y, Liu M, Zhao Q and Fu Y 2015 Recent applications of nanomaterials in prosthodontics *J. Nanomater.* **2** 1–11
[35] Ozak S T and Ozkan P 2013 Nanotechnology and dentistry *Eur. J. Dent.* **7** 145–51
[36] Rathee M and Bhoria M 2014 Nanodentistry: the emerging tiny tools—a review *Int. J. Biosci. Nanosci.* **1** 63–67
[37] Saravana K R and Vijaylakshmi R 2006 Nanotechnology in dentistry *Indian J. Dent. Res.* **17** 62–65
[38] Kong L X, Peng Z, Li S and Mark P 2000 Nanotechnology and its role in the management of periodontal diseases *Periodontology* **40** 184–96
[39] Jesitus J Nanotechnology accelerates wound healing http://dermatologytimes.modernmedicine.com
[40] Priyadarsini S, Mukherjee S and Mishra M 2018 Nanoparticles used in dentistry: a review *J. Oral Biol. Craniofac. Res.* **8** 58–67
[41] Business Wire 2019 Global $36+ billion toothpaste market, 2024: growth, trends and forecast analysis from 2019 https://businesswire.com/news/home/20190509005698/en/Global-36-Billion-Toothpaste-Market-2024-Growth-Trends-and-Forecast-Analysis-from-2019---ResearchAndMarkets.com

[42] European Commission 2018 Catalogue of nanomaterials in cosmetic products placed on the market

[43] Florence C, Stephane V, Livia O, Cedric G and Denis B 2020 Review nanoparticles as anti-microbial, anti-inflammatory, and remineralizing agents in oral care cosmetics: a review of the current situation *Nanomaterials* **10** 1–32

[44] Loke K F, Mohammad M F, Armita F M and Mohadeseh S *et al* 2020 Applications of nano materials in diverse dentistry regimes *RSC Adv.* **10** 15430–60

[45] Resnik D B and Tinkle S S 2007 Ethics in nanomedicine *Nanomedicine* **2** 345–50

[46] Eman M O 2019 Environmental and health safety considerations of nanotechnology: nano safety *Biomed. J. Sci. Techn. Res. BJSTR* **19** 14501–15

[47] Hester K, Mullins M, Murphy F and Tofail S A M 2015 Anticipatory ethics and governance (AEG): towards a future care orientation around nanotechnology *Nano Ethics* **9** 123–36

[48] Brey P A 2012 Anticipatory ethics for emerging technologies *NanoEthics* **6** 1–13

[49] Khushf G 2006 An ethic for enhancing human performance through integrative technologies *Managing Nano-Bio-Info Cogno Innovations* (Berlin: Springer) pp 255–78

[50] Gupta N, Fischer A R and Frewer L J 2015 Ethics, risk and benefits associated with different applications of nanotechnology: a comparison of expert and consumer perceptions of drivers of societal acceptance *NanoEthics* **9** 93–108

[51] Kurzweil R 2006 The singularity is near: when humans transcend biology *Technol. Forecast. Soc. Change* **73** 95–127

[52] Macnaghten P, Kearnes M B and Wynne B 2005 Nanotechnology, governance, and public deliberation: what role for the social sciences *Sci. Commun.* **27** 268–91

[53] Erik F and Roop L M 2006 Contradictory intent? US federal legislation on integrating societal concerns into nanotechnology research and development *Sci. Public Policy* **33** 5–16

[54] Leo S and Louis T 2011 Environmental implications of nanotechnology—an update *Int. J. Environ. Res. Public Health* **8** 470–79

[55] Rawan N A 2018 The Implications and applications of nanotechnology in dentistry: a review *Saudi Dental J.* **30** 107–16

[56] Stone V, Hankin S, Aschberger R K, Anders B, Christensen F and Fernandes T *et al* 2010 Engineered nanoparticles: review of health and environmental safety (ENRHES) *Project Final Report, European Commission* http://nmi.jrc.ec.europa.eu/documents/pdf/ENRHES%20Review.pdf

[57] Jyoti G 2011 Review Article: dental biomaterials nanotechnology applications in medicine and dentistry *J. Investig. Clin. Dent.* **2** 81–88

[58] Karlsson H L, Cronholm P, Gustafsson J and Möller L 2008 Copper oxide nanoparticles are highly toxic: a comparison between metal oxide nanoparticles and carbon nanotubes *Chem. Res. Toxicol.* **21** 1726–32

[59] Rafal P, Krzysztof P and Agnieszka M 2018 Nanomaterials in dentistry: a cornerstone or a black box *Nanomedicine* **3** 639–67

[60] Kopera D, Soyer H P and Kopera H 1998 Nonprecious metal tattoos (amalgam tattoos) of the oral mucosa can be successfully removed with the Q-switched Ruby laser *Laser in Medicine* (Berlin: Springer) pp 453–55

[61] Buchner A 2004 Amalgam tattoo (amalgam pigmentation) of the oral mucosa: clinical manifestations, diagnosis and treatment *Refuat Hapeh Vehashinayim* **21** 25–8

[62] Medina C, Inkielewicz S I, Santos M J and Radomski M W 2014 Pharmacological and toxicological effects of co-exposure of human gingival fibroblasts to silver nanoparticles and sodium fluoride *Int. J. Nanomed.* **9** 1677

[63] Zielinska E, Tukaj C, Radomski M W and Inkielewicz-Stepniak I 2016 Molecular mechanism of silver nanoparticles-induced human osteoblast cell death: protective effect of inducible nitric oxide synthase inhibitor *PLoS One* **11** e0164137

[64] Gaillet S and Rouanet J M 2015 Silver nanoparticles: their potential toxic effects after oral exposure and underlying mechanisms—a review *Food Chem. Toxicol.* **77** 58–63
Gunduz N, Ceylan H, Guler M O and Tekinay A B 2017 Intracellular accumulation of gold nanoparticles leads to inhibition of macropinocytosis to reduce the endoplasmic reticulum stress *Sci. Rep.* **7** 1–10

[65] Jia G, Zhi A, Lai P F H, Wang G, XiaY, Xiong Z, Zhang H, Che N and Ai L 2018 The oral microbiota—a mechanistic role for systemic diseases *Br. Dent. J.* **224** 447–55

[66] Gunduz N, Ceylan H, Guler M O and Tekinay A B 2017 Intracellular accumulation of gold nanoparticles leads to inhibition of macropinocytosis to reduce the endoplasmic reticulum *Stress. Sci. Rep.* **7** 1–10

[67] Sung J H, Ji J H, Park J D, Song M Y, Song K S, Ryu H R, Yoon J U, Jeon K S, Jeong J and Han B S *et al* 2011 Subchronic inhalation toxicity of gold nanoparticles *Part. Fibre Toxicol.* **8** 16

[68] Agarwal H, Nakara A and Shanmugam V K 2019 Anti-inflammatory mechanism of various metal and metal oxide nanoparticles synthesized using plant extracts: a review *Biomed. Pharmacother. Biomed. Pharmacother.* **109** 2561–72

[69] Lynch E, Brauer D S, Karpukhina N, Gillam D G and Hill R G 2012 Multi-component bioactive glasses of varying fluoride content for treating dentin hypersensitivity *Dent. Mater.* **28** 168–78

[70] Kachoei M Y, Divband B, Tabriz F D, Helali Z N and Esmailzadeh 2018 A comparative study of antibacterial effects of mouthwashes containing Ag/ZnO or ZnO nanoparticles with chlorhexidine and investigation of their cytotoxicity *Nanomed. J.* **5** 102–10

[71] Karlsson H L, Cronholm P, Gustafsson J and Moller L 2008 Copper oxide nanoparticles are highly toxic: a comparison between metal oxide nanoparticles and carbon nanotubes *Chem. Res. Toxicol.* **21** 1726–32

[72] Geraets L, Oomen A G, Krystek P, Jacobsen N R, Wallin H, Laurentie M, Verharen H W, Brandon E F and de Jong W H 2014 Tissue distribution and elimination after oral and intravenous administrationof different titanium dioxide nanoparticles in rats *Part. Fibre Toxicol.* **11** 30

[73] Rompelberg C, Heringa M B, van Donkersgoed G, Drijvers J, Roos A, Westenbrink S, Peters R, vanBemmel G, Brand W and Oomen A G 2016 Oral intake of added titanium dioxide and its nanofraction from food products, food supplements and toothpaste by the Dutch population *Nanotoxicology* **10** 1404–14

[74] Baranowska-Wójcik E, Szwajgier D, Oleszczuk P and Winiarska-Mieczan A 2019 Effects of titanium dioxidenanoparticles exposure on human health—a review *Biol. Trace Elem. Res.* **193** 118–29

[75] Niska K, Knap N, Kedzia A, Jaskiewicz M, Kamysz W and Inkielewicz-Stepniak I 2016 Capping agent-dependent toxicity and antimicrobial activity of silver nanoparticles: an in vitro study, concerns about potential application in dental practice *Int. J. Med. Sci.* **13** 772–82

[76] Geng H, Poologasundarampillai G and Todd N *et al* 2017 Biotransformation of silver released from nanoparticle coated titanium implants revealed in regenerating bone *ACS Appl. Mater. Interfaces* **9** 21169–80

[77] Pokrowiecki R, Zareba T and Szaraniec B *et al* 2017 In vitro studies of nanosilver doped Ti-implants for oral and maxillofacial surgery *Int. J. Nanomed.* **12** 4285–97

[78] Pauksch L, Hartmann S and Rohnke M *et al* 2014 Biocompatibility of silver nanoparticles and silver ions in primary human mesenchymal stem cells and osteoblasts *Acta Biomater.* **10** 439–49

[79] Hsiao I L, Hsieh Y K, Wang C-F, Chen I C and Huang Y J 2015 Trojan-Horse mechanism in the cellular uptake of silver nanoparticles verified by direct intra- and extracellular silver speciation analysis *Environ. Sci. Technol.* **49** 3813–21

[80] Aruoja V, Dubourguier H C, Kasemets K and Kahru A 2009 Toxicity of nanoparticles of CuO, ZnO and TiO_2 to microalgae *Pseudokirchneriella Subcapitata Sci. Total Environ.* **407** 1461–68

[81] McShan D, Ray P C and Yu H 2014 Molecular toxicity mechanism of nanosilver *J. Food Drug Anal.* **22** 116–27

[82] Kaplan J H and Maryon E B 2016 How mammalian cells acquire copper: an essential butpotentially toxic metal *Biophys. J.* **110** 7–13

[83] Lyons T and Eide D 2007 Transport and storage of metal ions in biology *Biological Inorganic Chemistry—Structure and Reactivity* (Sausolito, CA: University Science Books) pp 57–78

[84] Park E J, Yi J, Kim Y, Choi K and Park K 2010 Silver nanoparticles induce cytotoxicity by a Trojan-horse type mechanism *Toxicol. in Vitro* **24** 872–78

[85] Jiravova J, Tomankova K B and Harvanova M *et al* 2016 The effect of silver nanoparticles and silver ions on mammalian and plant cells *in vitro Food Chem. Toxicol.* **96** 50–61

[86] Shang L, Nienhaus K and Nienhaus G U 2014 Engineered nanoparticles interacting with cells: size matter *J. Nanobiotechnol.* **12** 5

[87] Zhang X-F, Shen W and Gurunathan S 2016 Silver nanoparticle-mediated cellular responses in various cell lines: an *in vitro* model *Int. J. Mol. Sci.* **17** 1603

[88] Cheng X, Tian X and Wu A *et al* 2015 Protein corona influences cellular uptake of gold nanoparticles by phagocytic and nonphagocytic cells in a size-dependent manner *ACS Appl. Mater. Interfaces* **7** 20568–75

[89] Dobrovolskaia M A, Patri A K and Zheng J *et al* 2009 Interaction of colloidal gold nanoparticles with human blood: effects on particle size and analysis of plasma protein binding profiles *Nanomed. Nanotechnol. Biol. Med.* **5** 106–17

[90] Corbo C, Molinaro R, Parodi A, Toledano Furman N E, Salvatore F and Tasciotti E 2016 The impact of nanoparticle protein corona on cytotoxicity, immunotoxicity and target drug delivery *Nanomedicine* **11** 81–100

[91] Gaillet S and Rouanet J-M 2014 Silver nanoparticles: their potential toxic effects after oral exposure and underlying mechanisms—a review *Food Chem. Toxicol.* **77C** 58–63

[92] Konstantinova V, Ibrahim M and Lie S A *et al* 2017 Nano-TiO_2 penetration of oral mucosa: *in vitro* analysis using 3D organotypic human buccal mucosa models *J. Oral Pathol. Med.* **46** 214–22

[93] Walczak A P, Fokkink R and Peters R *et al* 2013 Behavior of silver nanoparticles and silver ions in an *in vitro* human gastrointestinal digestion model *Nanotoxicology* **7** 1198–210

[94] Roblegg E, Frohlich E, Meindl C, Teubl B, Zaversky M and Zimmer A 2012 Evaluation of a physiological *in vitro* system to study the transport of nanoparticles through the buccal mucosa *Nanotoxicology* **6** 399–413
[95] Laffleur F, Hintzen F, Shahnaz G, Rahmat D, Leithner K and Bernkop-Schnurch A 2014 Development and *in vitro* evaluation of slippery nanoparticles for enhanced diffusion through native mucus *Nanomedicine* **9** 387–96
[96] Teubl B J, Leitinger G and Schneider M *et al* 2014 The buccal mucosa as a route for TiO_2 nanoparticle uptake *Nanotoxicology* **5390** 1–9
[97] Fröhlich E and Roblegg E 2016 Oral uptake of nanoparticles: human relevance and the role of *in vitro* systems *Arch. Toxicol.* **90** 2297–314
[98] Wang B, Feng W and Wang M *et al* 2008 Acute toxicological impact of nano- and submicro-scaled zinc oxide powder on healthy adult mice *J. Nanoparticle Res.* **10** 263–76
[99] Ebabe Elle R, Gaillet S and Vide J *et al* 2013 Dietary exposure to silver nanoparticles in Sprague–Dawley rats: effects on oxidative stress and inflammation *Food Chem. Toxicol.* **60** 297–301
[100] Liu H, Ma L, Liu J, Zhao J, Yan J and Hong F 2010 Toxicity of nano-anatase TiO_2 to mice: liver injury, oxidative stress *Toxicol. Environ. Chem.* **92** 175–86
[101] Subhashree P, Sumit M and Monalisa M 2018 Review article nanoparticles used in dentistry: a review *J. Oral Biol. Craniofac. Res.* **8** 58–67
[102] Besinis A, De Peralta T, Tredwin C J and Handy R D 2015 A review of nanomaterials in dentistry: interactions with the oral microenvironment, clinical applications, hazards and benefits *ACS Nano* **9** 2255–89

Chapter 12

Self-assembly of teeth proteins and their applications in dentistry

Monalisa Mishra

Teeth are the hardest part of our body and structurally can be divided into enamel, dentin and pulp. The precursor cells for these different parts of the teeth vary from one region to another. Different types of cells which correspond to the variation of proteins and enzymes present within it are specific to that region. Each protein has a specific role in the development and maintenance of the teeth. A healthy tooth is as a function of well-coordinated proteins present within the teeth. Similarly protein disorder is also reported in many diseased teeth. Alternatively, a healthy tooth can be maintained properly by regulating the function of proteins associated with teeth. Thus, many studies were carried out to understand the role of proteins in the healthy maintenance of the teeth. The current chapter summarises the role of different proteins present within the teeth and how they can be used to treat teeth diseases.

12.1 Introduction

The teeth are the hardest part of our body. They have highly organized mineral tissues which are formed from many specialized cells. Each cell has different proteins which are required for the healthy maintenance of the teeth [1]. The outermost region of the teeth is called enamel and it is as hard as steel. The enamel is exposed towards the oral cavity. Thus it is challenged with many microbes and altered condition of pH. The microbes present in the oral cavity mix with our food whenever we eat and make the food contaminated [2]. Often microbes affect the teeth and form pores in the teeth resulting in infected teeth. Microbes can degrade sugar and other fermented material which has a deleterious effect on the teeth minerals [3]. Degradation of enamel causes a serious threat to oral hygiene. Thus, dental caries is a multifactorial disorder occurring due to bacteria, protein in the saliva and carbohydrates present in the food. Thus new approaches have been developed to treat teeth enamel via natural proteins to prevent dental cavities. Cavities can be

doi:10.1088/978-0-7503-3671-0ch12

regenerated by delivering peptides to the damaged enamel [4]. Like peptides, several proteins are also designed to rebuild damaged teeth and recover from caries [5]. An infected tooth makes a significant contribution towards an infected digestive system. Since all our neurons end up in the digestive system, a healthy gut helps to maintain a good physiological system.

12.2 Structure of teeth

The tooth has many structural components essential to maintaining its integrity. Teeth are enriched with many minerals and proteins which contribute towards their maintenance. Different parts of the teeth have different proteins. From a structural point of view a longitudinal section of a tooth has three different structures: crown, neck and root [1]. The crown is the part of the teeth which is visible to us. The other two reside within the jaw. In a transverse section the crown has three different layers: (1) enamel (2) dentin and (3) dental pulp [1]. Out of these three layers enamel is enriched with minerals followed by dentin. The dental pulp has blood vessels, nerves, and mesenchyme tissue. The root has dentin, cementum, and periodontal ligament. The colour of the teeth is associated with the enamel which changes with age, temperature and physiological conditions. The enamel has 96% hydroxyapatite (HaP) and the remaining 4% corresponds to organic substances and water [6]. HaP forms rod-like structures which are at right angle to the enamel and form a dentin–enamel junction. Distance amongst inter-rods is 0.1 μm. An enamel rod is encircled by organic substances and there exists space between crystals. Interprismatic space between enamel rods is 0.1 μm. The pores in the enamel are due to prism junction (6 nm) and intercrystalline pores (4–6 nm) [7]. Inter-rod space also affects the light scattering ability of the teeth. In yellow teeth the inter-rod space is occupied with fine granular structures [7]. Phosphate is produced during demineralization and remineralisation. Calcium phosphate is composed of octacalcium phosphate (OCP) and Whitlockite (WH), HaP, Brushite (DCPD). These phosphates also contribute for healthy maintenance of the teeth.

12.2.1 Formation of enamel

Enamel is the hardest part that covers the dentin. The tooth enamel development has four different stages: (1) presecretory, (2) secretory, (3) short transition and (4) maturation stage. During the maturation stage, apoptosis of ameloblast cells occurs resulting in eruption of teeth to the oral cavity [8]. Ameloblast alters the morphology of the top part and enters into a transition stage from smooth to ruffle ended [9]. This alternative morphology is a hallmark for the maturation stage which is due to cleaved enamel matrix proteins to mineral crystals. As a consequence the protein present decreases and mineral content increases [4]. The structural enamel matrix protein, ion transportation and regulation of pH are requisites for correct enamel mineralization. The elements required for the teeth are present in the tooth enamel. Different trace elements present within the teeth help to prevent or protect the teeth from caries formation [10]. Atomic absorption spectroscopy and plasma resonance mass spectroscopy are also used to detect the biomass present within the teeth [11].

Besides, this protein present within the teeth has a key role in the development, regeneration and mineralization of the teeth. Gel-based assays are also developed to determine the proteins present within the teeth [12]. Iron, cobalt, magnesium and cadmium are most abundant in the enamel. Among these four, magnesium concentration is found to be highest, i.e. 5400–6000 μg gm^{-1} [13]. Elements like aluminum, copper and zinc are also present in the teeth. The zinc is acquired by HaP and it competes for calcium. Since Zn helps in the remineralisation of enamel,a higher amount of Zn makes the teeth susceptible to caries [13].

12.2.2 Dentin

This is the main structure which provides flexibility to the teeth. It is a highly mineralised tissue that covers pulp and supports the underlying enamel. It has predentin as an unmineralised portion of the teeth. Predentin is present amongst odontoblast and the mineralised dentin remains throughout life. The dentinal tubule runs amongst the pulp and dentin [14]. Dentin has 70% mineral and 20% glycoprotein and phospholipid and 10% water [15].

12.2.3 Cementum, dental pulp and enamel pellicle

The cementum is formed of special cells known as cementoblast. It covers the dental root and anchors the tooth to the alveolar bone. It is avascular in nature. It is non-innervated mineralised tissue having special morphology and matrix [16].

Dental pulp is the soft part of the teeth situated between the coronal pulp chamber and root canal. The pulp region is filled with connective tissue, lymphs and blood vessels [17]. The odontoblast layer of the pulp has an excess of sensory neurons that help to transmit the sensory stimuli from the dental pulp complex to protect the cell from pathogens [18].

The enamel pellicle is a layer on the enamel surface of 0.5–1 μm. It is formed of mucin, protein and glycoprotein [19]. Pellicle has important role in crystal growth and homeostasis [20].

12.3 Proteins present in different parts of the teeth

Teeth are enriched with inorganic salts and proteins [12]. Although the proteins are present at lower level, they are essential to conduct the cellular and molecular activities of the teeth [12]. Based on their structure the proteins are divided in to (1) structured and (2) unstructured proteins. Earlier it was thought that only the structured proteins regulate most of the activity. Recently, the role of unstructured proteins in cellular activities is also elucidated. Such proteins are labelled as intrinsically disordered proteins (IDPs). IDPs have affinity towards polar or charged amino acids (Lys, Arg, Glu, Gln, Ser) and no reaction towards hydrophobic amino acids (Try, TRP, Phe, Leu, Ile). These IDPs are flexible and help in the biomineralisation process. IDPs do not form α-helix and β-helix. After post translational modification or when bound with other proteins IDPs become more folded in their structure. Many proteins involved in biomineralisation of hydroxyapatite are IDPs. These help to perform several functions.

12.3.1 Protein from enamel

Variations were observed in different parts of the teeth with respect to the proteins. The enamel is produced from specialized cells known as ameloblast [21]. Enamel formation follows a secretory stage followed by a maturation stage. The secretory stage has more than 30% of protein. In the maturation stage the matrix gets degraded by protease. 95% is mineral and the proteins present in enamel are known as ameloblastin. 90% of the enamel proteins and enzymes are known as amelogenin. It has a low molecular weight. The remaining 10% are nonamilogenins protein. It includes enamelin and ameloblastins and tuftelin [21]. Amelogenins are hydrophobic in nature and enriched with proline, histidine, and glutamine. They belong to the extracellular matrix (ECM) family and are abundant in the enamel region. Amelogenins are regulated by several genes. Amilogenin controls morphology, organization and direction of the apatite crystals. Amilogenin promotes development of cementum. Nonamilogenin undergoes rapid extracellular processing and they do not accumulate in the enamel. Enamelin is essential for normal enamel tissue and highly conserved across species. Enamelin interacts with amelogenin and changes its confirmation. It stabilizes oligomers and dissociates amilogenin. One of the nonamilogenic proteins is tuftelin. It is an acidic phosphorylated glycoprotein. It promotes the mineralization process during the development of teeth. It is expressed in the enamel–dentin junction prior to the formation of amelogenin [22]. All these proteins have a role in the development of enamel. Besides, two other matrix proteins are essential for proper enamel formation. They are matrix metallo proteinase-20 (MMP-20) and kallikrein-4(KLK-4). These two proteases remove matrix protein and provide space for growth of apatite crystal. Ameloblast secretes KLK-4 from secretory to maturation stage. Absence of KlK-4 does not impact enamel thickness and microstructure of teeth.

12.3.2 Enamel matrix protein

Odontoblast specific gene products express highly mineralized ECM of dentin. HaP is the soul inorganic part, whereas the organic part is formed of type I collagen [23]. The noncollagenous proteins are decorin, biglycan and osteonectin. Osteopontin, and bone sialoprotein which are detected in the dentin. Two other ECM proteins specific to dentin are dentin sialoprotein (DSP) and dentin silaophosphoprotein (DSPP) [23]. Enamelysin (MMP20) is a member of the MMP family secreting marixprotein. Serin proteinase family member Kallikrein4 (KLK4) is a bulk digestive enzyme. The product of ameloblast gets secreted to the enamel and has a role in crystal formation. Two other secretory proteins are amelotin and odontogenic ameloblast-associated matrix (ODAM) [24]. Both are produced from matured ameloblast. Amelotin and ODAM both reside on the surface of the enamel layer and help in the maturation of enamel. Both the proteins are found in the junctional epithelium and are essential to maintain the integrity of the teeth. ECM is the major component of enamel and has unmineralised pre-enamel layers such as osteoid and predentin. Morphologically the enamel appears in a uniformly decalcified manner. The nonamilogenin molecules create interaction with the cell membrane for elongation of crystal. Amilogenin and amiloblastin both are secreted from secretory granules. Amilogenin is a multifunctional protein that participates in epithelial-mesenchymal events. Amilogenin helps in the growth and thickness of crystals. The role of amyloblast products is evaluated in mice models. Phosphophoryn's other name is dentin. It is the product of dentin sialophosphoprotein (DSPP) and dentin phosphoprotein (DPP) [25]. DSPP is the most dominant protein of dentin and enamel. An abnormal level of defective enamelin also affects the nonamilogenin. In Mmp20 and KlK4 mutant the crystal growth is affected. Disruption of laminin 332 causes animal hypoplasia. Mineralization of the tooth is found in enamel, dentin and cementum regions. In the enamel 90% amelogenin protein was detected [3]. The calcification of the tooth was tightly regulated by the proteins present in the ECM. The role of unstructured protein in teeth mineralization was first detected in 1977 from the study of presence of calcium. Besides this, dentin is a reservoir of growth factors like TGFB, bone morphogenetic proteins (BMP) and fibroblast growth factor (FGF) [26].

12.3.3 Dentin protein

Dentin includes hollow tubules surrounded by peritubular dentin embedded in low-mineralized intertubular dentin matrix. The tubules develop from pulp and reach the dentin and enamel [27]. Dentin stimulates inflammatory pathogens through dentinal tubules [18]. The predentin-odontoblast barrier is the first line of defense against pathogens. This helps to repair the tissue after removal of decayed dentin. Odontoblast has a key role in organization of ECM. During maturation additional mineralization occurs when changes in the organic phase of the ECM occur [28]. Dentin disorders are hereditary and linked with collagen and proteins. They are classified into five different types: dentinogenesis imperfect (three types) and dentin dysplasia (two types) [4, 23]. Dentin has 80 to 90% of protein [29]. A sum total of 27 types of proteins are found in dentin, which is under the regulation of 42 different types of genes [30]. Collagen of dentin is closely related with MMP2, MMP9, MMP20 [31, 32]. In predentin, fibronectin is found to be associated with collagen fibrils [33]. The major non-collagen protein is DSPP. The DSPP is responsible for dentin dysplasia Type II and dentinogenesis imperfect Type II and III [15, 34]. DSPP cleaves to denin silaoprotein, dentin glycoprotein and dentin phosphoprotein [25, 35]. DSPP is required for the healthy maintenance of periodontium [36]. Sialic acid rich proteins are present in the dentin and it is noncollagenous in nature. It includes dentin silaoprotein, osteopontin, osteonectin, osteocalcin, dentalmatrix protein I and integrin binding silaoprotein [37]. These proteins help in the dentin mineralization [38]. Dental matrix protein I is required for differentiation of phosphate homeostasis and differentiation of odontoblast and osteoclast [38, 39]. In the presence of calcium self-assembly occurs to form filaments. Periostin, [40] osteocalcin and osteonectin [41] have a defined role in the mineralization of dentin. Peptidoglycan was identified in predentin and dentin [42]. Biglycan and decorin binds to collagen fibril and hydroxyapatite crystals [43]. Osteoadherin, Lumican and fibromodulin help in the process of odontogenesis [44]. 10% dentin protein shows an immunological response. Thus, dentin proteins help to prevent dental caries.

12.3.4 Cementum protein

Dental cementum is produced by cementoblast cells. The ectomesenchymal dental follicles have positional osteoblasts [45]. Dental cementum is a bone-like tissue having morphological similarities and matrix composition. Collagen type I is the main protein and the rest are noncollagenous proteins. The noncollagenous proteins are versican, decorin and biglycan. Glycoproteins like osteonectin, arginine-glycine-asparatic-acid and integrin binding proteins are also detected. Matrix protein, protein S and osteocalcin are involved in cell metabolism, mineralization and matrix deposition [46].

12.3.5 Dental pulp and pellicle protein

Dental pulp has dendritic, macrophages, lymphocytes and various epithelial and mesenchymal cells. They have a role in detecting foreign pathogens. The presence of dental pulp proteins was detected in western blot [47]. Pulp fibroblast and odontoblast express desmoplakin, junction plakoglobin and intracellular junction proteins are linked to adjacent cells [48]. Leucin rich peptidoglycans such as biglycan, lumican and mimecan were detected in dental pulp. The glycosoamino-glycan on biglycan helps to bind calcium and in the process of mineralization [49].

Protein analysis of pellicle detected 130 different types of proteins [50] categorized under three different groups depending on the ability to bind to different molecules. The first group has the ability to bind calcium (18%), the second group to bind phosphate (15%) and the third group has the ability to bind to 28% proteins [50]. Five proteins were detected having a role in enamel pellicle formation and a role in biofilm formation [51].

12.4 Proteomic analysis of teeth

Proteome analysis reveals the presence of different types of proteins present in the teeth. Teeth proteins are analysed using mass spectrometry, circular dichohorism and FTIR techniques. Analysis of teeth proteins using mass spectrometry detects the presence of 152 different types of proteins. These proteins include: serrotransferrin, fibrillin-1, collagen 1 and components of immunoglobin. Serrotransferrein and fibrillin 1 are immunoglobin and collagen subtypes are abundant in peptide hits. Serotransferrein is a glycoprotein present in the enamel pellicle and has a role in tooth morphology. Fibrillin-1 and collagens are ECM proteins. Collagen is a dentin protein present in the bone. Fibrilins are cysteine rich glycolproteins providing structural support to the teeth. Noncollagenous proteins like versican, vitronectin, hyaluronan, and proteoglycan are also identified in the teeth extract [52]. Proteins like Tenasci, Thrombospondin, L fibronectin, Dickkopf related protein and vibro-nectin are also identified [53]. Proteins are also found in the saliva. They include antileukoproteinase, lactotransferrin and serotransferrin. These proteins are adsorbed in enamel pellicle. 33 common proteins including 66 kDa heat shock protein were detected from this analysis [54]. Proteome analysis was also carried out using LC-MS. Different parts of the teeth like crown, enamel, dentin, dentin enamel junction (DEJ) and adjacent enamel organic matrix (EOM) were analysed to detect specific proteins. The different parts of the teeth were dissected by laser micro-dissection and analysed using LC-MS. Serpin family members like antithrombin-III and alpha–1-antitrypsin were detected in dentin but not in DEJ + EOM. Teeth extracts were also analysed by 12% SDS-PAGE followed by silver staining [12]. Proteins like BMP regulate the crystal formation during teeth mineralization. The teeth extracts were also analysed by using circular dichorism (CD). The CD analysis of the teeth suggest the presence of amide I region. FTIR can detect the secondary structure of proteins and peptides present in the teeth. FTIR of all the four teeth proteins falls in the mid-infrared spectra region (1700–1600 cm^{-1}). Amide I vibration belongs to C=O, amide II range from 1600–1500 cm^{-1} and amide III is at 1320 cm^{-1}. Apart from amide vibration protein extracts also undergo post-translation modification like glycosylation. Carbohydrate has a band at 1200–900 cm^{-1}. The bands observed at 1180 and 1044 cm^{-1} correspond to C–O and C–C. Most of the proteins detected from proteome analysis were IDPs suggesting its role in teeth mineralization. SDS PAGE and mass spectroscopy have identified 233 proteins from dentin. Similarly, 147 ethylene diamine tetraacetic acid soluble proteins were identified from dentin. nLC-MS/Ms identified 289 proteins out of which 90 are already reported. Dental cementum was analysed by LC-MS/MS for protein analysis. A total of 213 proteins were identified which is specific to cementum matrix 3 [55]. Superoxide dismutase 3 was also detected in cementum and cementum associated cells [55].

12.5 Proteins during pathological conditions

During pathological conditions including hypomineralization the enamel changes its colour to opaque, yellow or brown. Under such circumstances the amount of protein increases up to 8- to 21-fold. Serum albumin, antitrypsin and antithrombin were detected in the teeth. Elevated serum albumin inhibits the growth of apatite crystal [56].

During caries the odontoblast is encouraged towards reactionary dentin. This inspiration could be due to numerous signaling molecules generated from the dentin during the demineralization process. In the disease stage deep cavity formation occurs and the osteoblast disintegrates. The new odontoblast cells originated from dental pulp fibroblast promote restoration of dentin. The dentinogenesis during normal and pathological conditions has some similar pathways for repair and development of dental tissue. Thus, normal and injured tissues were collected for the identification of molecules involved in the formation of teeth. Notch signaling has a key role in the cell fate specification and cell proliferation. During odontogenesis *Delta1* and *Notch* have complementary expression patterns. *Delta1* is articulated in distinguishing odontoblast, whereas *Notch* is restricted to subodontogenic cells. In adult dental tissue both *Delta1* and *Notch* are absent. Nestin is an outstanding marker for differentiated odontoblast to study the mechanism of generation of odontoblast-like cells under pathological conditions. Nestin is expressed in the developing nervous tissue and muscle. Throughout dentinogenesis, Nestin is detected in odontoblast and the pulp fibroblast region. Nestin is present up to the enamel–denin junction. During dental carries decalcification and proteolysis occurs. Dentin decalcification of the odontoblast is stimulated resulting in reactionary dentin. With the increase of irritation the odontoblast disintegrates. Pulp cells replace missing odontoblast which differentiate to odontoblast-like cells and start the secretion and deposition of restorative dentin. The pulp volume is reduced with the elaboration of reactionary and reparative dentin. With the progression of caries, blood vessels dilate and inflammatory cells are evident in the pulp. During the inflammation nestin/Notch is found to be up-regulated. BMP and epidermal growth factor (EGF) released from the matrix and diffusing to adjacent cells during dentin decalcification. Dental pulp with response to BMP4 differentiates to odontoblst-like cells and starts the secretion and deposition of reparative dentin matrix. Nestin is expressed in the cell bodies and it has role in the innervation of odontoblast. The process of pulp healing and regeneration is stimulated by BMP2- and BMP4. Dental pulp undergoes calcification with respect to caries, trauma and retransplantation [57]. Denervated teeth have defective dentin repair and pulp survival occurring after injury. In recent days dental pulp stem cells are used to repair the pulp–dentin complex. These cells differentiate to chondroblast, odontoblast, adipocytes, neural cells and osteoblasts [58]. The dental stem cells are pluripotent in nature [59] and capable of forming mineralized tissue which can interact with biomaterials. A few studies have described the composition of dental pulp [47]. 2D gel electrophoresis followed by MS/MS detects 96 proteins in the dental pulp. Variations were detected in the dental pulp of healthy and carious teeth with respect to proteins [60]. In carious teeth antimicrobial peptide like S100A8, S100A9, S100A13 were detected [61]. Pulpal tissues have two immune response proteins like alpha 1-acid glycoprotein, and interleukin 11 [62]. Upregulation of five proteins was detected in the dental pulp. Nine different proteins were detected from dental pulp cells. Vimentin has the ability of regeneration of pulp. Asporin regulates calcium deposition in the predentin layer. Preamiloblast induces differentiation of dental pulp stem cells. 21 different types of proteins were detected in preamiloblast. All these bioactive proteins help in the regeneration of tissue [63]. In carious saliva a greater amount of lipid and proteins were detected. A variation of proteins was detected in the saliva with respect to sex and age [64]. Lectins are upregulated in both female and male saliva.

12.6 Protein analysis and its application in dentistry

12.6.1 Proteins to treat carries

During tooth development ameloblasts regulate the formation of enamel and it undergoes apoptosis as the mature enamel is formed. The formation of enamel is known as amelogenesis. For any reason if the enamel is damaged it cannot be repaired. The enamel forming cells, the ameloblasts secrete ECM, enamel proteins and proteases that regulate shape, arrangement, and mineralization of calcium phosphate crystals. An alternative approach to repair the enamel is by acellular mineralization. Fluoride and calcium phosphate nanocystals help to remineralise the eroded enamel matrix. It can inhibit the demineralization by fluoride incorporation and lower the ability of enamel to protect the teeth up to 30 μm. Osteogenic protein of transforming growth factor β (TGF- β) controls the bone formation. Bower's examined the role of osteogenin (bone-morphogenetic protein-3) in bone regeneration. Osteogenin along with bone allograft stimulates the bone regeneration. Grafting materials like autograft, allograft and xenografts are used for periodontal regeneration. Graft retention is critical for periodontal regeneration.

Several proteins are designed to rebuild the teeth and recover the caries. Small proteins like amilogenin can be used to construct a mineral layer on damaged enamel. Treatment of teeth with peptides allows 10–50 μm of deposition of enamel over the teeth. This method is widely used to repair the damaged enamel. Peptides help in the remineralization of thick dense 10 μm HaP mineral. Peptide guided remineralization is the future development of biomimetic products to treat health care. Peptide technology allows 10–50 μm of enamel on teeth. These peptides are used as biomimetic toothpaste or gel as a safe approach to treat dental disorders.

12.6.2 Protein and peptide based treatment

Proteins and peptides have the ability to stimulate events like chemotaxis, differentiation and matrix synthesis. Autologous serum is a natural source of proteins and signaling peptides. Platelet rich plasma is used to assist with replacement of bone grafts in periodontal regenerative therapy. Protein derived therapeutic substances are used for teeth regeneration. Emdogain is the first enamel matrix derivative (EMD) from procine amelogenin [65]. EMD enhances gene expression, production of protein, proliferation of cells, differentiation of periodontal ligament and osteoblastic cells. Emdogain is used topically on the tooth surface to treat the periodontal intrabony defects. EMD can treat flap debridement in tissue regeneration in intraosseous defects. EMD application in coronary advanced flap increases whole root coverage in Miller Class I and II gingival recession defects [66]. EMD along with grafts helps to treat intraosseous defects. PepGen P-15 is a bone derived hydroxyapatite, containing peptides of 15 aminoacids that mimic putative cell binding domain Type-I collagen which enhance osteogenesis [67]. An organic bone matrix resounds the cell-binding peptide P-15 surges the degree of cell attachment and endorses anchorage dependent cell survival. PepGen P-15 is used in clinical trials to replace grafts in intrabony periodontal defects [68].

GEM 2IS is a purified human platelet derived recombinant growth factor-BB (rhPDGF-BB) and it is a β-Tricalcium phosphate (β-TCP) scaffold. GEM 2IS is the principal bone grafting material provided by FDA along with recombinant growth factor. PDGF-BB stimulates angiogenesis and augments cell employment and multiplying of bone and periodontal ligament [69]. GEM-2IS is under clinical trial. Periodontal defects spliced with βTCP can heal long junctional epithelium. Growth/differentiation factor 5 (GDF-5) is recognized as a cartilage derived morphogenetic protein and is affiliated with TGF-β superfamily. It is localized at the dental follicle and promotes root development in the periodontal tissue. Nakamura *et al* promote a dose dependent propagation of human periodontal ligament and amalgamation of sulfated glycosaminoglycan after exposure of GDF-5 [70]. rhPDG5 helps in the regeneration of periodontal ligament fibroblast migration and proliferation within the periodontal ligament. GDF-5 along with β-TCP or poly-lactate-coglycolide-acid

(PLGA) composite are used as a transport device in periodontal defects. Rh-GDF-5 is used in combination with β-TCP for open flap debridement in acute periodontitis [71]. rhGDF-5/β-TCP treated defects were compared with control sites. This suggests repairing of growth factor for periodontal regeneration. FGF is a large polypeptide involved in embryogeneis, wound repair and angiogenesis. FGF-2 endorses differentiation, proliferation and migration of periodontal ligament. As wound healing progresses FGF-2 contributes for proliferation, migration and differentiation of periodontal ligament. Teriparatide is a drug used to treat osteoporosis and has recombinant 1–34 N-terminal fragment of human parathyroid hormone (rhPTH1–34) [72]. Teriparatide peptide improves cortical and cancellous bone in women with osteoporesis. Administration of teriparatide peptide modulates the process of wound healing. Protein and peptide based treatment can promote the renewal of periodontal tissue, and as a consequence is used in many clinical trials for regeneration purposes. The use of mammalian scaffolds enhances the process of regeneration. Biocompatible delivery of sensitivity with spatial and temporal requirement is essential for the process of wound healing [72].

Both structured and unstructured proteins present within the teeth play a role in their development and maintenance. The most dominant type of protein present within the teeth is amelogenin. In developing enamel, ameloblastins, enamelins, and tuftelins proteins are also found. The amelogenins help to organize the rods present in the enamel. They regulate the growth and initiation of hydroxyapatite crystals during enamel formation. Amelogenins further help to develop the cementum from the surface of the root. At physiological pH, ionic strength, and temperature, amelogenin stimulates the nucleation of hydroxyapatite. The nucleated nanocrystallites, nanospheres and amelogenin oligomers are present at less concentration. The formation of nanorods makes the critical nuclei at calcium phosphate crystallization. Nanorod formation is a synergistic communication between flexible amelogenin protein and inelastic calcium phosphate nanocrystallites. The intermediate structures share chemical and morphological similarities with the enamel. Amelogenin promotes apatite crystallization and organization. Establishment of organized bundles in amelogeninapatite composites is guided by self-assembly of amelogenin nanochain. The self-assembly of amelogenin affects the structure of developing enamel crystals at the early stage of nucleation. As a consequence, it can be used to repair enamel and treat caries *in vivo*. Amelogenin is regulated by two genes: AMELX and AMELY. AMELX is specific only to the X chromosome and AMELY is specific only to the Y chromosome. AMELY(58–64) peptide is found only in males. AMELX(44–52) is found in both males and females. The expression of AMELY is only 10% of AMELX. Hydroxyapatite nucleation can promote enamel repair. Application of anionic peptide to caries simulates intraoral conditions of pH cycling. This suggests peptide treatment can significantly increase the gain of minerals in the carious lesions. The self-assembling peptides help to modulate the mineral behavior in dental tissue engineering.

12.7 Conclusions

Proteomics is an evolving field of science from a biological, medical and analytical point of view. Proteins present in the teeth vary with age, sex and cleanliness of the teeth.

These proteins play a role in the onset of disease and dental caries. Thus, understanding the role of each protein is an alternative approach to treat teeth disease. Hence identification of new proteins and understanding their function in teeth is an alternative approach towards the treatment of many diseases associated with teeth.

Acknowledgements

MM lab is supported by the Department of Biotechnology, Ministry of Science and Technology (SERB/EMR/2017/003054, BT/PR21857/NNT/28/1238/2017, and Odisha DBT 3325/ST (BIO)–02/2017) for financial assistance.

References

[1] Rabilloud T, Chevallet M, Luche S and Leong C 2010 Two-dimensional gel electrophoresis in proteomics: past, present and future *J. Proteomics* **73** 2064–77

[2] Fabian T, Hermann P, Beck A, Fejerdy P and Fabian G 2012 Salivary defense proteins: their network and role in innate and acquired oral immunity *Int. J. Mol. Sci.* **13** 4295–320

[3] Garcia-Godoy F and Hicks M 2008 Maintaining the integrity of the enamel surface. The role of dental biofilm, saliva and preventive agents in enamel demineralization and remineralization *J. Am. Dent. Assoc.* **139** 25S–34S

[4] Paine M and Snead M 2005 Tooth developmental biology: disruptions to enamel-matrix assembly and its impact on biomineralization *Orthod Craniofac. Res.* **8** 239–51

[5] Vitorino R, Guedes S, Manadas B, Ferreira R and Amado F 2012 Toward a standardized saliva proteome analysis methodology *J. Proteom.* **75** 5140–65

[6] He L and Swain M 2008 Understanding the mechanical behavior of human enamel from its structural and compositionalcharacteristics *J. Mech. Behav. Biomed. Mater.* **1** 18–29

[7] Zenobio M, Tavres M, Zenobio E and Silva T 2011 Elemental composition of dental biologic tissues: study by means of different analytical techniques *J. Radioanal. Nucl. Chem.* **289** 161–6

[8] Hu J, Chun Y, A L, Hazzazzi T and Simmer J 2007 Enamel formation and amelogenesis imperfecta *Cells Tissues Organs* **186** 78–85

[9] Lee J, Lee D, Choung H, Shon W, Seo B, Lee E, Cho J and Park J 2011c Odontogenic differentiation of human dental pulp stem cells induced by preameloblast-derived factors *Biomaterials* **32** 9696–706

[10] Dusevich V, Xu C, Wang Y, Walker M and Gorski J 2012 Identification of a protein-containing enamel matrix layer which bridges with the dentine-enamel junction of adult human teeth *Arch. Oral Biol.* **57** 1585–94

[11] Webb E, Amarasiriwardena D, Tauch S, Green E, Jones J and Goodman A 2005 Inductively coupled plasma-mass (ICP-MS) and atomic emission spectrometry (ICP-AES): versatile analytical techniques to identify the archived elemental information in human teeth *Microchem. J.* **81** 201–8

[12] Sharma V, Srinivasan A, Roychoudhury A, Rani K, Tyagi M, Dev K, Nikolaje F and Kumar S 2019 Characterization of protein extracts from dierent types of human teeth and insight in biomineralization *Sci. Rep.* **9** 9314

[13] Cook J, Layrisse M, Martinez-Torres C, Walker R, Monsen E and Finch C 1972 Food iron absorption measured by an extrinsic tag *J. Clin. Invest.* **51** 805–15

[14] Jágr M, Eckhardt A, Pataridis S and Miksik I 2012 A comprehensive proteomics analysis of human dentin *Eur. J. Oral Sci.* **120** 259–68

[15] Lee K, Kang H, Yoo S, Lee J, Hwang Y, Nam K, Kim J S, Park J and Kim J 2011 Novel dentin phosphoprotein frameshift mutations in dentinogenesis imperfecta type II *Clin. Genet.* **79** 378–84

[16] Foster B *et al* 2011 The progressive ankylosis protein regulates cementum apposition and extracellular matrix composition *Cells Tissues Organs* **194** 382–405

[17] Byers M and Taylor P 1993 Effect of sensory denervation on the response of rat molar pulp to exposure injury *J. Dent. Res.* **72** 613–8

[18] Couve E, Osorio R and Schmachtenberg O 2013 The amazing odontoblast: activity, autophagy, and aging *J. Dent. Res.* **92** 765–72

[19] Hannig C, Hannig M and Attin T 2005 Enzymes in the acquired enamel pellicle *Eur. J. Oral Sci.* **113** 2–13

[20] Fabian T, Fejerdy P and Csermely P 2008 Salivary genomics, transcriptomics and proteomics: the emerging content of the oral ecosystem and their use in the early diagnosis of cancer and other diseases *Curr. Genom.* **9** 11–21

[21] Hubbard M and Kon J 2002 Proteomic analysis of dental tissues *J. Chromatogr.* B **771** 211–20

[22] Paine M and Snead M 2005 Tooth developmental biology: disruptions to enamel-matrix assembly and its impact on biomineralization *Orthodon. Craniofac. Res.* **8** 239–51

[23] Butler W, Brunn J and Qin C 2003 Dentin extracellular matrix (ECM) proteins: comparison to bone ECM and contribution to dynamics of dentinogenesis *Connect Tissue Res.* **44** 171–8

[24] Hu J, Yamakoshi Y, Yamakoshi F, Krebsbah P and Simmer J 2005 Proteomics and genetics of dental enamel *Cells Tissues Organs* **181** 219–31

[25] Tsuchiya S, Simmer J, Hu J, Richarddson A, Yamakoshi F and Yamakoshi Y 2011 Astacin proteases cleave dentin sialophosphoprotein (Dspp) to generate dentin phosphoprotein (DPP) *J. Bone Miner. Res.* **26** 220–8

[26] Smith A, Scheven B, Takahashi Y, Ferracane J, Shelton R and Cooper P 2012 Dentine as a bioactive extracellular matrix *Arch. Oral Biol.* **57** 109–21

[27] Bar-On B and Wagner H 2013 Structural motifs and elastic properties of hierarchical biological tissues—a review *J. Struct. Biol.* **183** 149–64

[28] Mckee M, Hoac B, Addison W, Barros N, Millan J and Chaussain C 2013 Extracellular matrix mineralization in periodontal tissues: noncollagenous matrix proteins, enzymes, and relationship to hypophosphatasia and x-linked hypophosphatemia *Periodontology* **2000** 102–22

[29] Linde A 1989 Dentin matrix proteins, composition and possible functions in calcification *Anatom. Rec.* **224** 154–66

[30] Myllyharju J and Kivirikko K 2004 Collagens, modifying enzymes and their mutations in humans, flies and worms *Trends Gen.* **20** 33–43

[31] Yamakishi Y, Hu J, Iwata T, Kobayashi K, Fuake M and Simmer J 2006 Dentin sialophosphoprotein is processed by MMP-2 and MMP-20 *in vitro* and *in vivo J. Biol. Chem.* **281** 38235–43

[32] Mazzoni A, Pashley D, Ay F, Gobbi P, Orsisni G, Ruggeri A, Carrilho M, Tjaderhane L, Dilenarda R and Breschi L 2009 Immunohistochemical identification of MMP-2 and MMP-9 in human dentin,correlative FEI-SEM/TEM analysis *J. Biomed. Mater. Res.* A **88** 697–703

[33] Park E, Cho H, Kwon T, Jang S, Lee S, An C, Shin H, Kim J and Cho J 2009 Proteomics analysis of human dentin reveals distinct protein expression profiles *J. Proteome Res.* **8** 1338–46

[34] Lee S-K, Lee K-E, Song J S, Hyun H-K, Lee S-H and Kim J-W 2013 A DSPP mutation causing dentinogenesis imperfecta and characterization of the mutational effect *Biomed. Res. Int.* **2013** 948181

[35] Qin C, Brunn J, Baba O, Wygant J, Mcintyre B and Butler W 2003 Dentin sialoprotein isoforms, detection and characterization of a high molecular weight dentin sialoprotein *Eur. J. Oral Sci.* **111** 235–42

[36] Gibson M, Zhu Q, Liu Q, D′souza R, Feng J and Qin C 2013 Loss of dentin sialophosphoprotein leads to periodontal diseases in mice *J. Periodontal Res.* **48** 221–7

[37] Qin C, Brunn J, Jones J, George A, Ramachandran A, Gorski J and Butler W 2001 A comparative study of sialic acid-rich proteins in rat bone and dentin *Eur. J. Oral Sci.* **109** 133–41

[38] Suzuki S, Sreenath T, Haruyama N, Honeycutt C, Terse A, Cho A, Kohler T, Muller R, Goldberg M and Kulkarni A 2009 Dentin sialoprotein and dentin phosphoprotein have distinct roles in dentin mineralization *Matrix Biol.* **28** 221–9

[39] Beniash E, Deshpande A, Fang P, Lieb N, Zhang X and Sfeir C 2011 Possible role of DMP1 in dentin mineralization *J. Struct. Biol.* **174** 100–6

[40] Ma D, Zhang R, Sun Y, Rios H, Haruyama N, Han X, Kulkarni A, Qin C and Feng J 2011 A novel role of periostin in postnatal tooth formation and mineralization *J. Biol. Chem.* **286** 4302–9

[41] Papagerakis P, Berdal A, Mesbah M, Peucmaur M, Malaval L, Nydegger J, Simmer J and Macdougall M 2002 Investigation of osteocalcin, osteonectin, and dentin sialophosphoprotein in developing human teeth *Bone* **30** 377–85

[42] Embery G, Hall R, Waddington R, Septier D and Goldberg M 2001 Proteoglycans in dentinogenesis *Critic. Rev. Oral Biol. Med.* **12** 331–49

[43] Haruyama N, Sreenath T, Suzuki S, Yao X, Wang Y, Honeycutt C, Iozzo R, Young M and Kulkarni A 2009 Genetic evidence for key roles of decorin and biglycan in dentin mineralization *Matrix Biol.* **28** 129–36

[44] Petersson U, Hultenby K and Wendel M 2003 Identification, distribution and expression of osteoadherin during toothformation *Eur. J. Oral Sci.* **111** 128–36

[45] Bosshardt D and Selvig K 1997 Dental cementum: the dynamic tissue covering of the root *Periodontology* **13** 41–75

[46] Nanci A 1999 Content and distribution of noncollagenous matrix proteins in bone and cementum: relationship to speed of formation and collagen packing density *J. Struct. Biol.* **126** 256–69

[47] Wei X, Wu L, Ling J, Liu L, Liu S, Liu W, Li M and Xiao Y 2008 Differentially expressed protein profile of human dental pulp cells in the early process of odontoblast-like differentiation *in vitro J. Endodontol.* **34** 1077–84

[48] Sawa Y, Kuroshima S, Yamaoka Y and Yoshida S 2005 Intracellular distribution of desmoplakin in human odontoblasts *J. Histochem. Cytochem.* **53** 1099–108

[49] Sandanayake N *et al* A combination of serum leucine-rich alpha-2-glycoprotein 1, CA 19-9 and interleukin-6 differentiate biliary tract cancer from benign biliary structures *Br. J. Cancer* **105** 1370–8

[50] Siqueira W, Zhang W, Helmerhorst E, Gygi S and Oppenheim F 2007 Identification of protein components in *in vivo* human acquired enamel pellicle using LC-ESI-MS/MS *J. Proteom. Res.* **6** 2152–60

[51] Lee Y, Zimmerman J, Custodio W, Xiao Y, Basiri T, Hatibovic-Kofman S and Siqueira L 2013 Proteomic evaluation of acquired enamel pellicle during *in vivo* formation *PLoS One* **8** e67919

[52] Sahlberg C, Aukhil I and I T 2001 Tenascin-C in developing mouse teeth: Expression of splice variants and stimulation by TGF and FGF *Int. Eur. J. Oral Sci.* **109** 114–24

[53] Tucker R and Degen M 2019 The expression and possible functions of Tenascin-W during development and disease *Front. Cell Develop. Biol.* **7** 53

[54] Kaur J and Reinhardt D 2015 Extracellular Matrix (ECM) molecules *Stem Cell Biology and Tissue Engineering in Dental Sciences* (Amsterdam: Elsevier)

[55] Salmon C, Tomazela D, Rui K, Foster B, Leme A, Sallum E, Somerman M and Nociti F 2013 Proteomic analysis of human dental cementum and alveolar bone *J. Proteomics* **91** 544–55

[56] Farah R, Monk B, Swain M and Drummond B Protein content of molar-incisor hypomineralization enamel *J. Dent* **38** 591–6

[57] Ohshima H, Nakakura K, Yamamoto H and Haeda T 2001 Alteration in the expression of heat shock protein (Hsp) 25-immunoreactivity in the dental pulp of rat molars following tooth replantation *Arch. Histol. Cytol.* **64** 425–37

[58] Iohara K, Zheng L, Ito M, Tomokiyo A, Matsushita K and Nakashima M 2006 Side population cells isolated from porcine dental pulp tissue with self-renewal and multipotency for dentinogenesis, chondrogenesis, adipogenesis, and neurogenesis *Stem Cells* **24** 2493–503

[59] Paakkonen V and Tjaderhane L 2010 High-throughput gene and protein expression analysis in pulp biologic research: review *J. Endodontol.* **36** 179–89

[60] Paakkonen V, Ohelmeir S, Bergmann U, Larmas M, Salo T and Tjäderhane L 2005 Analysis of gene and protein expression in healthy and carious tooth pulp with cDNA microarray and two-dimensional gel electrophoresis *Eur. J. Oral Sci.* **113** 369–79

[61] Mclachlan J, Sloan A, Smith A, Landini G and Cooper P 2004 S100 and cytokine expression in caries *Infect. Immun.* **72** 4102–8

[62] Mclachlan J, Smith A, Bujalska I and Cooper P 2005 Gene expression profiling of pulpal tissue reveals the molecular complexity of dental caries *Biochim. Biophys. Acta* **1741** 271–81

[63] Smith A, Scheven B, Takahashi Y, Ferracane J, Shelon R and Cooper P 2012 Dentine as a bioactive extracellular matrix *Arch. Oral Biol.* **57** 109–21

[64] Qin Y, Zhong Y, Zhu M, Dang L, Yu H, Chen Z, Chen W, Wang X, Zhang H and Li Z 2013 Age- and sexassociated differences in the glycopatterns of human salivary glycoproteins and their roles against influenza A virus *J. Proteome Res.* **12** 2742–54

[65] Grandin H, Gemperli A and Dard M 2012 Enamel matrix derivative: a review of cellular effects *in vitro* and a model of molecular arrangement and functioning *Tissue Eng. Part B Rev.* **18** 181–202

[66] Cairo F, Pagliaro U and Nieri M 2008 Treatment of gingival recession with coronally advanced flap procedures: a systematic review *J. Clin. Periodontol.* **35** 136–62

[67] Lallier T, Yukna R, St Marie S and Moses R 2001 The putative collagen binding peptide hastens periodontal ligament cell attachment to bone replacement graft materials *J. Periodontol.* **72** 990–7

[68] Yukna R, Salinas T and Carr R 2002 Periodontal regeneration following use of ABM/P-15: a case report *Int. J. Periodontics Restorative Dent.* **22** 146–55

[69] Hollinger J, Hart C, Hirsch S, Lynch S and Friedlaender G 2008 Recombinant human platelet-derived growth factor: biology and clinical applications *J. Bone Joint Surg. Am.* **90** 48–54

[70] Morotome Y, Goseki-Sone M, Ishikawa I and S O 1998 Gene expression of growth and differentiation factors-5, -6, and -7 in developing bovine tooth at the root forming stage *Biochem. Biophys. Res. Commun.* **244** 85–90

[71] Stavropoulos A, Windisch P, Gera I, Capsius B, Sculean A and Wikesjo U 2011 A phase IIa randomized controlled clinical and histological pilot study evaluating rhGDF-5/beta-TCP for periodontal regeneration *J. Clin. Periodontol.* **38** 1044–54

[72] Aspenberg P, Genant H, Johansson T, Nino A, See K and Krohn K 2010 Teriparatide for acceleration of fracture repair in humans: a prospective, randomized, double-blind study of 102 postmenopausal women with distal radial fractures *J. Bone Miner. Res.* **25** 404–14

Chapter 13

Neurotoxicity of nanomaterials used in the field of dentistry

Monalisa Mishra

Nanomaterials replace many conventional materials used in the field of dentistry. The desired size, chemical, and physical properties allow us to use them for several dental applications. With the extensive applications, nanomaterials are also channelized into our physiological system. Being foreign material, our body also activates immune pathways and redox signaling against them. The over-activated product of immune pathways and redox signaling also results in stress within our bodies. Stress can be measured by many signaling pathways and markers present within our body. More stress is harmful and affects the structure and functioning of the nervous systems. Our nervous system is enriched with polyunsaturated fatty acid which makes it more susceptible to stress. Although our central nervous system is well secured by the blood–brain barrier (BBB), many of the nanoparticles (NPs) can cross it and harm the central nervous system. A malformed nervous system results in abnormal behaviour in the organism. As a consequence, it is necessary to check the toxicity of neurons *in vitro*, *in vivo,* and *ex vivo*. The current chapter summarises the toxicity of several nanomaterials used in the field of dentistry and their toxicity on the nervous system carried out from several studies.

13.1 Introduction

Nanomaterials have several advantages over their precursor materials. Their properties allows them to be used in quite a few day-to-day activities. Their size varies from 1 to 100 nm. Using different synthesis methods the desired shape of the nanomaterials can be obtained. Similarly, by altering the source of synthesis, nanomaterials of preferred chemical properties can be obtained. All these properties help to increase the demand for nanomaterials by several branches of science including medicine and dentistry [1, 2]. Different nanomaterials are synthesized to be

doi:10.1088/978-0-7503-3671-0ch13

used in numerous applications in dentistry. In dentistry, nanomaterials are used as sealing agents, mouth wash, coating material, implants, and polymerizing agents [2].

With the wide application of nanomaterials, safety concerns associated with their toxicity cannot be overlooked [3]. The application of nanomaterials is also known to cause many complications in physiological systems including the nervous system [4]. This opens up a new field of nanotoxicology which is associated with neuronal and behavioral defects [5]. Reports from several studies deduce a common mechanism of NPs induced neurotoxicity. All these findings suggest oxidative stress, apoptosis, autophagy, neuroinflammation, and alteration of the structure of the BBB occur after NP treatment [5, 6]. These defects are considered a hallmark of neurotoxicity [7, 8]. Moreover, the neurotoxic effect after NP treatment can be altered by altering the structure and function of neurons or by altering the glia–neuron interaction. Often, the conventional methods are not efficient enough to detect the toxicity associated with neurons. Thus, novel methods were developed to detect the toxicity of neurons using *in vitro*, *in vivo,* and *ex vivo* techniques. Since the toxicity of nanomaterials depends on size, shape, and the precursor materials used to synthesize the nanomaterials, it is better to check the toxicity of individual nanomaterials before they are used for dental applications [9].

13.2 Types of nanomaterials used in dentistry

Many NPs used in dentistry are under clinical trials. Nanomaterials have advantages over the conventional methods of treatments used in the field of dentistry. The advantages of nanomaterials include: (1) better antibacterial and mechanical properties; (2) strength to the dentine; (3) more stability towards temperature; (4) regenerating ability of the teeth tissue; (5) antifungal properties; (6) adhesive properties; (7) enhanced osseointegration ability; (8) improved supermagnetic and optical properties; and (9) biocompatibility [4]. All these unique properties allow the NPs to be used as a dental adhesive, for root canal filling, repair of the damaged area, as an implant, for delivering the drug to the defective area, and to image the tumor region. To perform such a function many NPs like nano-hydroxyapatite (HAp), porous ZrO_2/HAp, nanoTiO_2, nanoZnO, nanoCeO, nanosilver, nanoalumina, nano zirconia, nano ZnO, nano-silica, magnetic iron oxide, are used [2]. Among polymeric NPs nano resin, polyethylene glycol 1000 succinate, liposome nanocarriers are used. The application of individual NPs in dentistry is described in this section. The toxicity of nanomaterials towards the structure and function of neurons is described in two different headings.

13.2.1 Nano hydroxyapatite and silver

Hydroxypatite NP (NanoHAp) is widely used for filling root canals since it shares compositional similarities with bone tissue. Its small size prevents damage to the root canal and seals the damaged tissue. Its antibacterial, as well as bacteriostatic properties, help to heal the root canal without any further side effects. It is biocompatible and can induce the healing of ruptured teeth. [10]. Porous ZrO_2/HAp provides excellent mechanical properties [11]. NanoHAp increases the bioactivity of the dental implant material [12].

Nanosilver (AgNP) is well known to prevent the growth of mouth bacteria *Streptococcus mutans* [13]. AgNP of 17 nm can prevent biofilm and decrease the mineralization of the enamel after orthodontic treatment [14]. AgNP can efficiently prevent the biofilm formation of *S. mutans*. Coverage of AgNP on human dentine prevents the growth of the bacteria thus inhibiting the biofilm formation. The release of silver ions from AgNP affects the exopolysaccharide (EPS) of the bacteria thus preventing biofilm formation [15]. AgNP coated brackets are used to prevent the lesions caused in the teeth during orthodontic treatment [16].

13.2.2 NanoTiO_2

NanoTiO_2 is widely used in toothpaste and mouthwashes due to its antibacterial properties. The structure and surface property of the implant plays a key role in tissue integration and healing. The porous surface allows it to be biocompatible for the drug loading process. TiO_2NTs having 60 nm diameter and 10 μm length loaded with BMP2 help in osseointegration [17]. TiO_2 nanotubes (NTs) are better biocompatible with blood [18]. TiO_2NTs help in the immobilization of some peptide-like GLY-Arg-Gly-Asp-Ser which is needed for osseointegration [19]. TiO_2NPs behave differently with the platelet with respect to the shape and size of the NP. Nano TiO_2/HA composite bioceramic improves the nerve regeneration around the implant [20]. The coating of TiO_2NPs can prevent the colonization of *S. mutans* over the teeth [21]. Glassionomer cement incorporated with TiO_2NPs has better antibacterial property and is long-lasting. Nitrogen-doped TiO_2 can efficiently prevent oral pathogenic bacteria [22].

13.2.3 NanoZnO and CeO

NanoZnO is widely used in toothpaste and mouth rinse to eliminate bacterial infection present in the mouth. ZnONPs can kill Gram-negative bacteria like *Escherichia coli* [23], *Vibrio cholerae* [24], and *Campylobacter jejuni* [25]. Among Gram-positive bacteria, ZnONPs can kill *Pseudomonas aeruginosa* [26] and *Streptococcus mutans* [27]. ZnONPs have recently been used for filling teeth since they promote cement mineralization. ZnONPs prevent the growth of *Porphyromonas gingivalis* and *Actinomyces naeslundii* and stimulate the mineralization of the cementum. ZnONP prevents biofilm formation near the dentine area. Zn doped NPs can promote the remineralization of the dentine present in the root of the teeth [28].

CeO_2 can stimulate the natural fluorescence of dental enamel [29]. Its high atomic number helps in the attenuation of the dental x-ray beam [30]. Incorporation of CeO_2 in adhesive helps in the adhesion of dental composite and antimicrobial agents [31, 32]. CeO_2 could be an alternative filler for biopolymer. CeO_2 is also well known for its antimicrobial property [30]. CeO_2NPs can alter the radiopacity of dental adhesiveness [30].

13.2.4 Nanoalumina and SiO_2

Alumina nanoparticle (Al_2O_3NP) is added to dental material to enhance mechanical strength [33]. The addition of Al_2O_3NP to polymethylmethacrylate (PMMA) increases the mechanical strength and decreases the hydrophilic property [34, 35]. An orthodontic bracket having Al_2O_3NP enhances the strength and reduces mechanical strength [36].

SiO_2NP is used in dentistry due to its less toxicity, less absorption, less density, and biocompatible property [37]. It is used as a dental filler to improve mechanical strength. SiO_2NP can clean the plaque that has developed on the teeth. Thus, it is used in many dental applications for the cleaning of teeth. Mesoporous SiO_2 is used to block dentinal tubules which are responsible for hypersensitivity of the teeth [38]. A combination of SiO_2NPs and HA can heal the demineralized dentin up to a significant level [39].

13.2.5 Nanogold and zirconia

Gold NP (AuNP) is a novel metal and is well known for its biocompatibility, antibacterial and inert nature. Titania gold nanocomposite acts as an osteoinductive agent when incorporated as a dental implant. ZrNPs of 20–50 nm are biocompatible, osteoconductive, and well known to prevent plaque formation [40]. ZrO_2NPs are also used for filling teeth due to their mechanical property [39]. ZrO_2 improves strength and reduces the cracks present in the matrix of cement [41]. The addition of ZrO_2 to dentures improves strength, colour, electrical and thermal conductivity [42].

13.2.6 Nano ZnO and CuO

ZnONPs also prevent bacterial biofilm formation. ZnO-coated brackets can prevent the biofilm formation of *S. mutans and Lactobacillus* [43]. ZnONPs further significantly prevent the friction of the brackets. These NPs prevent anchorage and reduce the desorption risk. ZnONPs release oxygen radicals and Zn^{+2} ions that upsurge oxidative stress and obstruct the growth of the microorganisms [44]. Cellulose nanocrystal/ZnONP hybrid can augment the mechanical and antibacterial properties of dental resin composite.

CuONP is as effective as AgNP. However, it is cheaper and stable both physically and chemically. CuONP has been incorporated into adhesive material while its biofilm-forming ability remains unaltered [45]. It is also used as a nanofiller to improve the bond strength of the material. CuONP also prevents the growth of mouth bacteria *S. mutans* upto 70 to 80%[46]. Cu tends to produce hydroxyl radical which causes redox imbalance and results in the death of the cell [47]. CuONP generates free-radicals which interact with nucleic acid and protein synthesis of the microorganism [48].

13.2.7 Graphene NP

Graphene oxide (GO) has a less toxic effect on human dental follicle cells [49]. Graphene nanosheets can prevent the growth of mouth bacteria like *P. gingivalis*,

S. mutans, *F. nucleatum* [50], and *E. ludwigii* [51]. Zn graphene oxide nanocomposite can decrease the EPS synthesis and thus reduce the biofilm-forming ability of *S. mutans* [52]. The osseointegration ability of graphene makes it an excellent implant [53]. Graphene coated with titanium prevents the adhesion of *S. sanguinis* and *S. mutans*. GO thin film with titania and silver prevents the growth of *S. mutans* and *P. gingivalis*. Incorporation of GO into dental cement prevents pore formation. Graphene with glass ionomer (GI) enhances the physicomechanical properties of GI [54]. Reinforcement of graphene gold NPs with resin improves the physicochemical properties of dental nanocomposite [55].

13.2.8 Magnetic iron oxide

The magnetic NP can transport drugs better than other NPs. In magnetic NPs, iron is at the core and silica-containing shells are at the surface. Magnetic NP-doped resin is used for demineralized dentine tissue. The application of a 5% magnetic NP did not affect the degree of conversion. In this context application of magnetic NPs at 2% by mass enhances dentine bond strength and 2% penetration into demineralized dentin [56]. Incorporation of NPs of amorphous calcium phosphate (NaCP) with magnetic NPs neutralizes the acid and enhances the calcium and phosphate concentration and pH of the biofilm [57].

13.2.9 Nanochitosan and nano resin

This is a polymeric NP, non-toxic, biocompatible, and has antimicrobial activity. It can prevent the growth of oral fungi and bacteria like *P. gingivalis*, *S. mutans*, *A. actinomycetemcomitans* [58]. Although its insoluble property does not allow it to be used in mouthwashes it is used in many oral cavities for other applications. It prevents biofilm by interaction with the lipopolysaccharide, metal ions, and polymer. Its low temperature and insoluble property allow it to reside within the oral cavity for a longer duration. Nano resin includes nanocluster and nanofiller for bulk properties [59]. Nanocluster fillers have better surface bulk properties. The nanofillers are 1–100 nm and the nanohybrid is from 0.4 to 5 μm.

13.3 Several routes via which NPs used in dentistry enter the nervous system

Nanomaterials are used in the field of dentistry to fill the root canal, in dental prosthesis, coating materials in teeth, as a dental implant, or during rinsing of the mouth [60]. The application of all these nanomaterials in the field of dentistry gives a chance to enter these nanomaterials into patients' physiological systems [61]. In patients, the dental material stays within the mouth and thus comes in direct contact with the oral cavity or the maxiofacillary region. Thus, there is a high chance of getting in contact with the lymph, oral mucosa, and several parts of the digestive system. The dental prosthetic contains many nanometals, ceramics, and resins. Nanomaterials generated from the prosthetic materials mix with the blood and are transported to different physiological systems of the body [62]. In comparison to other systems, the presence of the BBB prevents attack to the central nervous system easily [63]. By whatever means, if the nanomaterial can cross the BBB then it can reach the central nervous system. Nanomaterials can reach the brain employing nerves [64]. The BBB is well structured with endothelial cells which are interconnected via tight junctions [65]. NPs are transcytosed/endocytosed either via endothelial cells or via the tight junction by affecting the transmembrane of flux [66]. NPs may alter the permeability of the neural endothelial cells [3]. During development the BBB is incomplete in the fetus, thus, one is more susceptible to dental nanomaterials that enter the blood [67]. Nano silica and nano TiO_2 cause complications in pregnancy when injected into pregnant mice [68]. Nanomaterial even crosses the placental barriers of mice and causes neurotoxicity in the offspring [69]. Developmental neurotoxic reports coming from several studies suggest the need for thorough neurotoxic investigation of nanomaterials used in the field of dentistry.

13.4 *In vitro* neurotoxic assessment of nanomaterials used in dentistry

In vitro methods are the quick way to check the toxicity of nanomaterials. Using *in vitro* methods, toxicity can be checked swiftly, with less ethical concern [70]. Thus the fat of the nanomaterials can be checked using several cell lines associated with the nervous system. The cytotoxicity of nanomaterials can be accessed easily. The effect of nanomaterials on organelle, transportation, and DNA damage can be studied using cell lines. Within the brain, when the NP comes in contact with different types of cells it can alter their functionality. Types of brain-specific cell lines have BBB, neuroglia (microglia and macroglia), neurons having white matter, and lacking a myelin sheath are used to check the neuronal toxicity [6]. Small unmyelinated neurons are used for *in vitro* study since they are associated with several diseases [64]. Sensitive methods like fluorescence and chemiluminescence are used to detect the alteration of neurons caused due to nanomaterials [71]. Many nanomaterials result in apoptosis, which can be detected by nuclear degradation, chromatin fragmentation, and shrinkage of cells [72]. Generation of free radicals, metabolites, and glutathione levels is also measured as a marker of oxidative stress [73]. DNA damage is also detected while considering the toxicity of nanomaterials [71]. For BBB, hCMEC/D3 cells (a cell line isolated from humans) are used [74]. Also, microvascular endothelial cells and human cerebral endothelial cells are also used for cell and NP interaction [75]. Among neuroglial cells, astrocytes and microglial cells are widely used. Astrocytes have a significant role in the instruction and preservation of the integrity of the BBB [76]. As a consequence, astrocyte cells are cocultured with microvascular endothelial cells and human cerebral endothelial cells. In certain cases, astrocytes are replaced with microglia [77] and pericyte cells [78]. Many studies were accompanied to check the interaction of NPs with astrocytes. ZnONPs and super magnetic iron oxide NPs persuade apoptosis and activate JNK signaling pathways in astrocyte cells [79]. SiNP induces ROS and nitrogen oxide species (NOS) production which affects the microglia cells and the surrounding neurons [80]. The activation of microglia is an alarm for the brain.

Neuronal cell lines used for NP interaction comprise (1) rat PC12 neuronal cells [81], (2) a human SHSY5Y neuroblastoma cell line [82]. Primary culture cell lines include HCN-1A, N27, CA1, and CA3 [3]. Nano CuO on CA1 cell lines causes alteration of the potassium channel and results in neuronal firing [83]. This change affects the physiology, behaviour, and cognition ability of rats [84]. Internalization of NPs affects the endo-lysosomal pathways and alters the functioning of the BBB. Accumulation of TiO_2NP is detected in the endothelial cells [85]. The dysfunction of endothelial cells is considered the principal step for neurodegeneration. AgNP damages the membrane and causes necrosis in a cell [86]. Iron oxide NPs alter synaptic transmission, nerve conduction, neuroinflammation [87]. TiO_2 induced neurotoxicity is associated with neuroinflammation, synaptic plasticity, and altering signaling pathways [88]. SiNP alters microglia function resulting in an undesirable effect on the striatum and dopaminergic neurons [89]. Carbon nanotubes affect cell proliferation, apoptosis, and reduction in astrocyte function [90]. Although *in vitro* experiments are easy and do not need ethical clearance, the long-term effect of nanomaterials on an organism cannot be answered using the *in vitro* approach and hence need study from animal models. Moreover, the *in vitro* model cannot mimic the effect of a microenvironment and thus provides unreliable results [91]. Recently, organ-on-a-chip, organoid, and 3D printing-like techniques have developed to check the toxicity of nanomaterials [68].

13.5 *In vivo* neurotoxic assessment of nanomaterials used in dentistry

Nanomaterials used in dentistry may enter the body via various means. Nanomaterials used for bone repair, periodontal dressing, paste and coating of implants are allowed to pass through the blood. During the preparation of nanomaterials, they may enter our body through the nasal and olfactory routes. The toxicity of the nanomaterials depends on the size, material, charge, surface coating as well as the route via which they enter our body [92]. Several model organisms are used to answer the questions associated with the toxicity of nanomaterials which cannot be answered through *in vitro* experiments [69]. The structural and compositional similarity of the BBB in different organisms allows us to detect the toxicity of nanomaterials using model organisms [93]. Nanomaterials cannot be excreted from the body easily and hence cause damage to the brain and neurons [94]. The long-term effect of these nanoparticles can be accessed using model organisms [95]. Among model organisms, mice and Drosophila are extensively employed to check the noxiousness of nanomaterials.

13.5.1 Neurotoxic report from the mice model

Mice share many genetic similarities with human beings and offer themselves as a nice model to check the harmfulness of nanomaterials used in dentistry [96]. Intraperitoneal injection of NPs to C57B16 mice resulted from the accumulation of NPs in different regions of the brain [97]. Intravenous injection of ZnNPs was detected in the brain after 14 days of exposure [98]. Nasal intake of TiO_2NPs was identified in the brain and caused oxidative stress in the brain, proliferation of glial cells, and apoptosis [99]. AlNPs can alter the gene and protein expression profile of mitogen-activated protein kinase activity [100]. Oral intake of nano alumina, nanoZnO, nanoTiO_2 causes accumulation of ROS and inhibits stress biomarkers CuZnSOD and MnSOD [101]. TiO_2NP induced ROS in the mouse hippocampus and resulted in neuroinflammation and neurodegeneration [102]. NP-induced neurotoxicity also causes abnormality in mitochondria, apoptosis, and autophagy [103]. Chitosan NPs when injected into rat intravenously caused neuronal apoptosis, neuroinflammation, oxidative stress and resulted in loss of weight [104].

13.5.2 Neurotoxic reports from the fly model

Drosophila, although placed in a lower phylum in the evolutionary tree, shares many organizational and functional similarities with the vertebrate nervous system. For example, the presence of neurotransmitters, a well-developed central nervous system, and a neuronal network allow it to be used as a typical organism to check the neurotoxicity of nanomaterials used in dentistry [9]. AgNP causes depigmentation in the cuticle of Drosophila by altering the dopamine pathway. Dopamine is further associated with the production of ROS, the nervous system, and ultimately the lifespan of the fly [105, 106]. AgNP also affects the JNK signaling pathways in Drosophila [107]. Ingestion of AuNPs alters the development of the eye and mechanosensory organs like wing and thorax bristle of Drosophila [108]. Oral intake of AlNP causes malformed interneuronal activity in the antennae of Drosophila [109]. Oral intake of TiO_2 alters the development of the eye and mechanosensory organs of Drosophila [110]. Oral intake of CoNP causes defective bristles in Drosophila [111]. ZnONP causes single-wing formation, deformed thorax with lack of ventral nerve cord in several generations [112]. Oral intake of ZrO causes defective eye and mechanosensory organs in Drosophila. Defective eye and mechanosensory organs are also reported after the treatment of HApNP to Drosophila. SiNp affects the development of the eye and wing in Drosophila [113]. Graphene nanosheet affects the development of the eye and mechanosensory organs of Drosophila [114].

13.6 Behavioural neurotoxicity

Behavior is a complex phenomenon evolved from the reflex action of many physical and physiological functions. Behavioral defects are reported from living organisms after treatment with the nanoparticle. Below are the behavioral defects associated with NPs used in dentistry.

Drosophila: Oral intake of AlNP causes malformed climbing and tasting behaviors in flies [115]. Oral intake of TiO_2 resulted in defective climbing behavior in Drosophila [110]. ZrNP consumed through the digestive system caused defective larvae crawling and malformed climbing behavior in Drosophila [116]. Oral intake of HApNP caused defective grooming and climbing behavior in adult flies [117]. CNP causes alteration of grooming and climbing behavior in Drosophila [118]. Ingestion of graphene nanosheet resulted in defective climbing, crawling, and mouth probosci's reflex action in Drosophila [114].

Mice: Nanomaterials used in dentistry alter synaptic changes and cause alteration of long-term potential (LTP) which regulate cognition like learning and memory. NanoMnO_2 can alter the LTP and thus affect cognition [119]. Intracerebroventricular injection of TiO_2 affects the complex behaviour of rats [120]. TiO_2NP affects the development of the brain of the offspring and causes depression in adult rats. ROS generated from the nanomaterials causes apoptosis in the neurons of brain development [121]. ROS can affect the neuronal channels, transcriptional factors, and kinases [122, 123]. Nanomaterials affect the neurotransmitters involved in learning and memory [124]. AuNP accumulates in the brain and increases seizure, cognition, and astrogliosis [125]. Astrogliosis is associated with hypoxia and seizures [126]. CNT causes anxiety and neurobehavioural toxicity and depression [127]

13.7 Conclusion

Nanomaterials are widely used in the field of dentistry. With the wide applications, numerous studies are also reporting its toxic effect on the structure and function of the nervous system. Animal studies also suggest damaged microglia and defective behaviours associated with it. All these results warn us to check the nanomaterials thoroughly using different systems to nullify their toxicity to a significant extent.

Acknowledgements

MM lab is supported by Department of Biotechnology,Ministry of Science and Technology [SERB/EMR/2017/003054,BT/PR21857/NNT/28/1238/2017, and Odisha DBT 3325/ST(BIO)–02/2017] for financial assistance.

References

[1] Priyadarsini S, Mukherjee S and Mishra M 2018 Nanoparticles used in dentistry: a review *J. Oral Biol. Craniofac. Res.* **8** 58–67

[2] Priyadarsini S, Mukherjee S, Bag J, Nayak N and Mishra M 2020 *Nanoparticles in Medicine* (Berlin: Springer) pp 55–98

[3] Feng X, Chen A, Zhang Y, Wang J, Shao L and Wei L 2015 Application of dental nanomaterials: potential toxicity to the central nervous system *Int. J. Nanomed.* **10** 3547

[4] Shahi S, Özcan M, Maleki Dizaj S, Sharifi S, Al-Haj Husain N, Eftekhari A and Ahmadian E 2019 A review on potential toxicity of dental material and screening their biocompatibility *Toxicol. Mech. Methods* **29** 368–77

[5] Xiaoli F and Longquan S 2018 Neurotoxicity of nanomaterials *Emerging Nanotechnologies in Dentistry* ed K Subramani and W Ahmed (Amsterdam: Elsevier) ch 20 421–44

[6] Teleanu D M, Chircov C, Grumezescu A M and Teleanu R I 2019 Neurotoxicity of nanomaterials: an up-to-date overview *Nanomaterials* **9** 96

[7] Manzo L, Artigas F, Martínez E, Mutti A, Bergamaschi E, Nicotera P, Tonini M, Candura S M, Ray D E and Costa L G 1996 Biochemical markers of neurotoxicity. A review of mechanistic studies and applications *Human Exp. Toxicol.* **15** S20–35

[8] Nanda R, Panda P and Mishra M 2020 *Nanomaterials in Diagnostic Tools and Devices* (Amsterdam: Elsevier) pp 479–523

[9] Mishra M and Panda M 2021 Reactive oxygen species: the root cause of nanoparticle-induced toxicity in *Drosophila melanogaster Free Radical Res.* **55** 1–17

[10] Jang H J, Lee E C, Kwon G J and Seo Y K 2020 The effect of coated nano-hydroxyapatite concentration on scaffolds for osteogenesis *J. Biomater. Appl.* **34** 827–39

[11] Samanipour F, Bayati M, Golestani-Fard F, Zargar H, Mirhabibi A, Shoaei-Rad V and Abbasi S 2011 Innovative fabrication of ZrO_2–HAp–TiO_2 nano/micro-structured composites through MAO/EPD combined method *Mater. Lett.* **65** 926–8

[12] Soundrapandian C, Bharati S, Basu D and Datta S 2011 Studies on novel bioactive glasses and bioactive glass–nano-HAp composites suitable for coating on metallic implants *Ceram. Int.* **37** 759–69

[13] Teixeira J A, Santos Júnior V E d, Melo Júnior P C d, Arnaud M, Lima M G, Flores M A P, Stamford T C M, Dias Pereira J R, Ribeiro Targino A G and Galembeck A 2018 Effects of a new nano-silver fluoride-containing dentifrice on demineralization of enamel and *Streptococcus mutans* adhesion and acidogenicity *Int. J. Dent.* **2018** 1351925

[14] Bapat R A, Joshi C P, Bapat P, Chaubal T V, Pandurangappa R, Jnanendrappa N, Gorain B, Khurana S and Kesharwani P 2019 The use of nanoparticles as biomaterials in dentistry *Drug Discov. Today* **24** 85–98

[15] Galvez A M, Ramos K M, Teja A J and Baculi R 2021 Bacterial exopolysaccharide-mediated synthesis of silver nanoparticles and their application on bacterial biofilms *J. Microbiol. Biotechnol. Food Sci.* **2021** 970–8

[16] Espinosa-Cristóbal L F, López-Ruiz N, Cabada-Tarín D, Reyes-López S Y, Zaragoza-Contreras A, Constandse-Cortéz D, Donohué-Cornejo A, Tovar-Carrillo K, Cuevas-González J C and Kobayashi T 2018 Antiadherence and antimicrobial properties of silver nanoparticles against Streptococcus mutans on brackets and wires used for orthodontic treatments *J. Nanomater.* **2018** 9248527

[17] Spriano S, Yamaguchi S, Baino F and Ferraris S 2018 A critical review of multifunctional titanium surfaces: new frontiers for improving osseointegration and host response, avoiding bacteria contamination *Acta Biomater.* **79** 1–22

[18] Simon A P, Santos V A Q, Rodrigues A, Santos J S, Trivinho-Strixino F, Pereira B L, Lepienski C M and Sikora M d S 2019 Enhancement of mechanical properties and wettability of TiO_2NT arrays formed in simulated body fluid-based electrolyte *Adv. Eng. Mater.* **21** 1900813

[19] Netti P A 2014 *Regenerative Medicine Applications in Organ Transplantation* (Amsterdam: Elsevier) pp 137–50

[20] Yuan Q, He G, Tan Z, Gong P, Li X Y, Chen Z Q, Ban Y, Tang H, Wang F F and Man Y 2007 Biocompatibility of nano-TiO_2/HA bioceramic coating for nerve regeneration around dental implants *Key Engineering Materials* (College Park, MD: TransTech Publ) pp 1393–6

[21] Salehi P, Babanouri N, Roein-Peikar M and Zare F 2018 Long-term antimicrobial assessment of orthodontic brackets coated with nitrogen-doped titanium dioxide against Streptococcus mutans *Prog. Orthod.* **19** 1–6

[22] Cao B, Wang Y, Li N, Liu B and Zhang Y 2013 Preparation of an orthodontic bracket coated with an nitrogen-doped $TiO_{2-x}N_y$ thin film and examination of its antimicrobial performance *Dent. Mater. J.* **32** 311–6

[23] Jiang Y, Zhang L, Wen D and Ding Y 2016 Role of physical and chemical interactions in the antibacterial behavior of ZnO nanoparticles against E. coli *Mater. Sci. Eng.* **C69** 1361–6

[24] Jiang J, Pi J and Cai J 2018 The advancing of zinc oxide nanoparticles for biomedical applications *Bioinorg. Chem. Appl* **2018** 1062562

[25] Windiasti G, Feng J, Ma L, Hu Y, Hakeem M J, Amoako K, Delaquis P and Lu X 2019 Investigating the synergistic antimicrobial effect of carvacrol and zinc oxide nanoparticles against *Campylobacter jejuni Food Control* **96** 39–46

[26] Feris K, Otto C, Tinker J, Wingett D, Punnoose A, Thurber A, Kongara M, Sabetian M, Quinn B and Hanna C 2010 Electrostatic interactions affect nanoparticle-mediated toxicity to gram-negative bacterium *Pseudomonas aeruginosa* PAO1 *Langmuir* **26** 4429–36

[27] Wang S, Wu J, Yang H, Liu X, Huang Q and Lu Z 2017 Antibacterial activity and mechanism of Ag/ZnO nanocomposite against anaerobic oral pathogen *Streptococcus mutans J. Mater. Sci.: Mater. Med.* **28** 1–8

[28] Toledano M, Vallecillo-Rivas M, Aguilera F S, Osorio M T, Osorio E and Osorio R 2021 Polymeric zinc-doped nanoparticles for high performance in restorative dentistry *J. Dent.* **107** 103616

[29] Kraipok A, Mamanee T, Ruangsuriya J and Leenakul W 2021 Investigation of phase formation and mechanical properties of lithium disilicate glass-ceramic doped CeO_2 *J. Non-Cryst. Solids* **561** 120772

[30] Garcia I M, Leitune V C B, Takimi A S, Bergmann C P, Samuel S M W, Melo M A and Collares F M 2020 Cerium dioxide particles to tune radiopacity of dental adhesives: Microstructural and physico-chemical evaluation *J. Funct. Biomater.* **11** 7

[31] Augustine R, Hasan A, Patan N K, Dalvi Y B, Varghese R, Antony A, Unni R N, Sandhyarani N and Moustafa A-E A 2019 Cerium oxide nanoparticle incorporated electrospun poly (3-hydroxybutyrate-co-3-hydroxyvalerate) membranes for diabetic wound healing applications *ACS Biomater. Sci. Eng.* **6** 58–70

[32] Augustine R, Zahid A A, Hasan A, Dalvi Y B and Jacob J 2020 Cerium oxide nanoparticle-loaded gelatin methacryloyl hydrogel wound-Healing patch with free radical scavenging activity *ACS Biomater. Sci. Eng.* **7** 279–90

[33] Kaze R C, Beleuk à Moungam L M, Cannio M, Rosa R, Kamseu E, Melo U C and Leonelli C 2018 Microstructure and engineering properties of Fe_2O_3 (FeO)-Al_2O_3-SiO_2 based geopolymer composites *J. Clean. Prod.* **199** 849–59

[34] Gad M M and Abualsaud R 2019 Behavior of PMMA denture base materials containing titanium dioxide nanoparticles: a literature review *Int. J. Biomater.* **2019** 6190610

[35] Xu Z-l, Yu L-y and Han L-f 2009 Polymer-nanoinorganic particles composite membranes: a brief overview *Front. Chem. Sci. Eng. China* **3** 318–29

[36] Hossen M M, Chowdhury F, Gafur M, Hakim A A and Nasrin S 2014 Investigation of mechanical properties of $Al_2O_{3\text{–}20}$ wt% ZrO_2 composites as a function of sintering temperature *Eur. Sci. J.* **10**

[37] Sonin D L, Korolev D V, Postnov V N, Naumysheva E B, Pochkaeva E I, Vasyutina M L and Galagudza M M 2016 Silicon-containing nanocarriers for targeted drug delivery: synthesis, physicochemical properties and acute toxicity *Drug Deliv.* **23** 1747–56

[38] De Cremer K, Braem A, Gerits E, De Brucker K, Vandamme K, Martens J, Michiels J, Vleugels J, Cammue B and Thevissen K 2017 Controlled release of chlorhexidine from a mesoporous silica-containing macroporous titanium dental implant prevents microbial biofilm formation *Eur. Cells Mater.* **33** 13–27

[39] Barot T, Rawtani D and Kulkarni P 2021 Nanotechnology-based materials as emerging trends for dental applications *Rev. Adv. Mater. Sci.* **60** 173–89

[40] Gulati K, Maher S, Findlay D M and Losic D 2016 Titania nanotubes for orchestrating osteogenesis at the bone–implant interface *Nanomedicine* **11** 1847–64

[41] De Aza A, Chevalier J, Fantozzi G, Schehl M and Torrecillas R 2002 Crack growth resistance of alumina, zirconia and zirconia toughened alumina ceramics for joint prostheses *Biomaterials* **23** 937–45

[42] Gautam C, Joyner J, Gautam A, Rao J and Vajtai R 2016 Zirconia based dental ceramics: structure, mechanical properties, biocompatibility and applications *Dalton Trans.* **45** 19194–215

[43] Behnaz M, Davani S T, Abdi A, Azadi G and Goudarzi M 2018 Efficacy of antimicrobial agents in orthodontic adhesive systems and brackets: a narrative review *J. Dent. Sch.* **37** 112–7

[44] Ahmed B, Solanki B, Zaidi A, Khan M S and Musarrat J 2019 Bacterial toxicity of biomimetic green zinc oxide nanoantibiotic: insights into ZnONP uptake and nanocolloid–bacteria interface *Toxicol. Res.* **8** 246–61

[45] Pelling H, Nzakizwanayo J, Milo S, Denham E, MacFarlane W, Bock L, Sutton J and Jones B 2019 Bacterial biofilm formation on indwelling urethral catheters *Lett. Appl. Microbiol.* **68** 277–93

[46] Son N T, Oda M, Hayashi N, Yamaguchi D, Kawagishi Y, Takahashi F, Harada K, Cuong N M and Fukuyama Y 2018 Antimicrobial activity of the constituents of Dalbergia tonkinensis and structural-bioactive highlights *Nat. Prod. Commun.* **13**

[47] Fatfat M, Abou Merhi R, Rahal O, Stoyanovsky D A, Zaki A, Haidar H, Kagan V E, Gali-Muhtasib H and Machaca K 2014 Copper chelation selectively kills colon cancer cells through redox cycling and generation of reactive oxygen species *BMC Cancer* **14** 1–12

[48] Vimbela G V, Ngo S M, Fraze C, Yang L and Stout D A 2017 Antibacterial properties and toxicity from metallic nanomaterials *Int. J. Nanomed.* **12** 3941

[49] Olteanu D, Filip A, Socaci C, Biris A R, Filip X, Coros M, Rosu M C, Pogacean F, Alb C and Baldea I 2015 Cytotoxicity assessment of graphene-based nanomaterials on human dental follicle stem cells *Colloids Surf.* B **136** 791–8

[50] He J, Zhu X, Qi Z, Wang C, Mao X, Zhu C, He Z, Li M and Tang Z 2015 Killing dental pathogens using antibacterial graphene oxide *ACS Appl. Mater. Interfaces* **7** 5605–11

[51] Konar M, Nayak N, Priyadarsini S, Mishra M and Sahoo H 2019 Antimicrobial activity of nanoparticle-based dental fillers on novel chromogenic bacteria *Enterobacter ludwigii Mater. Res. Exp.* **6** 085407

[52] Kulshrestha S, Khan S, Meena R, Singh B R and Khan A U 2014 A graphene/zinc oxide nanocomposite film protects dental implant surfaces against cariogenic *Streptococcus mutans Biofouling* **30** 1281–94

[53] Wang C, Hu H, Li Z, Shen Y, Xu Y, Zhang G, Zeng X, Deng J, Zhao S and Ren T 2019 Enhanced osseointegration of titanium alloy implants with laser microgrooved surfaces and graphene oxide coating *ACS Appl. Mater. Interfaces* **11** 39470–83

[54] Ge Z, Yang L, Xiao F, Wu Y, Yu T, Chen J, Lin J and Zhang Y 2018 Graphene family nanomaterials: properties and potential applications in dentistry *Int. J. Biomater.* **2018** 1539678

[55] Sarosi C, Biris A R, Antoniac A, Boboia S, Alb C, Antoniac I and Moldovan M 2016 The nanofiller effect on properties of experimental graphene dental nanocomposites *J. Adhesion Sci. Technol.* **30** 1779–94

[56] Toledano M, Osorio R, Osorio E, Medina-Castillo A L, Toledano-Osorio M and Aguilera F S 2017 Ions-modified nanoparticles affect functional remineralization and energy dissipation through the resin-dentin interface *J. Mech. Behav. Biomed. Mater.* **68** 62–79

[57] Li Y, Hu X, Xia Y, Ji Y, Ruan J, Weir M D, Lin X, Nie Z, Gu N and Masri R 2018 Novel magnetic nanoparticle-containing adhesive with greater dentin bond strength and antibacterial and remineralizing capabilities *Dent. Mater.* **34** 1310–22

[58] Vitt A, Sofrata A, Slizen V, Sugars R, Gustafsson A, Gudkova E, Kazeko L, Ramberg P and Buhlin K 2015 Antimicrobial activity of polyhexamethylene guanidine phosphate in comparison to chlorhexidine using the quantitative suspension method *Ann. Clin. Microbiol. Antimicrob.* **14** 1–9

[59] Curtis A R, Palin W M, Fleming G J, Shortall A C and Marquis P M 2009 The mechanical properties of nanofilled resin-based composites: characterizing discrete filler particles and agglomerates using a micromanipulation technique *Dent. Mater.* **25** 180–7

[60] Satpathy A, Ranjan R, Priyadarsini S, Gupta S, Mathur P and Mishra M 2019 *Medical Imaging Methods* (Berlin: Springer) pp 59–95

[61] McCabe J F and Walls A W 2013 *Applied Dental Materials* (New York: Wiley)

[62] Buzea C, Pacheco I I and Robbie K 2007 Nanomaterials and nanoparticles: sources and toxicity *Biointerphases* **2** MR17–71

[63] Pachter J S, de Vries H E and Fabry Z 2003 The blood-brain barrier and its role in immune privilege in the central nervous system *J. Neuropathol. Exp. Neurol.* **62** 593–604

[64] Stadelmann C, Wegner C and Brück W 2011 Inflammation, demyelination, and degeneration—recent insights from MS pathology *Biochim. Biophys. Acta (BBA)-Molecular Basis of Disease* **1812** 275–82

[65] Wolburg H, Noell S, Mack A, Wolburg-Buchholz K and Fallier-Becker P 2009 Brain endothelial cells and the glio-vascular complex *Cell Tissue Res.* **335** 75–96

[66] Xiao G and Gan L-S 2013 Receptor-mediated endocytosis and brain delivery of therapeutic biologics *Int. J. Cell Biol.* **2013** 703545

[67] Curtis J, Greenberg M, Kester J, Phillips S and Krieger G 2006 Nanotechnology and nanotoxicology *Toxicol. Rev.* **25** 245–60

[68] Al-Lamki R S, Bradley J R and Pober J S 2017 Human organ culture: updating the approach to bridge the gap from in vitro to in vivo in inflammation, cancer, and stem cell biology *Front. Med.* **4** 148

[69] Li Y, Fan J and Ju D 2019 *Brain Targeted Drug Delivery System* (Amsterdam: Elsevier) pp 377–408

[70] Kumar V, Sharma N and Maitra S 2017 In vitro and in vivo toxicity assessment of nanoparticles *Int. Nano Lett.* **7** 243–56

[71] Figueroa-González G and Pérez-Plasencia C 2017 Strategies for the evaluation of DNA damage and repair mechanisms in cancer *Oncol. Lett.* **13** 3982–8

[72] Blanco A and Blanco G 2017 *Apoptosis Medical Biochemistry* (New York: Elsevier) ch 32 pp 791–6

[73] Lovisolo D, Dionisi M, Ruffinatti F A and Distasi C 2018 Nanoparticles and potential neurotoxicity: focus on molecular mechanisms *AIMS Molecul. Sci.* **5** 1–13

[74] Weksler B, Romero I A and Couraud P-O 2013 The hCMEC/D3 cell line as a model of the human blood brain barrier *Fluids Barriers CNS* **10** 1–10

[75] Greiffenberg L, Goebel W, Kim K S, Weiglein I, Bubert A, Engelbrecht F, Stins M and Kuhn M 1998 Interaction of Listeria monocytogenes with human brain microvascular

endothelial cells: InlB-dependent invasion, long-term intracellular growth, and spread from macrophages to endothelial cells *Infect. Immun.* **66** 5260–7

[76] Al Ahmad A, Gassmann M and Ogunshola O 2009 Maintaining blood–brain barrier integrity: pericytes perform better than astrocytes during prolonged oxygen deprivation *J. Cell. Physiol.* **218** 612–22

[77] Guttenplan K A and Liddelow S A 2019 Astrocytes and microglia: models and tools *J. Exp. Med.* **216** 71–83

[78] Nakagawa S, Deli M A, Kawaguchi H, Shimizudani T, Shimono T, Kittel A, Tanaka K and Niwa M 2009 A new blood–brain barrier model using primary rat brain endothelial cells, pericytes and astrocytes *Neurochem. Int.* **54** 253–63

[79] Arakha M, Roy J, Nayak P S, Mallick B and Jha S 2017 Zinc oxide nanoparticle energy band gap reduction triggers the oxidative stress resulting into autophagy-mediated apoptotic cell death *Free Radic. Biol. Med.* **110** 42–53

[80] Choi J, Zheng Q, Katz H E and Guilarte T R 2010 Silica-based nanoparticle uptake and cellular response by primary microglia *Environ. Health Persp.* **118** 589–95

[81] Larner S F, Wang J, Goodman J, O'Donoghue Altman M B, Xin M and Wang K K 2017 *In vitro* neurotoxicity resulting from exposure of cultured neural cells to several types of nanoparticles *J. Cell Death* **10** 1179670717694523

[82] Bhattamisra S K, Shak A T, Xi L W, Safian N H, Choudhury H, Lim W M, Shahzad N, Alhakamy N A, Anwer M K and Radhakrishnan A K 2020 Nose to brain delivery of rotigotine loaded chitosan nanoparticles in human SH-SY5Y neuroblastoma cells and animal model of Parkinson's disease *Int. J. Pharm.* **579** 119148

[83] Xu L J, Zhao J X, Zhang T, Ren G G and Yang Z 2009 *In vitro* study on influence of nano particles of CuO on CA1 pyramidal neurons of rat hippocampus potassium currents *Environ. Toxicol.* **24** 211–7

[84] Lapiz M and Morilak D 2006 Noradrenergic modulation of cognitive function in rat medial prefrontal cortex as measured by attentional set shifting capability *Neuroscience* **137** 1039–49

[85] Kreyling W G, Holzwarth U, Haberl N, Kozempel J, Hirn S, Wenk A, Schleh C, Schäffler M, Lipka J and Semmler-Behnke M 2017 Quantitative biokinetics of titanium dioxide nanoparticles after intravenous injection in rats: part 1 *Nanotoxicology* **11** 434–42

[86] Sun C, Yin N, Wen R, Liu W, Jia Y, Hu L, Zhou Q and Jiang G 2016 Silver nanoparticles induced neurotoxicity through oxidative stress in rat cerebral astrocytes is distinct from the effects of silver ions *Neurotoxicology* **52** 210–21

[87] Valdiglesias V, Fernández-Bertólez N, Kiliç G, Costa C, Costa S, Fraga S, Bessa M J, Pásaro E, Teixeira J P and Laffon B 2016 Are iron oxide nanoparticles safe? Current knowledge and future perspectives *J. Trace. Elem. Med. Biol.* **38** 53–63

[88] Song B, Zhang Y, Liu J, Feng X, Zhou T and Shao L 2016 Unraveling the neurotoxicity of titanium dioxide nanoparticles: focusing on molecular mechanisms *Beilstein J. Nanotechnol.* **7** 645–54

[89] Zhou M, Xie L, Fang C-J, Yang H, Wang Y-J, Zhen X-Y, Yan C-H, Wang Y, Zhao M and Peng S 2016 Implications for blood-brain-barrier permeability, in vitro oxidative stress and neurotoxicity potential induced by mesoporous silica nanoparticles: Effects of surface modification *RSC Adv.* **6** 2800–9

[90] Shi D, Mi G and Webster T 2017 *Neurotoxicity of Nanomaterials and Nanomedicine* (Amsterdam: Elsevier) pp 259–84

[91] Dusinska M, Rundén-Pran E, Schnekenburger J and Kanno J 2017 *Adverse Effects of Engineered Nanomaterials* (Amsterdam: Elsevier) pp 51–82

[92] Pappus S A and Mishra M 2018 A drosophila model to decipher the toxicity of nanoparticles taken through oral routes *Cell. Mol. Toxicol. Nanopart.* **1048** 311–22

[93] O'Brown N M, Pfau S J and Gu C 2018 Bridging barriers: a comparative look at the blood–brain barrier across organisms *Genes Devel.* **32** 466–78

[94] Song B, Zhang Y, Liu J, Feng X, Zhou T and Shao L 2016 Is neurotoxicity of metallic nanoparticles the cascades of oxidative stress? *Nanoscale Res. Lett.* **11** 1–11

[95] Barik B K and Mishra M 2019 Nanoparticles as a potential teratogen: a lesson learnt from fruit fly *Nanotoxicology* **13** 258–84

[96] Presland R B, Boggess D, Lewis S P, Hull C, Fleckman P and Sundberg J P 2000 Loss of normal profilaggrin and filaggrin in flaky tail (ft/ft) mice: an animal model for the filaggrin-deficient skin disease ichthyosis vulgaris *J. Investig. Dermatol.* **115** 1072–81

[97] Vilella A, Tosi G, Grabrucker A M, Ruozi B, Belletti D, Vandelli M A, Boeckers T M, Forni F and Zoli M 2014 Insight on the fate of CNS-targeted nanoparticles. Part I: Rab5-dependent cell-specific uptake and distribution *J. Control. Release.* **174** 195–201

[98] Amara S, Ben-Slama I, Mrad I, Rihane N, Jeljeli M, El-Mir L, Ben-Rhouma K, Rachidi W, Sève M and Abdelmelek H 2014 Acute exposure to zinc oxide nanoparticles does not affect the cognitive capacity and neurotransmitters levels in adult rats *Nanotoxicology* **8** 208–15

[99] Ze Y, Hu R, Wang X, Sang X, Ze X, Li B, Su J, Wang Y, Guan N and Zhao X 2014 Neurotoxicity and gene-expressed profile in brain-injured mice caused by exposure to titanium dioxide nanoparticles *J. Biomed. Mater. Res. Part A* **102** 470–8

[100] Marano F, Hussain S, Rodrigues-Lima F, Baeza-Squiban A and Boland S 2011 Nanoparticles: molecular targets and cell signalling *Arch. Toxicol.* **85** 733–41

[101] Shrivastava R, Raza S, Yadav A, Kushwaha P and Flora S J 2014 Effects of sub-acute exposure to TiO_2, ZnO and Al_2O_3 nanoparticles on oxidative stress and histological changes in mouse liver and brain *Drug Chem. Toxicol.* **37** 336–47

[102] Win-Shwe T-T and Fujimaki H 2011 Nanoparticles and neurotoxicity *Int. J. Mol. Sci.* **12** 6267–80

[103] Zhang Y, Yu C, Huang G, Wang C and Wen L 2010 Nano rare-earth oxides induced size-dependent vacuolization: an independent pathway from autophagy *Int. J. Nanomed.* **5** 601

[104] Yuan Z-Y, Hu Y-L and Gao J-Q 2015 Brain localization and neurotoxicity evaluation of polysorbate 80-modified chitosan nanoparticles in rats *PLoS One* **10** e0134722

[105] Raj A, Shah P and Agrawal N 2017 Dose-dependent effect of silver nanoparticles (AgNPs) on fertility and survival of Drosophila: an in-vivo study *PLoS One* **12** e0178051

[106] Perez R G and Hastings T G 2004 Could a loss of α-synuclein function put dopaminergic neurons at risk? *J. Neurochem.* **89** 1318–24

[107] Tian H, Eom H-J, Moon S, Lee J, Choi J and Chung Y D 2013 Development of biomarker for detecting silver nanoparticles exposure using a GAL4 enhancer trap screening in Drosophila *Environ. Toxicol. Pharmacol.* **36** 548–56

[108] Vecchio G, Galeone A, Brunetti V, Maiorano G, Rizzello L, Sabella S, Cingolani R and Pompa P P 2012 Mutagenic effects of gold nanoparticles induce aberrant phenotypes in Drosophila melanogaster *Nanomed.: Nanotechnol. Biol. Med.* **8** 1–7

[109] Huang N, Yan Y, Xu Y, Jin Y, Lei J, Zou X, Ran D, Zhang H, Luan S and Gu H 2013 Alumina nanoparticles alter rhythmic activities of local interneurons in the antennal lobe of Drosophila *Nanotoxicology* **7** 212–20

[110] Sabat D, Patnaik A, Ekka B, Dash P and Mishra M 2016 Investigation of titania nanoparticles on behaviour and mechanosensory organ of Drosophila melanogaster *Physiol. Behav.* **167** 76–85

[111] Vales G, Demir E, Kaya B, Creus A and Marcos R 2013 Genotoxicity of cobalt nanoparticles and ions in Drosophila *Nanotoxicology* **7** 462–8

[112] Carmona E R, Inostroza-Blancheteau C, Rubio L and Marcos R 2016 Genotoxic and oxidative stress potential of nanosized and bulk zinc oxide particles in Drosophila melanogaster *Toxicol. Indus. Health* **32** 1987–2001

[113] Mallick T, Karmakar A, Bag J, Sahu S, Mishra M and Begum N A 2020 Carbazole analog anchored fluorescent silica nanoparticle showing enhanced biocompatibility and selective sensing ability towards biomacromolecule *Dyes Pigment.* **173** 107994

[114] Priyadarsini S, Sahoo S K, Sahu S, Mukherjee S, Hota G and Mishra M 2019 Oral administration of graphene oxide nano-sheets induces oxidative stress, genotoxicity, and behavioral teratogenicity in Drosophila melanogaster *Environ. Sci. Pollut. Res.* **26** 19560–74

[115] Anand A S, Gahlot U, Prasad D N, Amitabh and Kohli E 2019 Aluminum oxide nanoparticles mediated toxicity, loss of appendages in progeny of Drosophila melanogaster on chronic exposure *Nanotoxicology* **13** 977–89

[116] Mishra M, Sabat D, Ekka B, Sahu S, Unnikannan P and Dash P 2017 Oral intake of zirconia nanoparticle alters neuronal development and behaviour of Drosophila melanogaster *J. Nanoparticle Res.* **19** 1–12

[117] Sahu S and Mishra M 2020 Hydroxyapatite nanoparticle causes sensory organ defects by targeting the retromer complex in Drosophila melanogaster *NanoImpact* **19** 100237

[118] Liu X, Vinson D, Abt D, Hurt R H and Rand D M 2009 Differential toxicity of carbon nanomaterials in Drosophila: larval dietary uptake is benign, but adult exposure causes locomotor impairment and mortality *Environ. Science Technol.* **43** 6357–63

[119] Li T, Shi T, Li X, Zeng S, Yin L and Pu Y 2014 Effects of nano-MnO_2 on dopaminergic neurons and the spatial learning capability of rats *Int. J. Environ. Res. Public Health* **11** 7918–30

[120] Kim E-M, Palmer P, Howard V, Elsaesser A, Taylor A, Staats G and O'Hare E 2013 Effect of Intracerebroventricular injection of TiO_2 nanoparticles on complex behaviour in the rat *J. Nanosci. Nanotechnol.* **13** 8325–30

[121] Sorce S and Krause K-H 2009 NOX enzymes in the central nervous system: from signaling to disease *Antioxid. Redox. Signal.* **11** 2481–504

[122] Schäppi M G, Jaquet V, Belli D C and Krause K-H 2008 Hyperinflammation in chronic granulomatous disease and anti-inflammatory role of the phagocyte NADPH oxidase *Seminars in Immunopathology* (Berlin: Springer) pp 255–71

[123] Thomas M P, Chartrand K, Reynolds A, Vitvitsky V, Banerjee R and Gendelman H E 2007 Ion channel blockade attenuates aggregated alpha synuclein induction of microglial reactive oxygen species: relevance for the pathogenesis of Parkinson's disease *J. Neurochem.* **100** 503–19

[124] Hu R, Gong X, Duan Y, Li N, Che Y, Cui Y, Zhou M, Liu C, Wang H and Hong F 2010 Neurotoxicological effects and the impairment of spatial recognition memory in mice caused by exposure to TiO_2 nanoparticles *Biomaterials* **31** 8043–50

[125] Flora S 2017 *Neurotoxicity of Nanomaterials and Nanomedicine* (Amsterdam: Elsevier) pp 179–203

[126] Das M, Ghosh M, Gharami K and Das S 2018 Thyroid hormone and astrocyte differentiation *Vitamins Hormones* **106** 283–312

[127] Gholamine B, Karimi I, Salimi A, Mazdarani P and Becker L A 2017 Neurobehavioral toxicity of carbon nanotubes in mice: focus on brain-derived neurotrophic factor messenger RNA and protein *Toxicol. Indus. Health* **33** 340–50

www.ingramcontent.com/pod-product-compliance
Ingram Content Group UK Ltd.
Pitfield, Milton Keynes, MK11 3LW, UK
UKHW051041120926
8556IPUK00030B/670

9 780750 336697